Atta-ur-Rahman · P.W. Le Quesne (Eds.)

Natural Products Chemistry III

With 104 Figures and 86 Schemes

Springer-Verlag
Berlin Heidelberg New York
London Paris Tokyo

Prof. Dr. ATTA-UR-RAHMAN
HEJ Research Institute of Chemistry
University of Karachi
Karachi-32
Pakistan

Prof. Dr. PHILIP WILLIAM LE QUESNE
Department of Chemistry
Northeastern University
360 Huntington Avenue
Boston, MA 02115
USA

ISBN-13:978-3-642-74019-0 e-ISBN-13:978-3-642-74017-6
DOI: 10.1007/978-3-642-74017-6

Library of Congress Cataloging-in-Publication Data. Natural products chemistry / Atta-ur-Rahman, Philip W. Le Quesne, eds. p. cm. Proceedings of the 3rd International Symposium and Pakistan-U.S. Binational Workshop on Natural Products Chemistry, held in Karachi, Pakistan, Jan. 1988. Includes bibliographies and index. 1. Natural products–Congresses. I. Rahman, Atta-ur-, 1942- II. Le Quesne, Philip W. III. International Symposium and Pakistan-U.S. Binational Workshop on Natural Products Chemistry (3rd. : 1988 : Karachi, Pakistan) QD415.A1N38 1988 547.7–dc 19 88-28145

Softcover reprint of the hardcover 1st edition 1988

2131/3130-543210 – Printed on acid-free paper

Preface

Natural product chemistry has undergone an explosive growth during the latter half of the current century. This has been brought about by a number of factors. One of these has been the growing number of substances from natural sources which display interesting pharmacological activities. These include antibiotics, anti-tumor agents, immunostimulants, drugs affecting the cardiovascular and central nervous systems, analgesics etc. Another factor has been the improvements made in the technology of isolation processes which includes the development of such techniques as high pressure liquid chromatography which has allowed the rapid isolation of substances which were previously difficult to obtain by classical procedures. The most important factor has been the development of new spectrosopic techniques which have opened up whole new vistas in this exciting field. Prominent in these advances has been the advent of powerful superconducting magnets with very stable magnetic fields, and pulse NMR in which the duration, direction and phases of pulses can be accurately controlled by means of minicomputers. These have heralded the advent of two-dimensional NMR spectroscopy which has now come to be routinely employed in unravelling complex structures. No less important, though somewhat less dramatic, have been the advances made in the field of mass spectroscopy where new ionization techniques such as positive and negative fast atom bombardment, field desorption, chemical ionization etc. have allowed the mass spectra of larger non-volatile substances to be recorded.

The field of organic synthesis has seen equally rapid advances. The emphasis has been in the development of new stereoselective methods for achieving specific synthetic transformations. The exciting synthetic approaches in the field of macrolide antibiotics, anti-tumor agents and other biologically important natural products serve to exemplify these developments.

The 3rd International Symposium and Pakistan - U.S. Binational Workshop on Natural Products Chemistry, which was organised in Karachi in January 1988 served to illustrate the frontiers of natural product chemistry as they stand today. The distinguished scientists who participated in the Symposium and presented Plenary lectures reflect the high standards of chemistry that was presented during this conference, as in the previous two conferences organised by us in Karachi in 1984 and 1986.

The present proceedings volume contains material which should be of interest to a large number of synthetic organic chemists, phytochemists and biochemists. The first few chapters are concerned with the use of modern spectroscopic techniques in natural products chemistry. These are followed by articles on stereoselective synthesis, biosynthesis and structure elucidation of natural products. In view of space restrictions, we have had to shorten the presentations by Pakistani scientists in this volume in order to include the articles from foreign scientists.

It would be apparent from the galaxy of eminent scientists who contributed to this conference that all in all it was a most exciting feast of chemistry which everyone thoroughly enjoyed. Of course all of it would not have happened, were it not for the staff and students of H.E.J. Research Institute of chemistry who worked so tirelessly to ensure its success - to them we owe our gratitude. But behind every major conference there are the sponsoring agencies who provide the financial support. In this case the two major sponsors were G.T.Z. (West Germany) and N.S.F. (U.S.A.). Other international and national agencies who contributed to the conference are UNESCO, ISESCO, IFS, PSF, British Council, International Seminar in Chemistry, Third World Academy of Sciences and Hamdard Foundation Pakistan.

One of us (Prof. Atta-ur-Rahman) wishes to thank Miss Khurshid Zaman for her assistance in the preparation of the index, and Mr. Mahmood Alam for secretarial help.

This volume is dedicated to Prof. Salimuzzaman Siddiqui on his 91st birthday.

Summer 1988

ATTA-UR-RAHMAN
Sc.D. (Cantab.)
PHILIP W. LE QUESNE

Contents

Contributors

You will find the addresses at the beginning of the respective contribution

Applications of NMR in Biochemistry and Biosynthesis

A. I. SCOTT
Texas A & M University, Chemistry Department,
College Station, TX 77843, USA

I. ^{1}H-NMR Spectroscopy of Catalysis

A splendid illustration of the use of 500 MHz ^{1}H NMR is provided by the water suppressed time course of porphyrin biosynthesis where porphobilinogen (PBG), the substrate for uro'gens I and III and Vitamin B_{12} is transformed into the apparent product uro'gen I which turns out to be a chemical artifact. Thus, during incubation of PBG with the enzyme, PBG deaminase ([D]; Fig. 1; M_r 33,000) the product released from the enzyme is pre-uro'gen, or hydroxymethyl bilane (HMB), whose characteristic spectrum (Fig. 1) ($t^1/_2$ = 180 sec/30°C) can be seen transiently when the reaction is carried out at 4°C [12a,b]. The NMR method also uniquely defines HMB as the substrate for the second enzyme, uro'gen III cosynthetase. In the absence of the latter enzyme, the cyclization of the head-to-tail tetrapyrrole HMB takes place spontaneously. The deep-seated intra-molecular rearrangement of HMB to uro'gen III can now be studied by this technique.

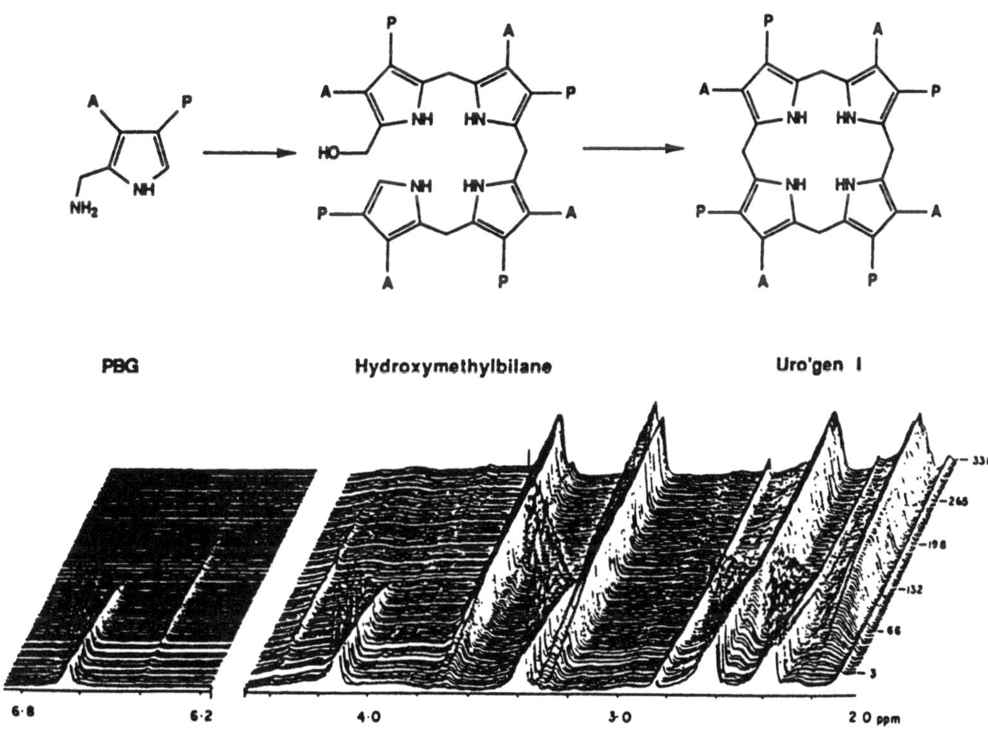

Figure 1: Formation of uro'gen 1 and HMB from PBG.

Perhaps this is indeed a unique case, for in our second example, Nature is not helpful in releasing the intermediate from the cyclizing enzyme for NMR viewing. Thus the enzyme iso-Pen synthetase, which converts the tripeptide ACV to isopenicillin N (Fig. 2) [3] forms both β-Lactam and thiazolidine rings synchronously on the NMR time scale without release from the enzyme. These examples bring us logically to the ultimate search for structural information during the catalytic cycle, by temporal resolution of the spectra of covalently bound intermediates.

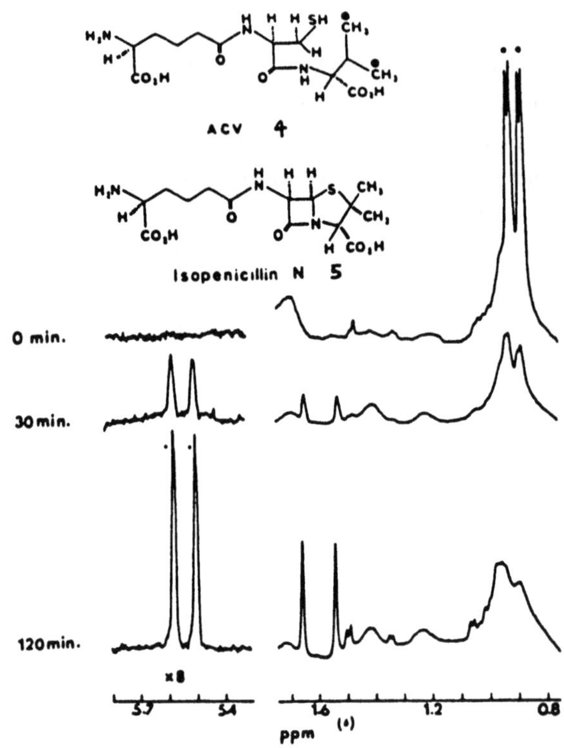

Figure 2: NMR Time Course of Penicillin Biosynthesis.

II. Studying Enzyme Mechanism by ^{13}C NMR.

Almost all enzymatic processes utilize multistep reactions during catalysis, and the characterization of each of these stages is essential if one is to understand enzyme mechanism at the molecular level. Earlier investigators have used spectrophotometric methods for characterizing enzyme-catalyzed reactions, whereas inhibitors that form stable "transition state analogues" have been examined by other techniques that require long periods of data accumulation, for example, X-ray analysis. However, the advent of NMR and its subsequent use in enzymology are beginning to provide a novel and penetrating probe for elucidating enzyme mechanism by directly characterizing intermediates formed in catalysis. The studies of "transition state analogues" allow access to the unique properties of enzymes that enable them to stabilize labile intermediates and therefore achieve their remarkable catalytic efficiency. We discuss first the structures of an enzyme-inhibitor adduct of the serine protease trypsin and compare the results of direct observation by ^{13}C NMR with the structural information inferred by more classical techniques. We then review the powerful combination of NMR and cryoenzymology. The marriage of the two techniques whereby enzyme-catalyzed reactions are studied at sub-zero temperatures in aqueous organic solvents (cryosolvents), allows such reactions to be slowed down and can prolong the lifetime of enzyme-substrate intermediates.

Fig. 3 shows the generally accepted mechanism for the hydrolysis of an amide function by a serine protease and is representative of the mechanism for all the hydrolyses. To confirm this mechanistic pathway, the visualization and rigorous characterization of productive tetrahedral intermediates and acyl enzymes by ^{13}C NMR is the ultimate goal, and we now discuss the progress made so far, particularly with the serine and thiol proteases.

In the domain of synthetic inhibitors, chloromethylketone derivatives of specific substrates are potent irreversible covalent inhibitors of the serine proteases, alkylating the active-center histidine at N-2. X-ray crystallographic studies led to the suggestion that, in addition to the above alkylation, there was also nucleophilic attack by the active-center serine hydroxyl to form a hemiketal, which is stereochemically analogous to the tetrahedral intermediate purported to occur during catalysis.

To test this suggestion, trypsin was inhibited with the highly specific reagent Na carbobenzyloxylysylchloromethylketone ($RCOCH_2Cl$), labeled in the ketone carbon with 90 percent ^{13}C, for it was predicted that a tetrahedral adduct, if formed, would be directly observable by this technique. In aqueous solution (Fig. 4) $RCOCH_2Cl$ exists as a mixture of the ketone (δ204.7 ppm) and its hydrate (δ = 95.4 ppm). At pH 3.2 no alkylation or inhibition of the enzyme is observed and the spectrum of [^{13}C = 0] $RCOCH_2Cl$ is unperturbed (Fig. 4, B and C), showing that at low pH there is no detectable binding or tetrahedral adduct formation prior to alkylation. At pH 6.9 there is rapid, irreversible inhibition, and resonances due to both [^{13}C = 0] $RCOCH_2Cl$ and its hydrate are replaced by a single resonance at 98.0 ppm (Fig. 4D), an indication that the ^{13}C enriched carbonyl of the inhibitor is tetrahedrally coordinated in the inhibitor-enzyme adduct. Denaturation of the trypsin led to reappearance of a carbonyl resonance (205.5 ppm) and a decrease in the intensity of the resonance at 98.0 ppm. This demonstrates that the tetrahedral adduct formed by the attack of the serine hydroxyl on the inhibitor carbonyl and characterized by the resonance at 90.0 ppm requires an intact trypsin structure.

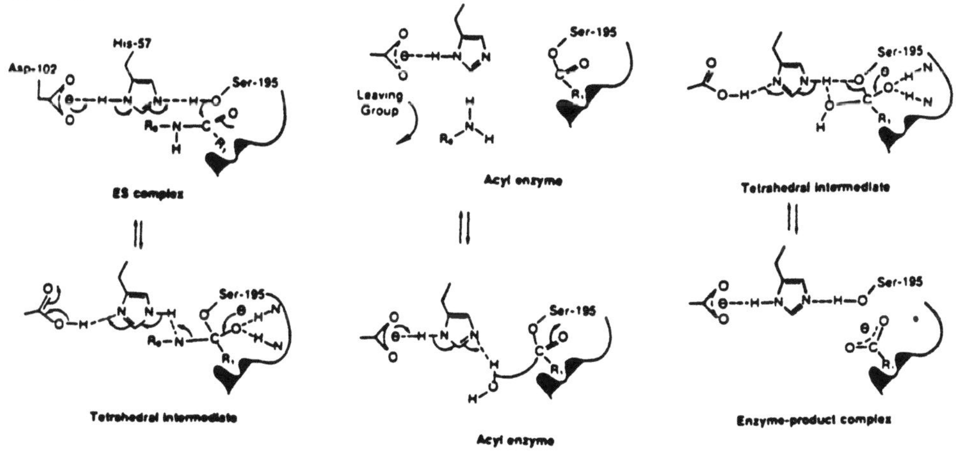

Figure 3: Serine protease mechanism.

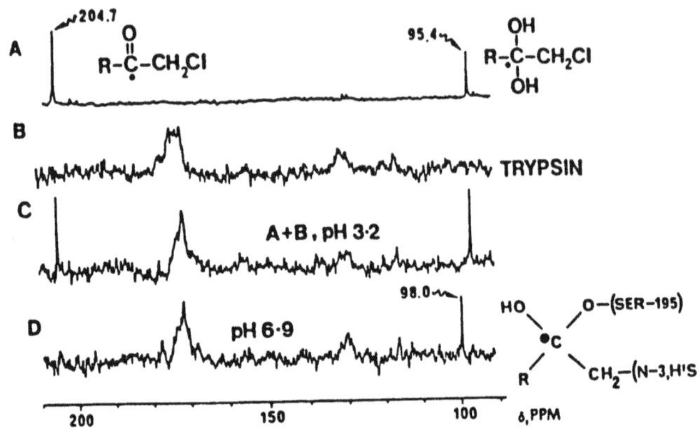

Figure 4: Formation of tetrahedral complex of trypsin with Z-Lys CMK.

In the above experiments with [$^{13}C = 0$] $RCOCH_2Cl$ and trypsin it is difficult to discount the possibility that the resonance at 98.0 ppm could result from hydration of the carbonyl of the covalently bound inhibitor. However, the use of ^{18}O isotopic shifts on the ^{13}C spectrum of the tetrahedral adduct has now shown that the adduct is indeed formed by nucleophilic attack of the Ser^{195}hydroxyl group (Fig. 5) a result completely confirmed by X-ray diffraction studies as portrayed in Fig. 6 [4].

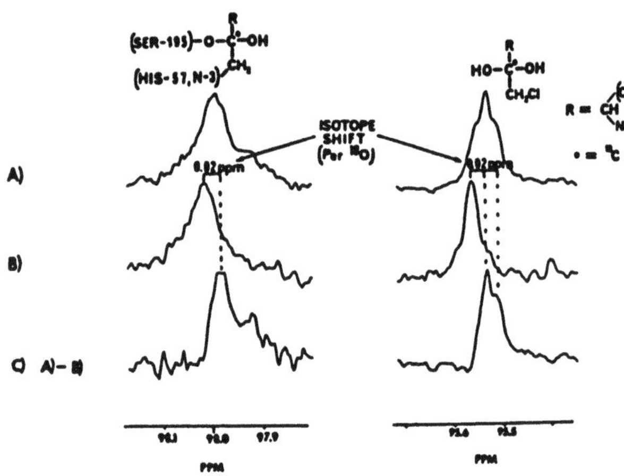

Figure 5: ^{18}O-shifted CMR spectra of Z-Lys CMK complex with trypsin.

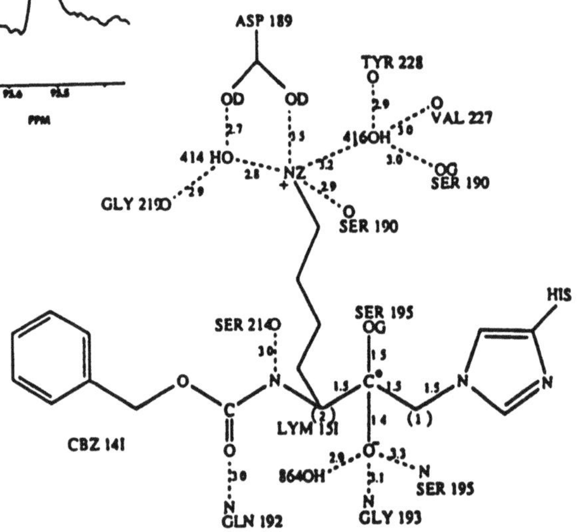

Figure 6: Bond distances in Z-Lys CMK with trypsin from X-ray data at 2.0 Å resolution.

III. The Quest for True Intermediate Structure: Cryoenzymology.

The detection and characterization of a productive acyl intermediate by ^{13}C NMR during catalysis with natural substrates at ambient temperatures is not possible at present, because there is an inherent lack of sensitivity in ^{13}C NMR spectroscopy. Therefore, the lifetimes of these intermediates must be extended into the domain of the NMR experiment. One approach is to utilize low-temperature cryoenzymological techniques to extend the lifetimes of such intermediates.

Since the ^{13}C NMR resonances of the carbonyl of thioesters are shifted (Δ 20 to 30 ppm), ^{13}C NMR spectroscopy should allow the direct monitoring of the formation and decay of a thioacyl intermediate. Using [^{13}C = O] N-benzoylimidazole (δ = 168.7 ppm), we were able to observe directly a thioacyl intermediate at δ = 195.9 ppm in the presence of papain under the cryoenzymological conditions of -t^{o} C in 25 percent aqueous dimethyl sulfoxide. Moreover, the thioacyl species is clearly a productive intermediate since the decrease in its signal intensity was accompanied by an increase in the product resonances and by release of free enzyme (half-life, 96 minutes) determined by titration of its thiol group. The line width of the resonance at 195.9 ppm was 25 ± Hz [5,6].

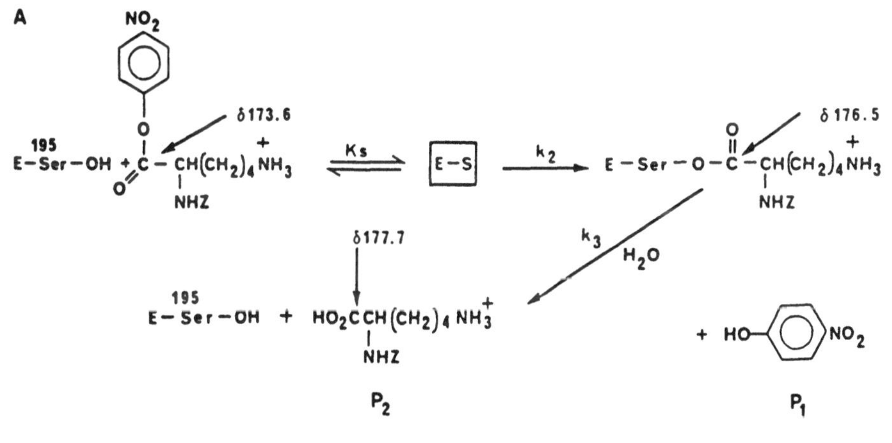

Figure 7: A) Trypsin-catalyzed hydrolysis of Z-Lys-PNP

The trypsin catalyzed hydrolysis (Fig. 7A) of the highly specific substrate Na carbobenzyloxy-L-lysine-p-nitrophenyl-ester (Z-lys-pNP) has been studied in detail under cryoenzymological conditions by both spectrophotometry and ^{13}C NMR spectroscopy. The kinetic data from both techniques confirm that the kinetics and mechanism under cryoenzymological conditions are essentially the same as those determined at ambient temperatures by rapid reaction techniques. The hydrolysis of [^{13}C]Z-lys-pNP[(S) in Fig. 7A, δ = 173.6 ppm] by trypsin was monitored by the decrease in intensity of this signal and the increase in the signal arising from the product (P_2 in Fig. 7A, δ = 177.7

ppm) at -21°C in 41 percent aqueous dimethyl sulfoxide. The continued formation of product (P_2 after all the substrate has been consumed (Fig. 7B) provides indirect evidence for an enzyme-bound

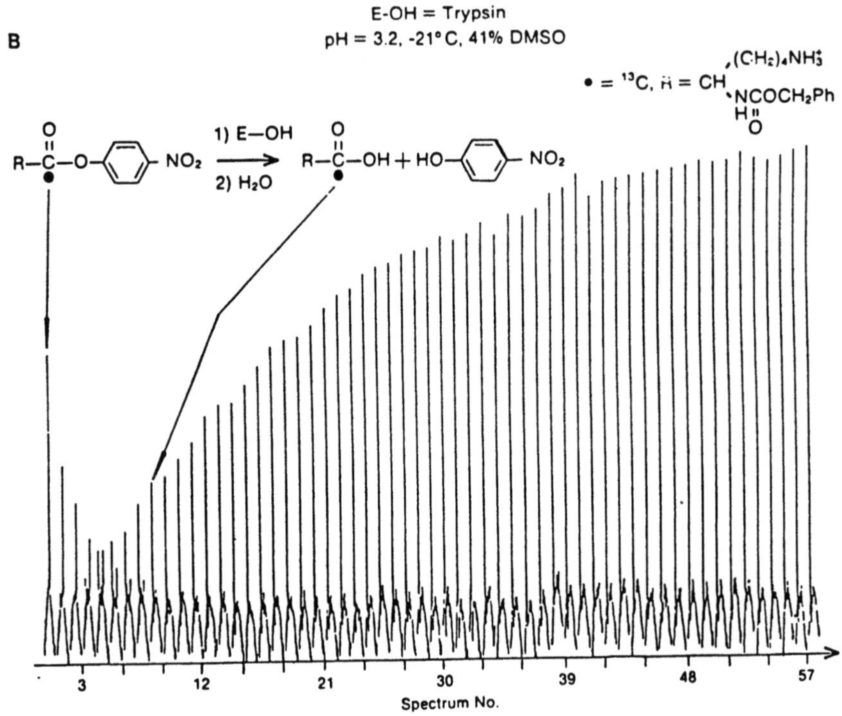

Figure 7B: Time course of hydrolysis of Z-Lys-PNP ($^{13}C = 0$) by trypsin.

intermediate whose line width is much greater than that of the free substrate or product and is therefor not directly observable in the individual ^{13}C NMR spectra of Fig. 7A. Improvement of th signal-to-noise ratio by summation of sets of the individual spectra in Fig. 7A made it possible to us difference techniques and allowed direct observation at -21°C of a resonance (d = 176.5 ppm) th: was assignable to the acyl enzyme on the basis of its chemical shift and the kinetics of it breakdown-formation. Experiments at -1.5°C (Fig. 8a) showed that the acyl intermediate was muc more readily detected at this higher temperature, the resonance being clearly resolved in the individu: spectra as a result of the smaller line width of the resonance at -1.5°C compared to that at -21°C (2 ±3) Hz and 100 ±10 Hz, respectively. This illustrates one of the main difficulties of a combine NMR-cryoenzymological approach in which the line width of enzyme-bound species increase dramatically as the cryosolvent viscosity increases on lowering the temperature. In recent studies witl chymotrypsin modified by replacement of His-57 by N-methyl His-57, stabilization of the acy intermediate was achieved at room temperature (Fig. 9).

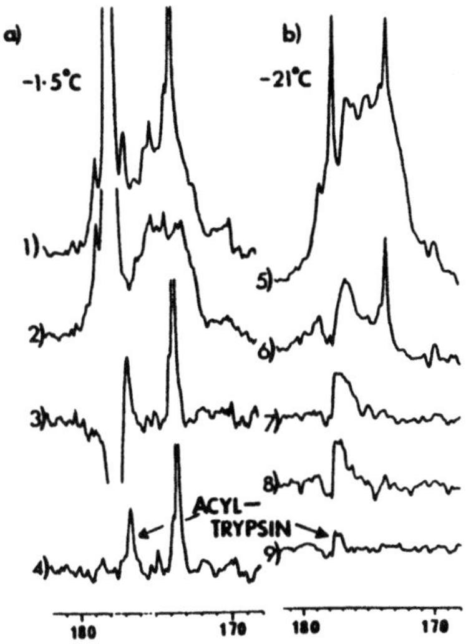

Figure 8: Difference NMR Spectra at (a) -1.5°C (b) -21°C showing signals for Acyl-trypsin.

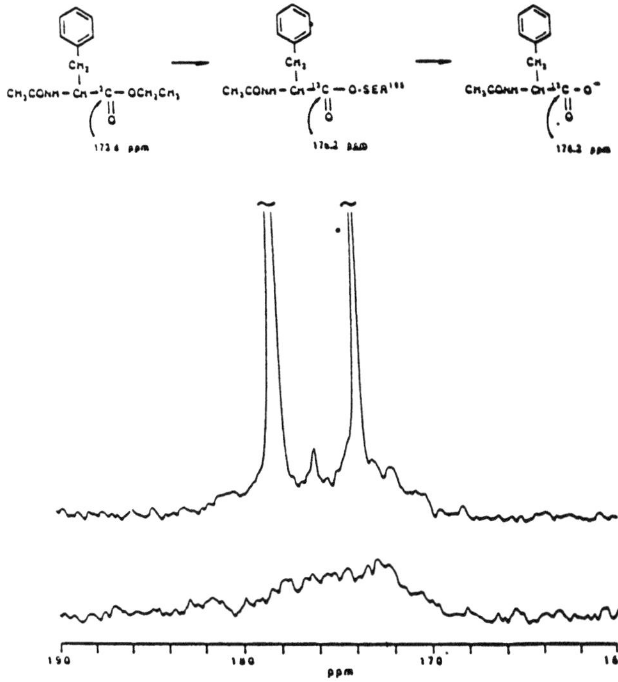

Figure 9: N-Me chymotrypsin.

Using a similar approach, the B_{12} coenzyme dependent isomerization of methyl malonyl CoA to succinyl CoA has been studied, as portrayed in Fig. 10, which shows that both substrate and product (^{13}C enriched) are observed in bound (protonated) and unbound (ionized) forms at pH 6.5 (-25°C/40% DMSO) [7]. Here the mutase enzyme has M_r = 160,000.

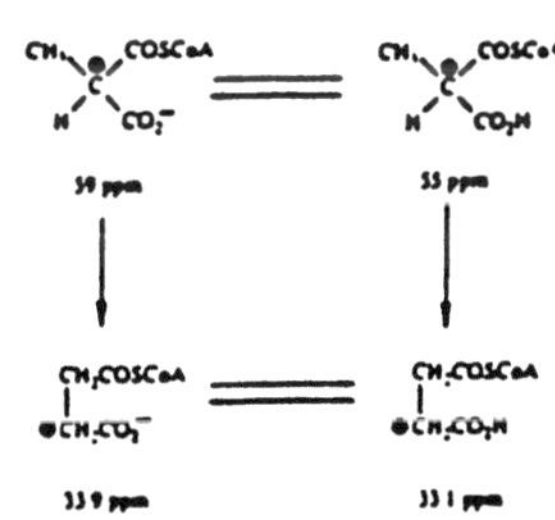

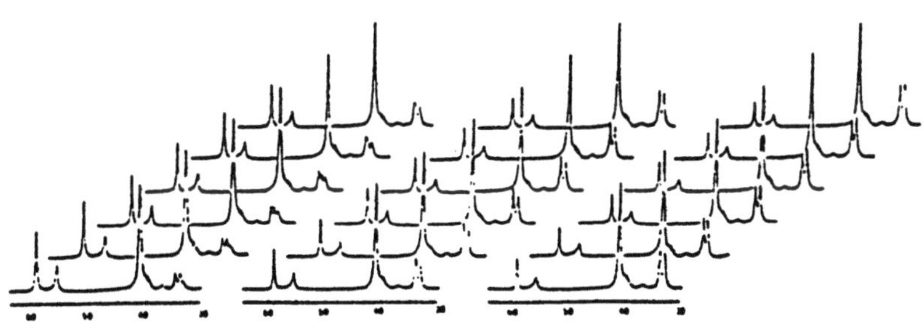

Figure 10: Time course of methyl malonyl CoA - succinyl CoA isomerization by MMCoA mutase/B_{12} coenzyme at -25°C in DMSO/Buffer.

IV. Solid State NMR as a Probe for Enzyme-Substrate Structure

The basis for the NMR studies with proteases has relied heavily on the crystal structure determination of the enzyme and/or enzyme complex at the active site. Since the phase change from solid to solution may well involve considerable conformational alteration, the correlation of the two physical methods (X-ray, NMR) in the same phase would be a desirable objective in furthering our understanding of enzyme mechanism. Our final theme concerns the application of solid state cross polarized magic angle spinning (CPMAS) NMR spectroscopy to the direct observation of the hydrolysis of a slow (pseudo) substrate catalyzed by an enzyme. Magic angle spinning (MAS) NMR has already been shown to be a powerful technique for the study of solids generally and has been used extensively in inorganic systems.

The crystal structures of several complexes of the metallo enzyme, carboxypeptidase A$^{\alpha}$ (CPA)(EC 3.4.17.1), have been examined in considerable detail. The structure of the complex with glycyl tryosine (Gly-Tyr) as been refined to 2.0 Å resolution and reveals inter alia interactions between the amide carbonyl oxygen and the catalytically essential zinc, and between the amide nitrogen and the hydroxyl of tryosine-248 (Tyr-248)(Fig. 11). The proposed mechanisms for hydrolysis of peptide and ester bonds by CPA have relied heavily on these crystal structures, but a clear distinction between the possible roles of glutamate-270 (Glu-270) in nucleophilic attack either by general base catalysis (Fig. 11A) or by covalent anyhydride formation (Fig. 11B) remains a major unresolved problem. Indeed, it is not yet certain whether esters and amides are hydrolyzed by CPA via identical mechanisms.

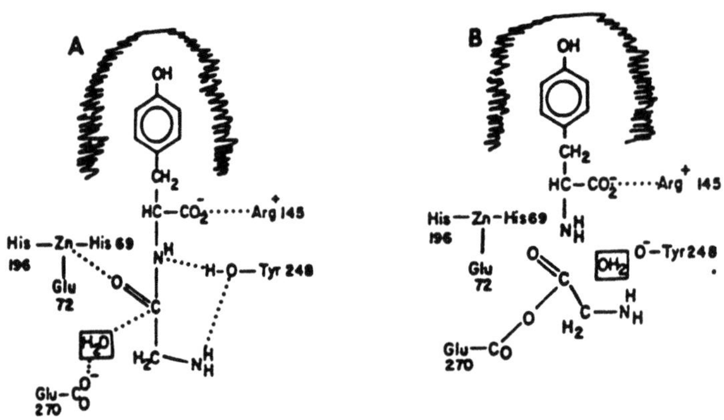

Figure 11: Two possible mechanisms (A & B) for the hydrolysis of Gly-Tyr by CPAse.

Gly-Tyr was synthesized as both the [^{13}C]-amido (90 atom % enrichment and amido-[^{13}C, ^{15}N] (90 atom % enrichment, respectively) isotopomers by known methods. CPA, purchased from Sigma as a suspension in toluene, was crosslinked and soaked with Gly-Tyr as described previously, and the washed, dried crystals examined by solid-state NMR. The ^{13}C CPMAS NMR spectrum of the ^{13}C-enriched Gly-Tyr (Fig. 12) displays residual ^{13}C-^{14}N quadrupolar coupling (δ = 176 ppm). However, the ^{13}C CPMAS difference spectrum of CPA and the CPA/Gly-Tyr complex (Fig. 12) displays a single resonance (δ = 178 ppm, $LW^{1}/_{2}$ 60 Hz) revealing that, under the experimental conditions, the peptide bond of this slow reacting substrate has been cleaved to the extent of > 90%.

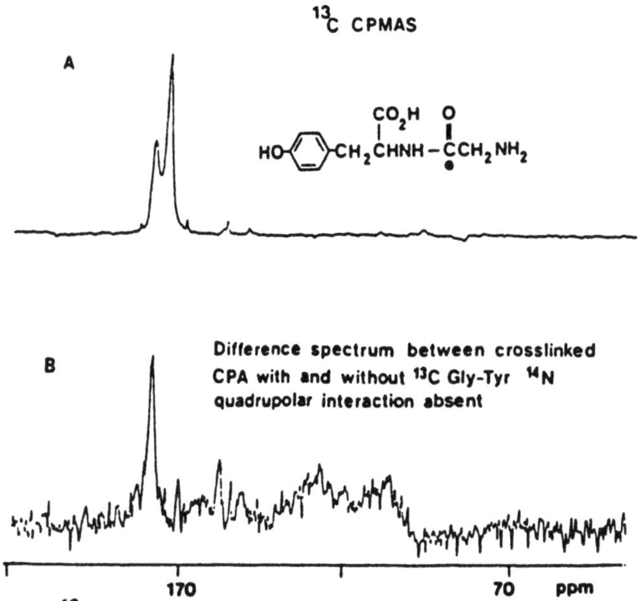

Figure 12: ^{13}C-CPMAS spectrum of Gly-Tyr showing disappearance of quadrupolar coupling (^{13}C-^{14}N) after reaction with CPAse.

In order to confirm this result, doubly labeled amido [^{13}C, ^{15}N] Gly-Tyr (not shown) was bound to CPA under identical conditions and the complex examined by ^{15}N resonance of the substrate (d 120 ppm) is sufficiently broad (LW$^{I}/_{2}$ 325 Hz) to conceal the ^{15}N-^{13}C one-bond scalar coupling (J = 16 Hz). In the enzyme complex (not shown), the observed ^{15}N resonances correspond to the amide region of CPA (δ122 LW$^{1}/_{2}$ = 400 Hz) and an amine (δ 42 ppm LW$^{1}/_{2}$ = 250 Hz), respectively. We infer from these experiments that we are observing either the first stable intermediate of the catalytic pathway, i.e. the anhyrdride of the cleaved glycine residue bound covalently to Glu-270 (Fig. 11), or the bound glycine accompanied in either base by the second product, tyrosine (Fig. 11) also bound to CPA, since in a separate experiment it could be shown that no unbound tyrosine or glycine was present in the complex. A distinction between the two glycine species (anhydride or acid) cannot be made at this stage since the ^{13}C chemical shifts of the carbonyl(s) in both the anhydride and the acid are within the observed chemical shift range. However, these preliminary experiments clearly indicate the feasibility of simultaneous X-ray and CPMAS NMR studies, expecially with slow substrates. In summary, a method of some generality for reaching the long-sought correlation between X-ray diffraction data, the resultant geometry of a substrate undergoing slow catalysis, and the corresponding NMR chemical shifts for selected atoms in the enriched substrate, obtained under identical conditions, is now available through the use of the CPMAS technique [8].

From the discussion presented above, it is clear that the stage is now set for intensive research into the mechanism of enzyme action with cryoenzymology. Although most of the enzymes discussed above lie in the molecular weight range of 20,000 to 35,000, an upper figure of 100,000 probably represents the maximum convenient size for ^{13}C NMR studies on currently available super-conducting instruments with substrates and inhibitors.

Even with high M.W. enzymes labeling coenzyme e.g. (^{13}CHO-PLP) can give excellent results. Thus the use of ^{13}C-labeled coenzyme in the reactions catalyzed by cytosolic aspartate amino transferase (EC 2.6.1.1) a dimeric protein (M_r 90,000), allows clear observation of both internal and external aldimine forms with inhibitors such as methyl aspartate as shown in Fig. 13 [9].

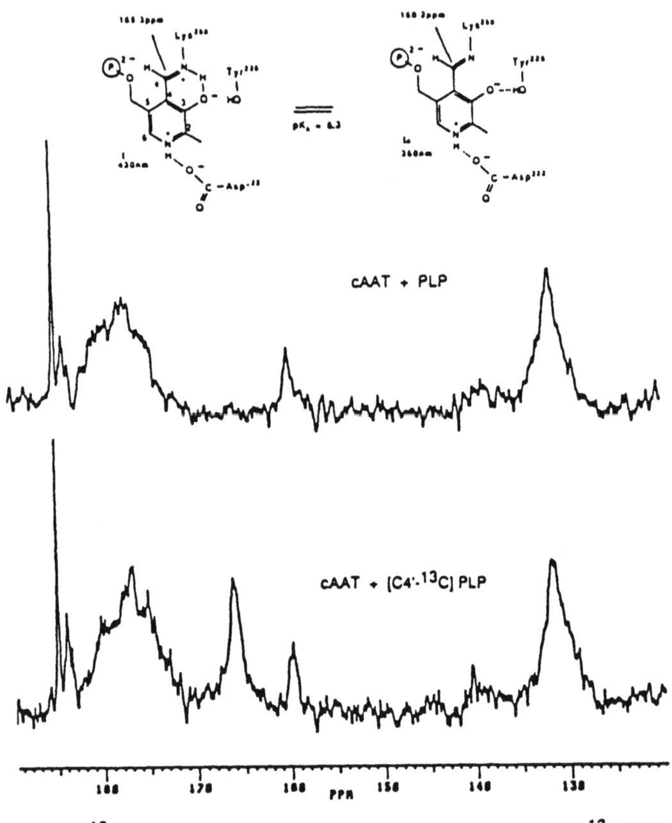

Figure 13: ^{13}C NMR spectrum of cytosolic AAT with (top) PL (bottom) [^{13}C-4] PLP.

V. <u>Concerning the Catalytic site of PBG Deaminase</u>

The availability of pure PBG deaminase has been enhanced at least 200 fold by molecular cloning experiments [10]. Purification by FPLC and concentration afforded a single enzyme (gel electrophoresis) M.W. 33,000 which on exposure to air (pH 3.5) underwent a color change (yellow → pink). Careful investigation revealed that a pyrromethene moiety is produced at the active site of the enzyme under oxidizing conditions. Activity can be restored with $NaBH_4$. That this remarkable dipyrrolic cofactor is responsible for the assembly process of PBG deaminase could readily be shown by isolation of the ^{13}C-enriched mono-PBG complex followed by NMR spectroscopy at pH 8 (better resolved at pH 12) where signals at 24.6 and 110 ppm were observed. The covalent attachment of <u>cofactor</u> to enzyme was uncovered in an NMR experiment using ^{13}C enriched deaminase grown on genetically engineered <u>*E. coli*</u> containing 5-^{13}C-ALA. In this way, all of the requisite signals for the pyrromethane including its attachment to a Cys-S-residue could be observed (Fig. 14). Most importantly, these enriched carbons are not transferred to the product of PBG deaminase, Uro'gen I, either as ^{14}C or ^{13}C isotopomers. The suggested mechanism for deaminase based on these recent results is shown in Fig. 15 [10]. This is in complete contrast to a claim [11] that a lysine residue mediates the assembly of the PBG units to HMB (see also Fig. 1).

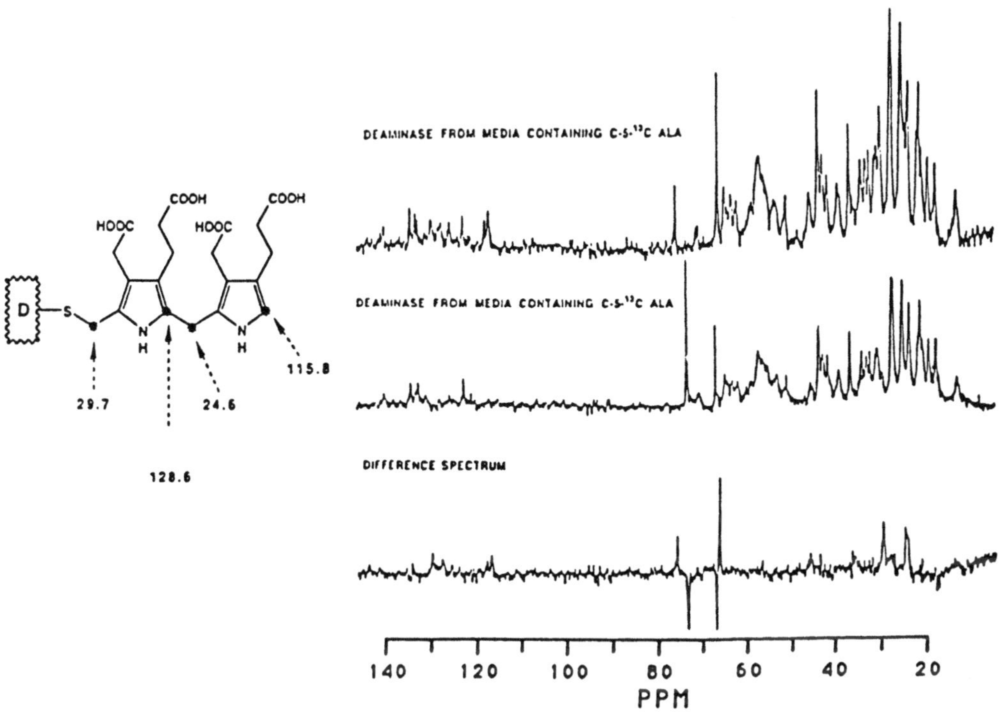

Figure 14: ^{13}C NMR spectrum (125 MHz) of enriched deaminase showing incorporation of 2 PBG units attached to a Cysteine residue in the enzyme, pH12, 20°C.

The proof that a cysteine thiol mediates the attachment of the growing oligopyrrole chain rationalizes our earlier observation that a thiol residue is responsible for the first covalent encounter of enzyme with substrate [26]. The two sets of results, far from being mutually exclusive suggest that indeed there are two sulfhydryl residues at the active site, one of which is used to guide the four moles of substrate successively onto the growing pyrrolic array. When the chain length reaches a total of six units, four are detached to form porphyrinogen, thus regenerating deaminase with the two original PBG units conserved as a dipyrromethane "arm". The processes can be viewed as a counterpart of the acetate-malonate condensation of the fatty acid synthase machinery. Similar results have just been repeated by Jordan and Warren [12]. Further experiments with this remarkable enzyme are in progress.

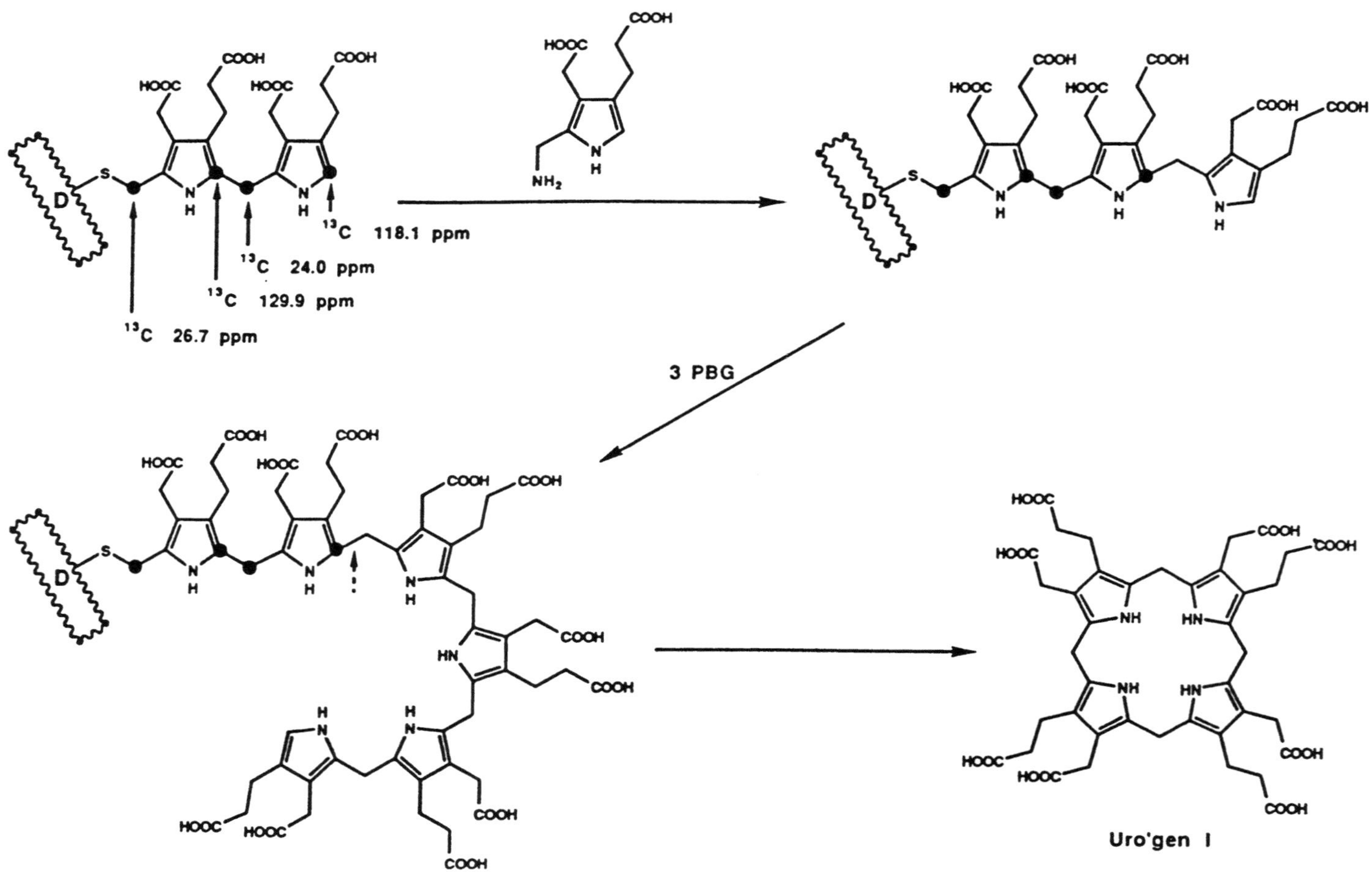

Figure 15: Postulated mechanism of action of Deaminase.

We thank the S.E.R.C. (U.K.), the National Institutes of Health, and the Robert A. Welch Foundation for generous support at Edinburgh and Texas A&M Universities.

References

1. A. I. Scott, M. Gonzalez, P. J. Santander, H. J. Williams and N. J. Stolowich, (1987), in preparation.
2. a) J. N. S. Evans, R. C. Davies, A. S. F. Boyd, I. Ichinose, N. E. Mackenzie, A. I. Scott, and R. L. Baxter (1986), Biosynthesis of Porphyrins and Corrins. I. Biochemistry 25, 896-904. b) J. N. S. Evans, G. Burton, P. E. Fagerness, N. E. Mackenzie, and A. I. Scott (1986), Biosynthesis of Porphyrins and Corrins II. Biochemistry 25, 905-912.
3. A. I. Scott, R. Shankaranarayan, G. Shrum (1987), unpublished work.
4. A. I. Scott, N. E. Mackenzie, J. P. G. Malthouse, W. V. Primrose, P. E. Fagerness, A. Brisson, Z. Q. Li, W. Bode, C. M. Carter, and I. J. Jang (1986), Inhibition of Trypsin, Tetrahedron 42, 3269-3276.
5. A. I. Scott, N. E. Mackenzie, and J. P. G. Malthouse (1984), NMR Studies of Enzyme Mechanism, Science 225, 883-889.
6. A. I. Scott, N. E. Mackenzie, and J. P. G. Malthouse (1984), Cryoenzyomology of Trypsin, Biochem. J., 219, 437-444.
7. A. I. Scott, and B. Aebischer (1987), Benzon Sypm. NMR Spectroscopy, Copenhagen 1987, to be published.
8. A. I. Scott, NMR in vitro and in vivo, Proc. R. A. Welch Foundation Conference XXVII. The 30th year of the Welch Foundation, Houston, Texas pp. 237-282, (1984).
9. A. I. Scott, N. J. Stolowich, and H. J. Williams (1987), Proc. International NMR Symposium Beijing, October 1987, (submitted for publication).
10. A. I. Scott, H. J. Williams, N. J. Stolowich, C. A. Roessner, and S. Grant, Proc. Nat. Acad. Sci., (USA), 1988, in press.
11. A. R. Battersby, C.J.R. Fookes, G. Hart, G.W.J. Matcham, and P. S. Pandey, J. Chem. Soc., Perkin Trans. I, 3041-3047.
12. P. M. Jordan and M. J. Warren, FEBS Letters, 1987, **225**, 87-92.

NMR Techniques for the Structure Elucidation and Conformational Analysis of Natural Products

GEOFFREY A. CORDELL*, GÁBOR BLASKÓ , MATTHIAS O. HAMBURGER ,
ZE-YUAN LUO , HELGA SHIEH, DAVID C. LANKIN, AND HILDEBERT WAGNER

* Program for Collaborative Research in the Pharmaceutical Sciences, College of Pharmacy, University of Illinois at Chicago, Chicago, Illinois 60612, U.S.A.

1. INTRODUCTION

For many years, we have been involved in the structure elucidation of biologically active natural products. As the availability of techniques for work in this area have changed, so has our ability to conduct detailed spectroscopic analysis on the isolated compounds. Several years ago it would not have been reasonable to determine the complete proton and carbon assignments for a complex, new natural product. The techniques currently available are such that it is now typically possible to prove all of the carbon and proton assignments, without relying on prior information or using chemical shift theory to assign carbon chemical shifts. Philosophically, this is an important distinction to make, for relatively few of the carbon-13 attributions in the literature could be regarded as proven unambiguously. Rather, reliance is usually placed either on model compounds or on additive or subtractive shifts based on the data for simple analogues. As a result, numerous proton and carbon-13 assignments must be either reconfirmed or revised, even in well-established natural products.

Advances in magnet design and technology, and the computer hardware and software for pulse programming, data acquisition and data processing have revolutionized both the quality of the available nmr spectra and the level of sophistication of the information which these data reveal. Perhaps the most important single development in the past ten years has been the ability to conduct correlation spectroscopy. Of the many variants of this technique that are available, three in particular have been of importance to the natural product chemist. These are, a) ^{1}H-^{1}H COSY in which either two bond or long range couplings may be emphasized, b) nOe COSY (NOESY) spectra in which proximate proton-proton relationships are displayed, and c) ^{1}H-^{13}C COSY (hetcor) spectra in which either one-bond or three-bond couplings can be shown.

The latter technique unfortunately has two substantial disadvantages, i) significant amounts of material (at least 30 mg) are normally required, which may be impossible for a new natural product, and ii) the assignment of carbon signals with close (δ 0.2 ppm) chemical shifts may be difficult because the digital resolution required cannot normally be achieved in a hetcor experiment. In addition, unless the hetcor experiment can be run under conditions where the J value emphasized is 4 - 8 Hz, no information concerning the assignment of quaternary carbons is possible.

Bax and his associates have succeeded in developing several innovative pulse programming sequences which are extremely useful for the assignment of complex spectra. Two of these techniques, the CSCM 1D (ref. 4) and the selective INEPT (ref. 5) have proven to be invaluable in our studies where sample size was a limiting factor (refs. 6-13). Before proceeding to some applications we will review the capabilities of these techniques, acknowledging that a well resolved, unambiguously assigned proton nmr spectrum obtained at moderate to high field (300 to 500 MHz) is of crucial importance for their success.

In the CSCM 1D technique (ref. 4) a carbon satellite either upfield or downfield of the proton is irradiated, magnetization is transfered to the attached carbon and a signal, either positive (upfield irradiation) or negative (downfield irradiation), is then observed. This might therefore be interpreted as a one dimensional technique corresponding to a ^{1}H-^{13}C hetcor experiment. The technique offers two principal advantages for the natural product chemist: sensitivity and selectivity, in part, because vast amounts of unwanted data are not acquired and because the digital resolution is the same as in the decoupled spectrum. The major limitation to the technique is that it is not useful in the instance of substantial proton overlap.

In the selective INEPT technique (ref. 5), the proton is irradiated with a soft pulse and the dwell time prior to observation is varied depending on the anticipated (or preferably known) coupling constant. Typically, the experiment is utilized for the selective enhancement of carbons three bonds away from the irradiated proton where J values in the range 6-8 Hz are used. There are two important exceptions to this range of J values. In the case of an aromatic carbon to which a methoxy group is attached, the $^3J_{CH}$ is of the order of 4 Hz, and we found that the technique was capable of picking up a three bond coupling through a heteroatom (either O or N) when they are the second or third atom in the sequence. It was this strategy which we used in work which led to a highly efficient technique for the determination of the regiochemistry on benzodioxane systems, such as coumarinolignans (ref. 6). Judicious interpretation of selective INEPT experiments corresponds to a carbon-carbon connectivity study.

In this brief review, the importance of difference nOe and selective INEPT spectroscopy in elucidating the structure of a novel naphthoquinone from Larrea tridentata and a red pigment from Dalbergia candenatensis will be presented and the powerful combination of CSCM 1D and selective INEPT spectroscopy will be used to show how the complete carbon-13 assignments of highly aromatic alkaloids, such as the protoberberines and the benzophenanthridines, have recently been established. The use of proton-proton and proton-carbon coupling constants, together with the observation of nOe effects will be discussed in relationship to a study of the conformation of savinin, whereby comparison could be made with the conformation determined through energy minimization calculations. Finally, the effectiveness of the selective INEPT technique to resolve problems with regioisomers will be demonstrated in the furanonaphthoquinone series.

2. CARBON-13 ASSIGNMENTS OF SANGUINARINE AND BERBERINE

The protoberberines and the benzo[c]phenanthridines are two of the most important and well-known classes of isoquinoline alkaloids. Recently, both groups of alkaloids have been the target of increased chemical and biological interest because some of their members have been found to possess potent anticancer activity (ref. 14). Moreover, benzo[c]phenanthridines are at the end of a long and fascinating biosynthetic pathway in which protoberberines serve as key intermediates derived directly from norcoclaurine, a universal precursor of a large number of different isoquinoline alkaloids (ref. 15). However, although several monographs (refs. 16, 17) and reviews (ref. 18) have reported detailed ^{13}C nmr studies on isoquinoline alkaloids, there are no previous ^{13}C nmr data on either berberine and related quaternary protoberberines or on any of the quaternary benzo[c]phenanthridines (ref. 19, 20). Our work (ref. 21) therefore constitutes the first unambiguous assignment of the proton and carbon-13 nmr spectra of the protoberberine alkaloid, berberine (1) and of the benzo[c]phenanthridine alkaloid sanguinarine (2), and is based on the integrated use of the CSCM 1D and the selective INEPT techniques. Initially, the ^1H nmr data of berberine (1) and sanguinarine (2) were established, and the results are summarized in Table 1.

The assignments are similar to those previously obtained at 60 MHz for berberine (1)(ref. 22) and sanguinarine (2) (ref. 23), but are now proven.

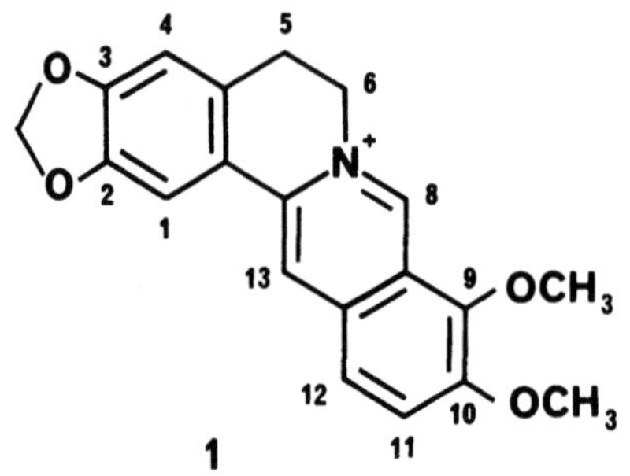

1

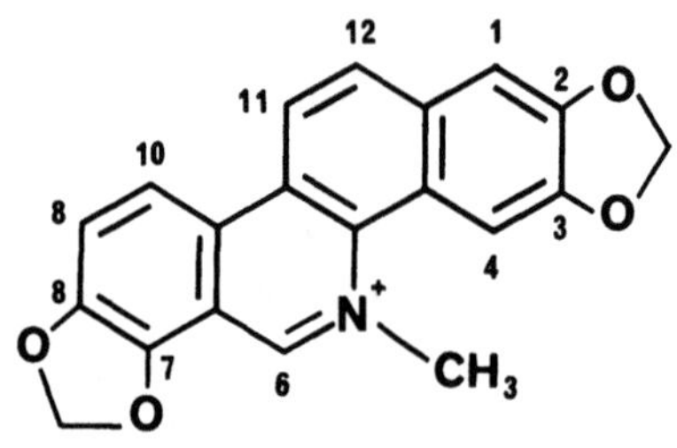

2

TABLE 1. ^{1}H NMR DATA OF BERBERINE (1) AND SANGUINARINE (2)[a]

Berberine (1)			Sanguinarine (2)		
Proton	Chemical shift		Proton	Chemical shift	
1	7.79 s		1	7.72 s	
4	7.09 s		4	8.24 s	
5	3.22 t	(5.6)	6	10.11 s	
6	4.95 t	(5.6)	9	8.70 d	(8.7)
8	9.91 s		10	8.09 d	(8.7)
11	8.20 d	(8.7)	11	8.56 d	(9.2)
12	8.01 d	(8.7)	12	8.23 d	(9.2)
13	8.96 s		5-CH_3	4.91 s	
2,3-OCH_2O	6.17 s		2,3-OCH_2O	6.35 s	
9-OCH_3	4.07 s		7,8-OCH_2O	6.61 s	
10-OCH_3	4.10 s				

[a] All spectra were recorded in DMSO-d_6 at 361.07 MHz. Chemical shifts (δ) are given in ppm using TMS as internal standard. Coupling constants are given in Hz in parentheses.

In its APT spectrum, berberine (1) shows the anticipated two methyl, three methylene, six methine and nine quaternary carbon atoms. The three methylene signals at δ 26.35, 55.19 and 102.07 were assigned to C-5, C-6 and the 2,3-methylenedioxy carbon, respectively. CSCM 1D irradiation of the ^{13}C satellites of H-1, H-4, H-8, H-11, H-12 and H-13 resulted in the magnetization transfer to their carbon atoms appearing at δ 105.43, 108.41, 145.42, 126.69, 123.53 and 120.18, respectively. The nine quaternary carbon atoms whose assignment was the focus of the study could be subdivided as

either ring-junction (δ 120.42, 121.37, 130.64, 132.94 and 137.41) or oxygen-substituted (δ 143.61, 147.63, 149.76 and 150.36).
Selective INEPT irradiation of H-8 enhanced the carbon signals at δ 55.19 (C-6), 132.94, 137.41 and 143.61, of which the latter could be assigned to C-9. The two other quaternary carbons, C-12a and C-13a, were distinguished when H-11 was irradiated, resulting in enhancements at δ 143.61 and 132.94, thereby confirming the attribution of C-9 and permitting the assignment of C-12a, respectively. Irradiation of H-12 enhanced a methine carbon signal at δ 120.18 (C-13) as well as an oxygenated quaternary carbon (δ 150.36) and a ring-junction carbon (δ 121.37), which were assigned to C-10 and C-8a, respectively. Polarization transfer from H-13 resulted in enhancement at δ 121.37 (C-8a), and δ 120.42, which should be the signal of C-13b. Irradiation of the protons in the methylenedioxy group enhanced two oxygenated quaternary carbon signals (δ 147.63 and 149.76). Which were distinguished by the following two experiments (Figure 1). Irradiation of H-1 enhanced the signals at δ 149.76, which must be C-3, at δ 137.41 (C-13a) and δ 130.64, which consequently should be assigned to C-4a. Finally, irradiation of H-4 permitted the assignment of C-2 and C-13b (δ 147.63 and 120.42, respectively). The complete assignment of the ^{13}C nmr spectrum of berberine (1) is shown in Table 2.

The benzo[c]phenanthridine alkaloid sanguinarine (2) has twenty carbon atoms, but in its ^{13}C nmr spectrum two of these overlapped. The ten quaternary carbons could be assigned as either oxygenated (δ 146.00, 147.27, 148.50, 148.52) or ring-junction (δ 109.15, 119.90, 125.28, 126.78, 131.00, 131.87) carbon atoms, respectively. The remaining carbon atoms were observed as seven methine carbons (δ 103.96, 105.54, 117.05, 118.53, 119.68, 131.00, 149.49) and a methyl group (δ 52.09). Initially, only the latter signal could be assigned unambiguously. CSCM 1D irradiation of the ^{13}C satellites of H-1, H-4, H-6, H-9 and H-10, and the two methylenedioxy groups permitted magnetization transfer to their respective carbon atoms. But it was not until the SINEPT irradiations of H-6 and the methylenedioxy protons were conducted that the proton assignments of the methylenedioxy groups could be firmly established through the attribution of C-7.

Two other signals, at δ 131.00 and 126.78, were also enhanced through polarization transfer on irradiation of H-6, and these should be C-4b and C-10a. They were distinguished when the quaternary methyl group was irradiated to enhance C-6 (δ 149.49) and C-4b (δ 131.00). Irradiation of H-10 enhanced the signals at δ 147.27, assigned to C-8, and the signals at δ 131.87 and δ 125.28. The latter two signals, C-6a and C-10b, were distinguished through the irradiation of H-12 (Figure 2). Irradiation of H-1 permitted the

Figure 1. Selective INEPT spectra of Berberine (1)

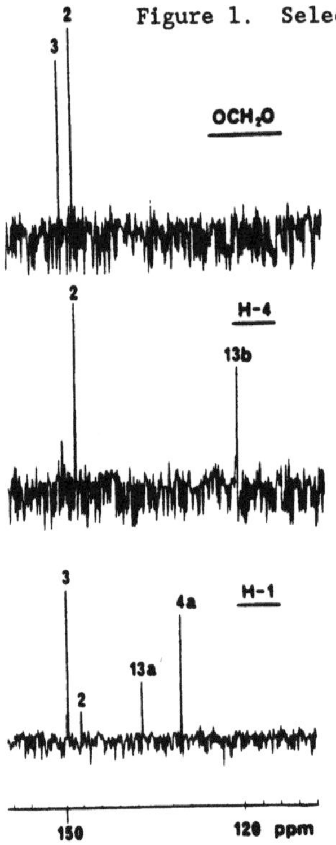

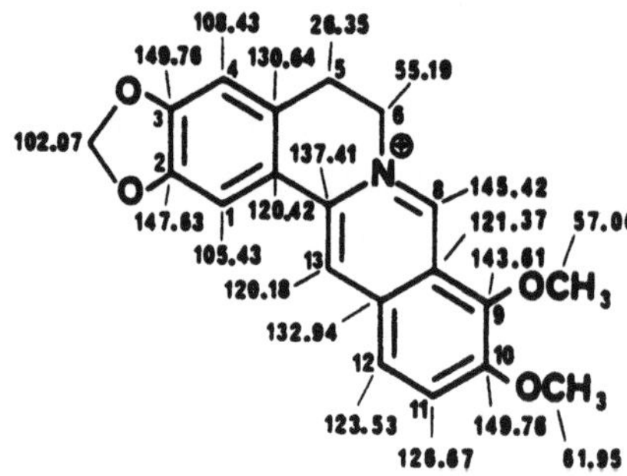

1

Figure 2. Selective INEPT spectra of Sanguinarine (2)

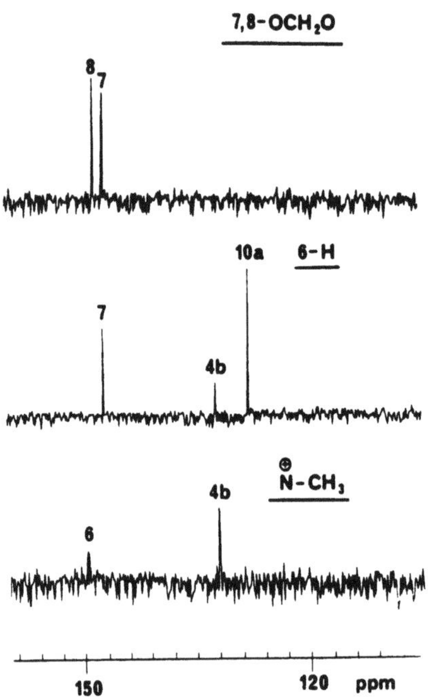

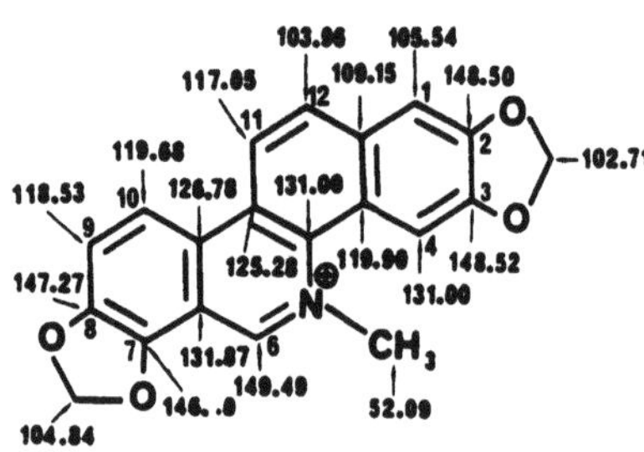

2

assignment of C-3 and C-4a (δ 148.52 and 119.90, respectively) and the remaining carbon, C-12a, was attributed through the irradiation of H-11. With this result, unambiquous assignment of all carbons of sanguinarine (**2**) was complete and is shown in Table 2.

TABLE 2. ^{13}C NMR DATA OF BERBERINE (1) AND SANGUINARINE (2)[a]

Berberine (1)		Sanguinarine (2)	
Carbon	Chemical shift	Carbon	Chemical shift
1	105.43	1	105.54
2	147.63	2	148.50
3	149.76	3	148.52
4	108.43	4	131.00
4a	130.64	4a	119.90
5	26.35	4b	131.00
6	55.19	6	149.49
8	145.42	6a	131.87
8a	121.37	7	146.00
9	143.61	8	147.27
10	150.36	9	118.53
11	126.67	10	119.68
12	123.53	10a	126.78
12a	132.94	10b	125.28
13	120.18	11	117.05
13a	137.41	12	103.96
13b	120.42	12a	109.15
2,3-OCH_2O	102.07	2,3-OCH_2O	102.71
OCH_3	57.06	7,8-OCH_2O	104.84
OCH_3	61.95	N-CH_3	52.09

[a] All spectra were recorded in DMSO-d_6 at 90.54 MHz. Chemical shifts (δ) are given in ppm using TMS as internal standard (δ_{TMS} = 0 ppm).

3. LARREANTIN

We have previously described the isolation of several new triterpenes (ref. 24) and lignans (refs. 25, 26) from the stems and leaves and the roots of the creosote bush, *Larrea tridentata* (DC) Coville (Zygophyllaceae), a plant under investigation for its fertility regulating principles. At the same time we also examined the methanol extract of the root of this plant for its cytotoxic potential in both the KB and P-388 assays (refs. 27, 28). Flash chromatography on Celite gave a cytotoxic fraction (P-388, ED_{50} 0.57 μg/ml), which was subjected to bioactivity-directed fractionation to afford larreantin (**3**) showing, in the P-388 assay, an ED_{50} value of 0.38 μg/ml.

Larreantin displayed a molecular ion at m/z 460 analyzing for $C_{27}H_{24}O_7$. Only the fragment ions at m/z 429 (M^+-31) and 137 were of any significance in the eims. The UV spectrum displayed maxima at 241.5, 259, 286 and 348 nm, and the ir spectrum showed hydroxyl group absorption at 3400 cm^{-1}, and with carbonyl bands at 1687 and 1649 cm^{-1}.

The ^{1}H-nmr spectrum indicated the presence of an aromatic methyl group (δ 2.024), a benzylic methylene (δ 4.057), three aromatic methoxy groups at δ 3.792, 3.818 and 3.837, and eight aromatic protons. Six of these protons were observed in two 1,2,4- (or 1,3,4-) trisubstituted aromatic systems, with two singlet aromatic protons at δ 6.083 and 7.983. The final two protons were observed as exchangeable phenolic protons at δ 5.697 and 5.781. The interrelationships of the aromatic protons were determined through a homonuclear COSY experiment and a COSY spectrum enhancing the long range couplings established that the three aromatic methoxyl groups were each coupled to different, single aromatic protons.

In the carbon-13 nmr spectrum all of the individual resonances were revealed and the APT spectrum permitted the observation of fourteen quaternary aromatic carbons, in addition to the protonated carbons anticipated from the ^{1}H-nmr spectrum. Two carbonyl carbons were observed at δ 185.78 and 180.44, and this information, together with an analysis of the carbon framework requirements, suggested that the central nucleus was a naphthoquinone, substituted by a methoxyl, a methyl, a 4-hydroxy-2(or 3)-methoxyphenyl and a 4-hydroxy-2(or 3)-methoxyphenylmethyl group. Crucial in defining the relative locations of the functional groups was the powerful combination of the nOe difference, CSCM 1D and the selective INEPT techniques.

The only assumption that was made, and later confirmed, was that the singlet aromatic proton at δ 7.983 was *peri* to one of the carbonyl groups. Selective INEPT irradiation of this proton with $^3J_{CH}$ = 6 Hz enhanced three quaternary carbons (δ 185.28, 143.49 and 127.09), which should be C-4, C-8a and C-7. The latter two signals could not be further assigned at this time. Irradiation of the methylene group at δ 4.057 led to the enhancement of six signals, δ 146.20, 143.49, 130.33, 127.23, 121.50 and 111.37. The signal at δ 127.23 was assigned to C-5, and the latter two to C-6" and C-2", respectively through appropriate CSCM 1D irradiations (Figure 3).

The irradiation of H-3 provided extremely valuable structure information. Under nOe conditions one of the methoxyl groups (δ 3.792) was enhanced as expected, and under selective INEPT conditions (Figure 4) it was apparent that the carbonyl carbon at δ 179.93 was also enhanced, thereby placing the methoxyl group at C-2. The COSY spectrum enhancing the long range couplings also revealed another important structure feature, namely, the coupling of the aromatic methyl group with both the benzylic methylene protons and H-5, suggesting that the methyl group was at C-7 and the benzylic group was at C-6 or C-8. A distinction between the latter two possibilities was achieved when H-5 was irradiated under selective INEPT conditions, where it was observed that the signal at δ 40.93 was also enhanced. The benzylic group is therefore at C-6.

The relationship of the hydroxy and methoxyl groups on the two phenolic rings were defined through nOe difference experiments. When the 7-CH_3 was irradiated, nOe effects were observed for the 2'- and the 6'- protons, indicating the presence of a 3,4-substituted system in ring C where the 3-substituent is a methoxyl group. For the D-ring, irradiation of either H-5 or the benzylic methylene group enhanced the <u>meta</u>-coupled protons (i.e. H-2" and H-6"), indicating that ring D has the same substitution as ring C. On this basis the methoxyl group is placed at C-2, the substituted benzyl group is placed at C-6, the methyl group at C-7 and the substituted phenyl moiety at C-8. Larreantin therefore has the structure 3.

Figure 3. CSCM 1D spectra of Larreantin (3)

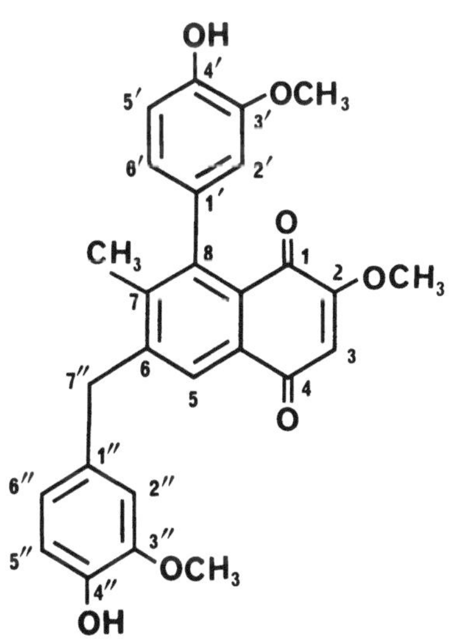

3

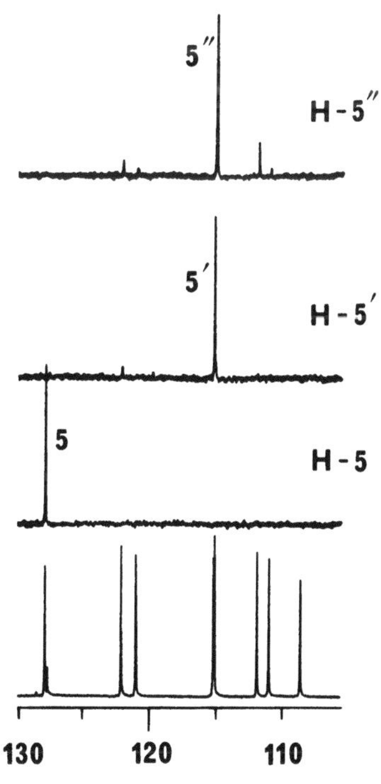

Selective INEPT experiments provided internally consistent confirmatory evidence for this structure proposal. Thus irradiation of the 7-CH_3 led to the enhancement of the quaternary carbon signals at δ 146.20, 143.57 and 143.49. Irradiation of the 2'- or 6'- protons also enhanced the resonance at δ 143.57 indicating that it should be C-8. Irradiation of H-5 enhanced the signals at δ 143.49 and 127.09 (C-8a), and consequently the signal at δ 143.49 could be attributed to C-7, whilst the signal at δ 146.20 should be C-6. The resonance for C-7 was also enhanced through the irradiation of the 7"-H_2.

All of the protonated carbons, including the three methoxyl carbons, could be assigned through the use of the CSCM 1D technique (Figure 3). But there remained several quaternary carbons in the phenyl rings to be assigned.

TABLE 3. PROTON AND CARBON-13 NMR DATA FOR LARREANTIN (3)

CARBON	δ_C	δ_H	Multiplicity, J (Hz)	
1	179.93	-		
2	160.75	-		
3	108.04	6.083	s	
4	185.28	-		
4a	130.63	-		
5	127.23	7.983	s	
6	146.20	-		
7	143.49	-		
8	143.57	-		
8a	127.09	-		
1'	132.12	-		
2'	110.46	6.539	d	2.1
3'	146.86	-		
4'	144.62	-		
5'	114.52	6.954	d	7.7
6'	120.33	6.499	dd	2.1, 7.7
1"	130.33	-		
2"	111.37	6.680	d	2.1
3"	146.74	-		
4"	144.29	-		
5"	114.59	6.839	d	7.7
6"	121.50	6.620	dd	2.1, 7.7
7"	40.35	4.057	s	
2-OCH_3	55.90	3.792	s	
3'-OCH_3	55.95	3.816	s	
3"-OCH_3	56.34	3.837	s	
7-CH_3	17.29	2.024	s	
4'-OH	-	5.781	s	
4"-OH	-	5.697	s	

[a] Recorded at 360 MHz in $CDCl_3$, δ_{TMS} = 0 ppm.
[b] Recorded at 90.54 MHz in $CDCl_3$, δ_{TMS} = 0 ppm.

Figure 4. Selective INEPT experiments on Larreantin (3) for location of substituents.

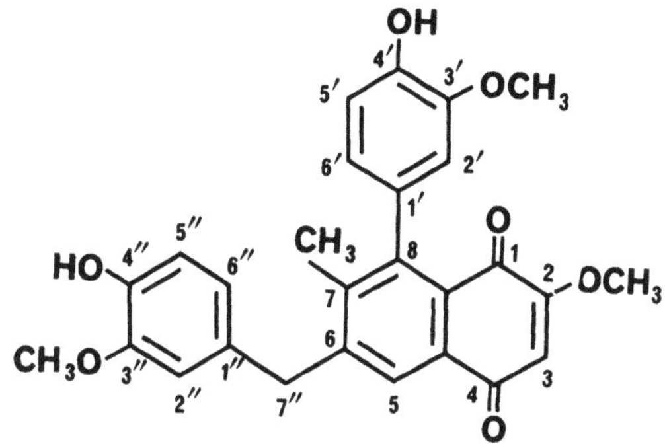

3

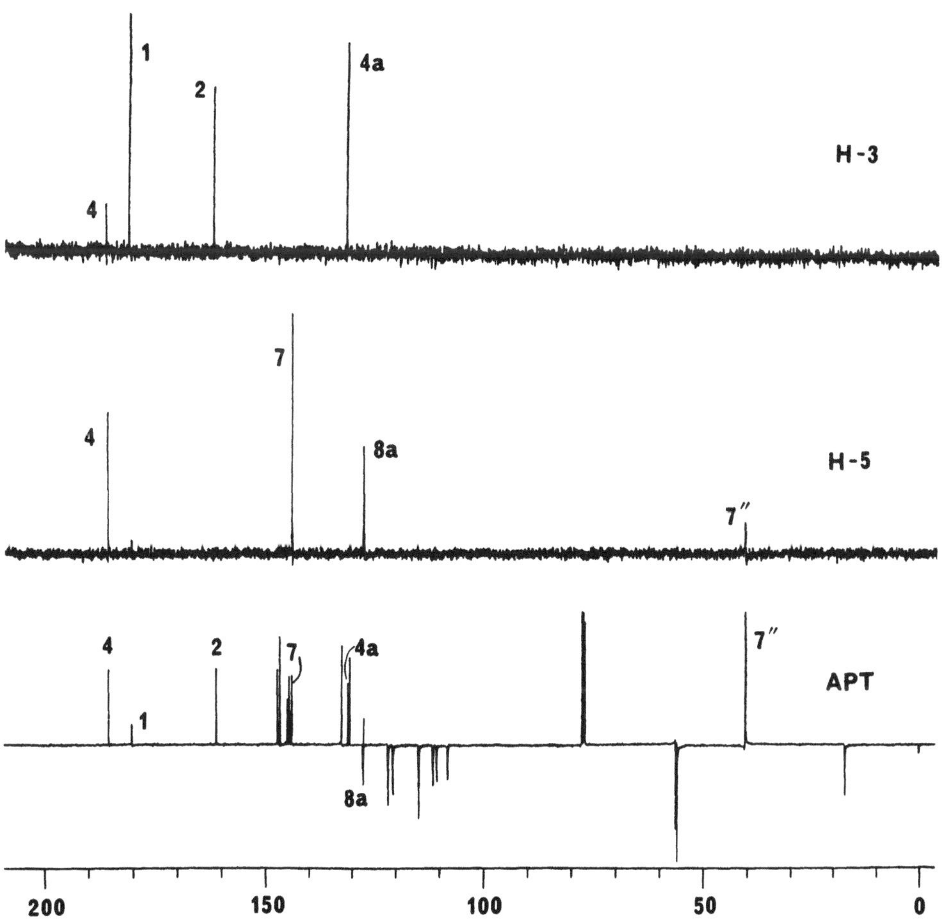

For example, the signals at δ 130.33 and 132.12 should be C-1' and C-1", and those at δ 144.29, 144.59, 146.74 and 146.86 should be C-3', C-3", C-4' and C-4". Selective INEPT irradiation of H-5" (δ 6.839) enhanced the signals at δ 130.33, 144.29 and 146.74, permitting a distinction between the two ring systems, and irradiation in the methoxyl group region led to enhancement of the signals at δ 160.75 (C-2), 146.86 and 146.74. The latter signals can therefore be assigned to C-3' and C-3", respectively, whilst those at δ 144.59 and 144.29 are C-4' and C-4", respectively. In this way all of the carbon atoms in larreantin could be attributed unambiguously (Table 3).

Larreantin represents a new class of natural product in which it would appear biogenetically that two isomeric phenylpropene units have combined with a preformed benzoquinone to afford an intermediate which can undergo oxidation to the naphthoquinone nucleus.

4. CANDENATONE

The heartwood of Dalbergia candenatensis Prain (Leguminosae) is used in Thailand as an antibacterial and also as a red dyestuff. We had previously identified the principal compounds responsible for the antimicrobial and cytotoxic activity (ref. 29) and were very intrigued by the intensely red color of the plant material. However, a literature search yielded no clues concerning the nature of the red pigments of D. candenatensis.

Preliminary TLC assays with a prepurified MeOH extract revealed a complex mixture of orange and purple pigments, all present in relatively small concentrations. Attempts to purify these compounds by column chromatography over silica gel failed. However, concerted use of liquid-liquid partitioning, followed by flash chromatography, droplet counter-current chromatography (DCCC) and gel permeation chromatography finally led to 15 mg of a pure purple pigment, to which the name candenatone (**4a**) was given.

The UV spectrum of candenatone (**4a**) showed absorption maxima at 558 nm and 527 nm, and a shoulder at 498 nm. A bathochromic shift was observed on addition of NaOMe, and a strong hypsochromic shift of both bands on addition of acid. A molecular weight of 522 daltons was deduced from the D/CI MS (m/z 523 $(M+H)^+$), corresponding to the molecular formula $C_{32}H_{26}O_7$ as determined by high resolution FAB MS.

During preliminary ^1H-NMR experiments, a significant change in the aromatic region was observed within a few hours after the dissolution of the sample, resulting in an increasingly complex spectrum. Addition of DCl to the sample also simplified and stabilized the spectrum, and consequently most of the ^{1}H-NMR studies were conducted in acidic media. The coupling pattern of the five aliphatic protons and the long-range coupling of the benzylic methylene group to an aromatic proton in the peri-position were typical for isoflavans with the dihydropyran ring in a half-chair conformation. At that point it became through biogenetic analysis that the pigment was an isoflavan with an additional C_{15} unit attached.

In the aromatic region of the spectrum obtained in the presence of acid, the signals for 10 protons belonging to three aromatic rings were observed as a p-substituted ring and two 1,2,4-tri- substituted rings. The substituents were identified as two methoxy groups (δ 3.68 and 4.02) and three hydroxyls appearing at δ 9.88, 10.77 and 11.62. In the COSY spectrum, the downfield

Figure 5. Delayed COSY spectrum of Candenatone (**4a**)

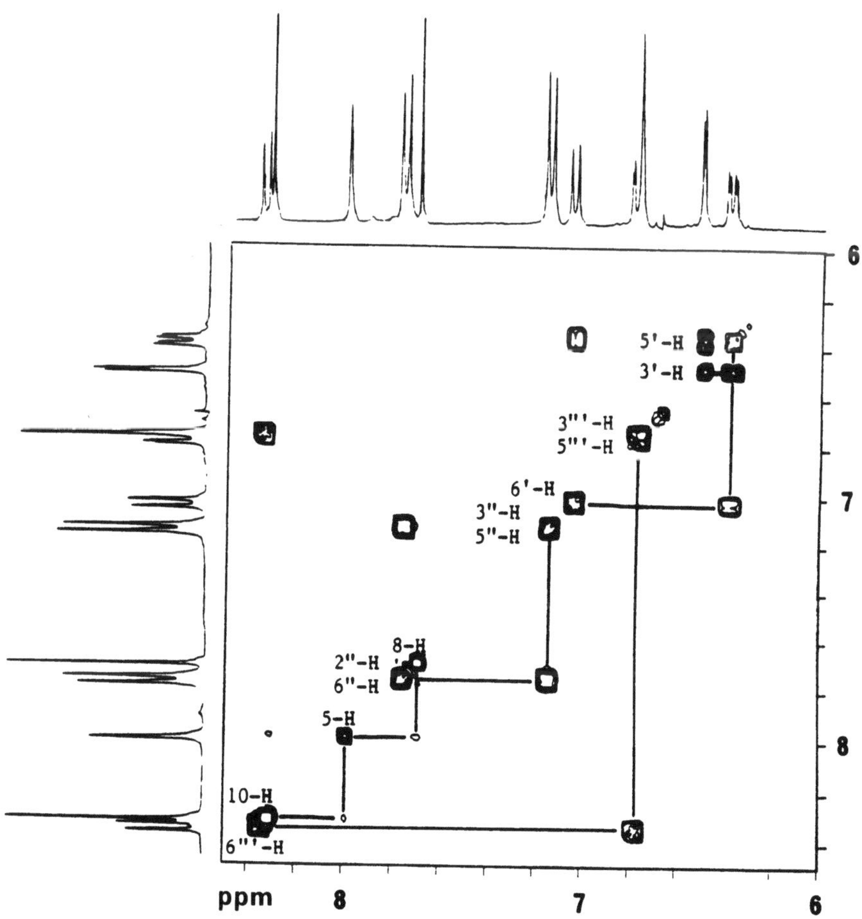

methoxy singlet displayed a cross-peak with the signal at δ 5.77, which was therefore attached at the 2- position of the ring. In a delayed COSY experiment emphasizing long-range coupling, the methoxy singlet at δ 3.68 displayed cross peaks with the signals at δ 6.38 and 6.49, placing it at the 4-position on the second trisubstituted aromatic ring. The singlet at δ 8.01 exhibited cross peaks with both singlets at δ 7.73 and 8.35, but no long-range coupling was observed between the latter two signals (Figure 5).

The ^{1}H NMR spectrum of a freshly prepared sample of candenatone **4a** in DMSO-$\underline{d}_6$ (Figure 6) showed some striking differences when compared with the spectra recorded in acidic solution. With the exception of H-3′, H-5′, H-6′ and MeO-4′, all of the other aromatic protons appeared significantly upfield, and from the coupling constants a quinone-type ring system was suggested. Furthermore, only two phenolic protons were observed instead of three for candenatone in acidic solution, leading to the conclusion that, i) all the

Figure 6. ^{1}H-NMR spectra of Candenantone (**4a**)

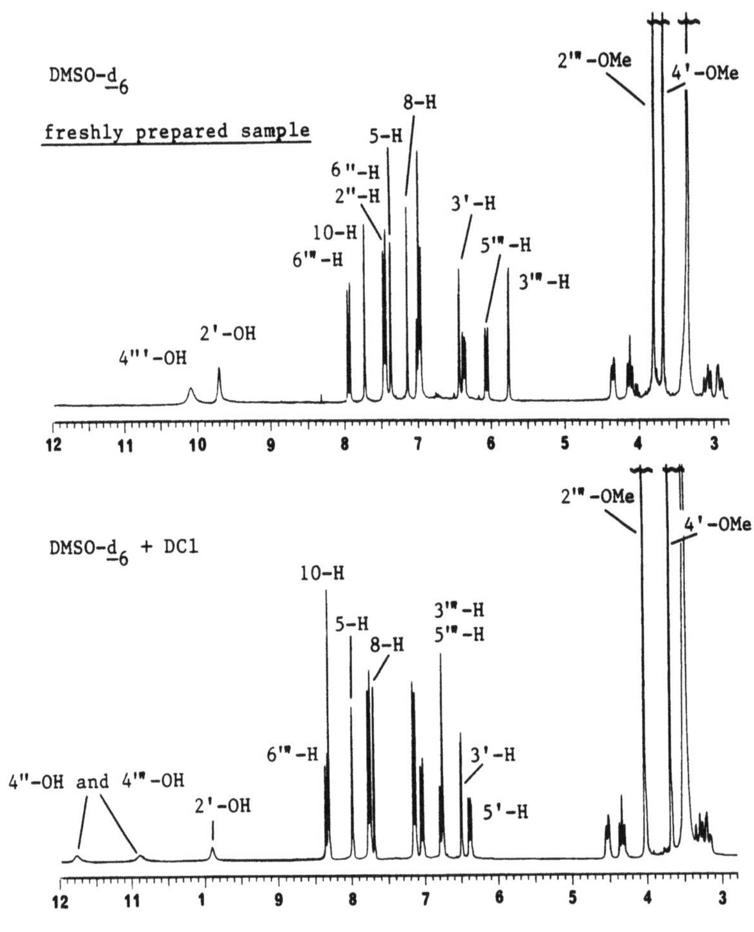

aromatic protons with the exception of H-3′, H-5′, H-6′ and MeO-4′ had to be located in the chromophore of the molecule, and ii) the quinone system observed in neutral solution became aromatic in acidic media. This conclusion was supported by the ^{13}C NMR data of **4a** and 5. A carbonyl carbon was observed at δ 183.08 for **4a**, but was missing in the spectrum of 5. The single carbonyl signal, the trisubstituted 6-membered ring and a carbonyl band at 1635 cm^{-1} in the IR spectrum were only compatible with the partial structure of a p-quinone methide bearing a methoxy group in the 2-position.

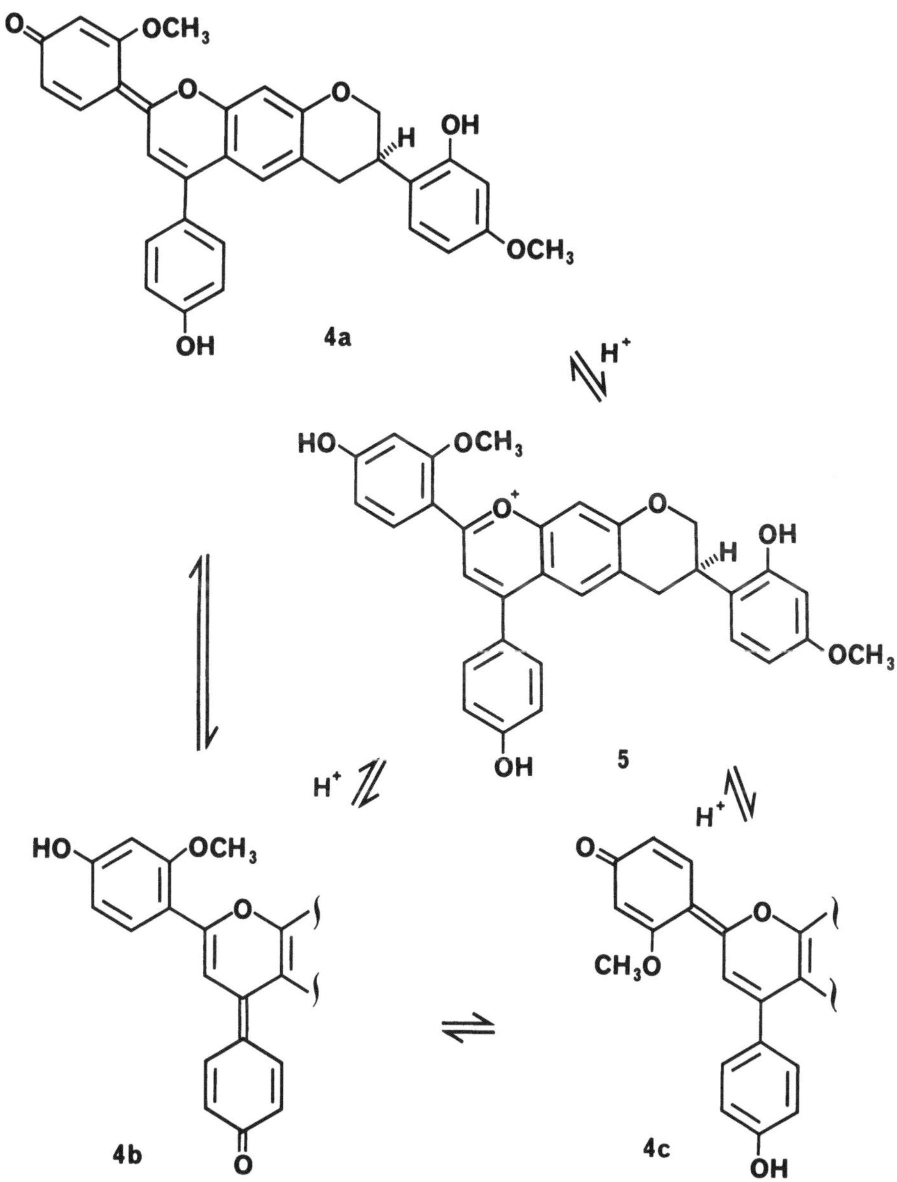

The partial structure of the chromophore was substantiated by nOe difference spectroscopy under acidic conditions, which also revealed the location of the three aromatic rings in 5. Irradiation of H-3 led to a negative nOe for H-6′ (ref. 30), whereas irradiation of H-5 led to the enhancement of H-2″ and H-6″ (Figure 7a). The p-substituted ring was therefore attached to C-6.

Figure 7. NOe measurements of Candenatone (4a)

7a

7b

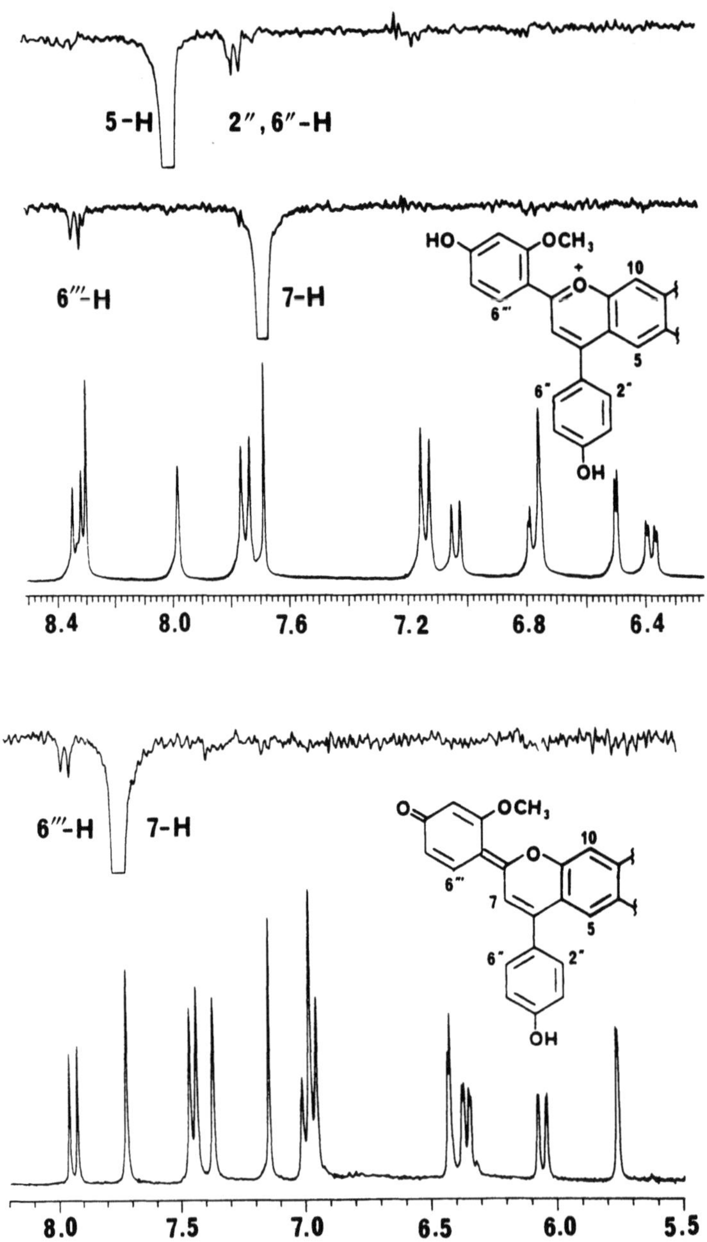

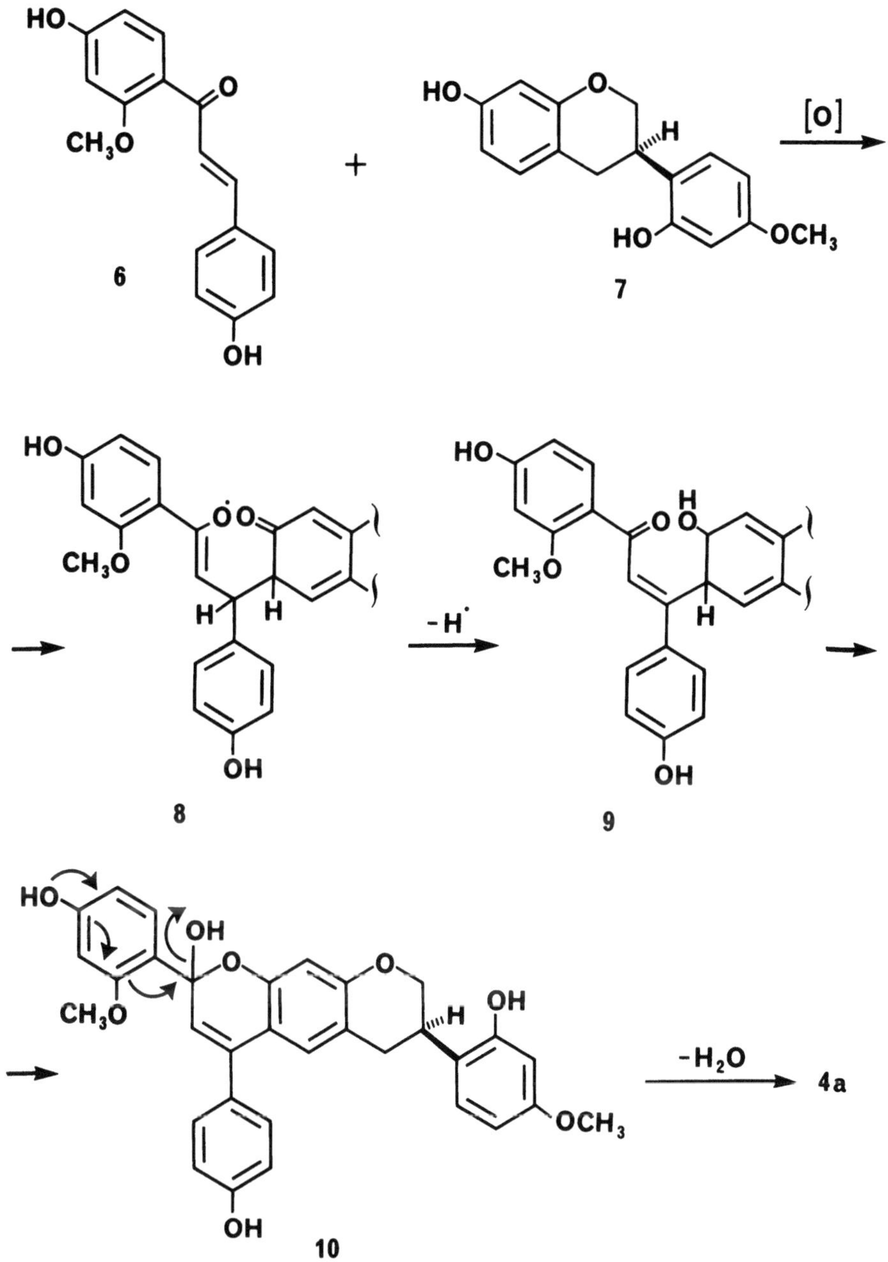

Scheme 1. Suggested biogenesis of candenatone (4a)

For the establishment of the stereochemistry of the quinone substituent in **4a**, an nOe difference experiment was performed on a freshly prepared sample of candenatone. Irradiation of H-7 produced an enhancement of H-6" (Figure 7b) which is only compatible with the tautomer **4a**. The structure of the other major tautomer **4b** present in DMSO-$\underline{d}_6$ was investigated through the ^{1}H NMR spectrum of an equilibrated sample. Comparison of the CD spectra of a partially saturated derivative of candenatone and (-)-vestitol (**7**) (ref. 31) led to the conclusion that the absolute configuration of **4a** at C-3 is R.

Candenatone **4a** represents the first example of a new type of fused isoflavan. A biogenesis of the compound is outlined in Scheme 1 in which a radical coupling of the chalcone **6** and the preformed isoflavan vestitol (**7**) could lead to the radical species **8**. Elimination of a H radical and tautomerization would give the intermediate **9**. Candenatone would be obtained through cyclization to **10** and subsequent elimination of water.

5. CARBON-13 ASSIGNMENTS AND CONFORMATION OF SAVININ

The conformational analysis of small flexible molecules with biological activity is presently an area of great interest. For our initial studies in this area we chose savinin (**11**), an α-arylidene-γ-lactone lignan found in numerous plant extracts (refs. 32-36), and isolated by us as a cytotoxic constituent of <u>Aristolochia</u> <u>indica</u> roots (ref. 37). The present study of savinin was initiated to i) establish the proton and carbon-13 nmr assignments, and ii) provide information regarding the solution conformation. In conducting this study a broad combination of complimentary 1D and 2D nmr techniques, including a variety of homonuclear shift correlations (COSY, Relayed COSY), two dimensional nuclear Overhauser experiments (NOESY), heteronuclear shift correlation (hetcor) emphasizing both one-bond and long-range coupling were employed. As a result complete and unambiguous

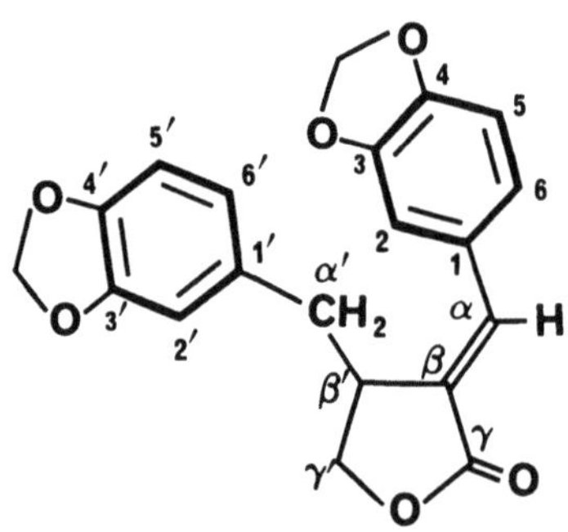

11

assignments of both the carbon and proton chemical shifts of savinin including several revisions to our previous assignments have been achieved. The NOESY spectrum, using a mix time of 3 sec (Figure 8) revealed that the α'-protons were not proximate to the β'- or γ'-protons. But the β'-H was proximate to the 2- and 6-protons, and the 2-H also showed a nOe cross peak with the 2'-H (Figure 9). Since the γ'-proton was not coupled to one of the α'-H these can be attributed stereotopically. A weak nOe was also observed between the methylenedioxy protons but this still has not yet afforded a precise conformation of savinin in solution. Our next step will be to conduct quantitative nOe experiments in order to establish interproton distance ranges and then compare those data with the results of molecular force field calculations which have led to the energy-minimzed conformation shown in 12.

Figure 8. NOESY spectra of Savinin (11)

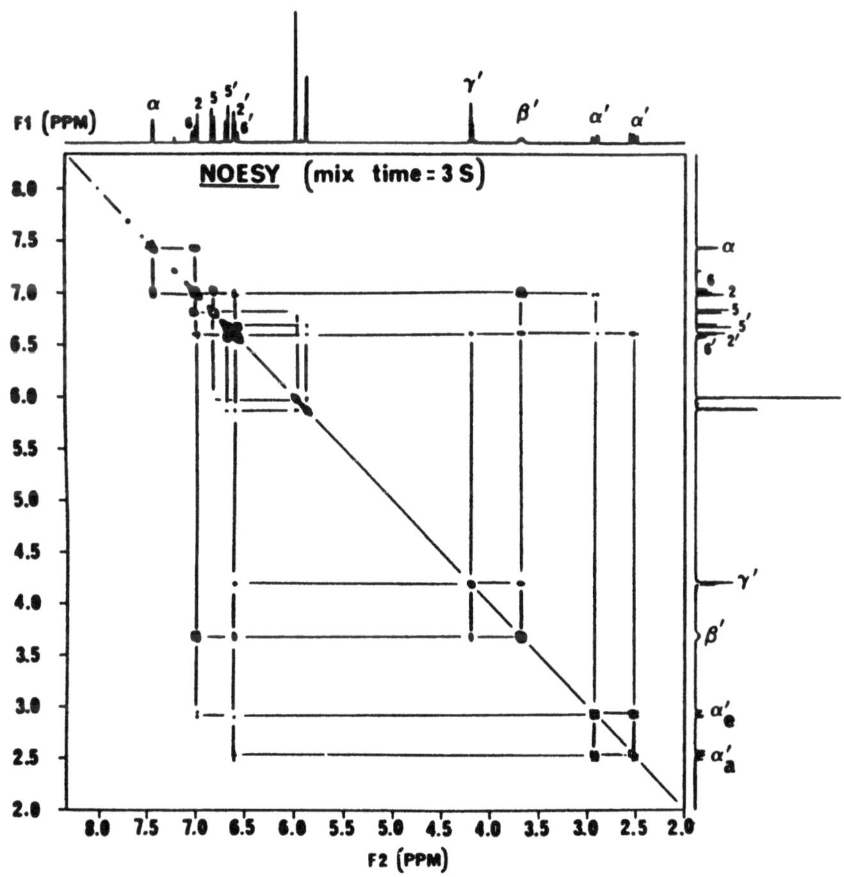

Figure 9. Enlarged NOESY spectra of Savinin (11)

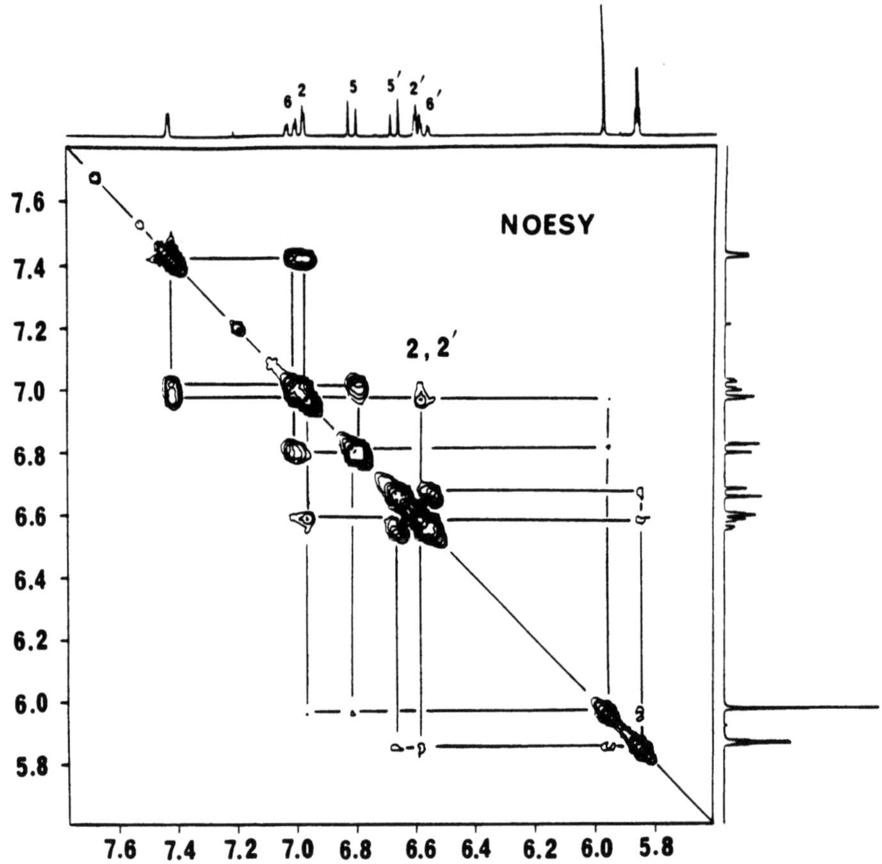

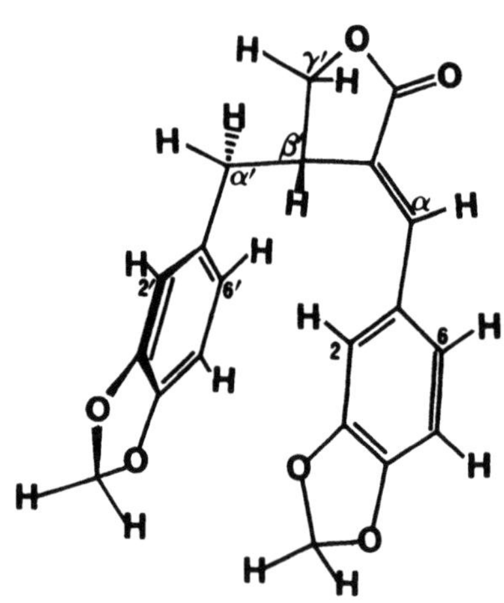

12

6. A FURANONAPHTHOQUINONE STRUCTURE THROUGH SELECTIVE INEPT SPECTROSCOPY

Finally, we would like to discuss what can be regarded as a classic example of a structure elucidation where the selective INEPT technique provided rapid resolution to an otherwise difficult problem (ref. 38). Wagner and co-workers (ref. 39) isolated a mixture of the furanonaphthoquinones **13** and **14** from Tabebuia avellandeae (Bignoniaceae), and succeeded in separating one of the isomers by HPLC only with great difficulty. It was felt that a distinction between the two possible structures could be made on the basis of the irradiation of the furan 3-proton and the aromatic proton *peri* to the carbonyl group. The theory was that in isomer **13** irradiation of these two protons would lead to the same carbonyl carbons being enhanced, whereas in isomer **14** two different carbonyl carbons would be enhanced. In practice the experiment proved somewhat more difficult than anticipated because only 4.5 mg of the compound was available and because although the coupling constant between the *peri* proton and the carbonyl was correctly estimated as about 4 Hz, we were surprised to find that the coupling constant between the furan 3-proton and the carbonyl carbon was only 1.5 Hz (ref. 40).

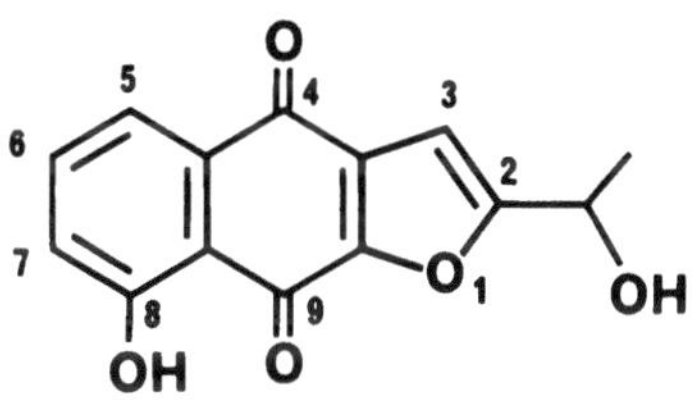

13

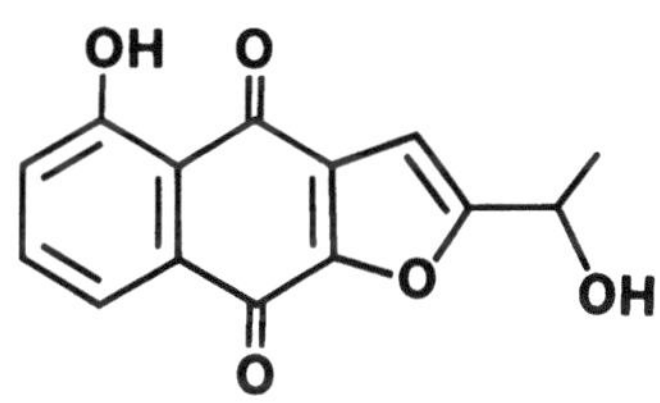

14

Figure 10. SINEPT irradiations and complete ^{13}C NMR assignments of Furanonaphthoquinone (**14**).

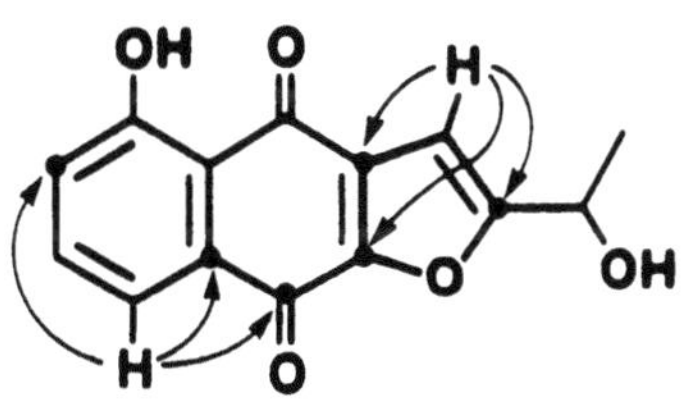

SINEPT experiments

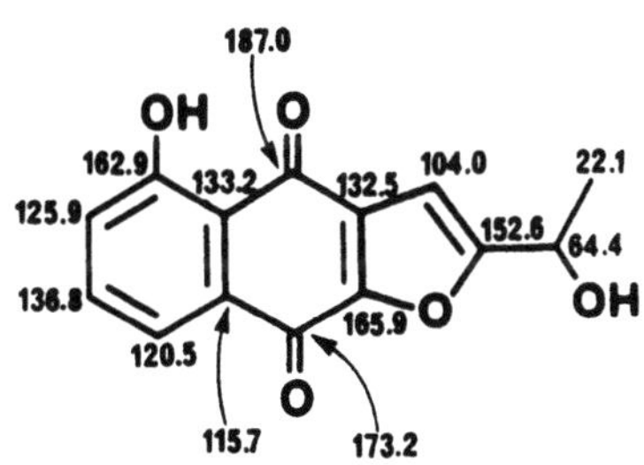

14

Thus initially we estimated the ^{1}H-^{13}C coupling constant between H-3 and the carbonyl to be of the order 6 Hz. However, this gave rise to enhancement of C-9a and when the delay time was adjusted to correspond to a $\underline{J}$ value of 4 Hz, C-3a was enhanced. A similar result was obtained using $\underline{J}$ = 3 Hz. It was only when the delay was set for $\underline{J}$ = 1.5 Hz that the carbonyl group at δ 187.01 was selectively enhanced. In the case of the irradiation of H-5(8), the delay was set for $\underline{J}$ = 7 Hz with the result that three carbons were enhanced, C-6, C-8a and C-9.

The results of the irradiation experiments indicated that the two protons were three-bond coupled to two different carbonyl groups. Consequently, the pure isolate has the structure **14** (Figure 10). During the course of this work, the complete carbon-13 assignments of the isolate were also determined through the concurrent use of the CSCM 1D technique. Like our earlier work with the coumarinolignans (ref. 6), the selective INEPT technique provides a facile solution to what would otherwise be a difficult problem without resorting to synthesis. In addition, the combination with the CSCM1D technique permits unambiguous carbon assignment when sample size is limited.

7. SUMMARY

In this presentation we have summarized our recent work on the carbon-13 nmr assignments of quaternary isoquinoline alkaloids, e.g. berberine (**1**) and sanguinarine (**2**), and described the structure elucidation of the first representatives of two new classes of natural product; larreantin (**3**), a highly substituted naphthoquinone **14**, and candenatone (**4a**), a new, intensely colored type of fused isoflavan. We have also shown how the use of a broad range of new NMR techniques can lead to an analysis of the conformation of molecular in solution and how that information can be related to computer generated, energy minimized structures. Finally, the powerful combination of the selective INEPT and the CSCM 1D techniques was used to show how regioisomers may be distinguished and complete carbon assignments made when sample availability is limited. The importance of the integration of information from the difference nOe, CSCM 1D and selective INEPT techniques in the determination of new natural product structures has been demonstrated.

8. ACKNOWLEDGEMENTS

This work was supported, in part, by a grant from the Division of Cancer of Treatment, National Cancer Institute, Bethesda, MD. One of us (M.O.H.) was the recipient of a grant from the Swiss National Science Foundation.

9. REFERENCES

1. On leave from the Central Research Institute for Chemistry of the Hungarian Academy of Sciences, Budapest, Hungary.
2. Present Address: Institut de Pharmacognosie et Phytocimie, Universite de Lausanne, CH-1005 Lausanne, Switzerland.
3. Present Address: Sichuan Institute of Chinese Materia Medica, Chongqing, People's Republic of China.
4. S.K. Sarkar and A. Bax, A simple and sensitive one-dimensional NMR technique for correlation of proton and carbon chemical shifts, J. Magn. Reson., 62 (1985) 109-112.
5. A. Bax, Structure determination and spectral assignment by pulsed polarization transfer via long range ^{1}H-^{13}C couplings, J. Magn. Reson., 57 (1984) 314-318.
6. L.-J. Lin and G.A. Cordell, Application of the SINEPT pulse programme in the strucutre elucidation of coumarinolignans, J. Chem. Soc., Chem. Commun., (1986) 377-379.
7. Y. Schun and G.A. Cordell, 21-Oxogelsevirine, a new indole alkaloid from Gelsemium rankinii, J. Nat. Prod., 49 (1986) 483-487.
8. Y. Schun and G.A. Cordell, Rankinidine, a new oxindole alkloid from Gelsemium rankinii, J. Nat. Prod., 49 (1986) 806-808.
9. Y. Schun and G.A. Cordell, Cytotoxic Steroids of Gelsemium sempervirens, J. Nat. Prod., 50 (1987) 195-198.
10. L.-J. Lin, D. Meksuriyen, G.A. Cordell, W.S. Woo, and C.K. Lee, The structure of americanin A, Kor. J. Pharmacog., 18 (1987) 94-98.
11. L. Xun, G. Blaskó and G.A. Cordell, 2'-Hydroxyflavone from Daphnopsis sellowiana, J. Nat. Prod., in press.
12. D. Meksuriyen, G.A. Cordell, N. Ruangrungsi and P. Tantivatana, Traditional medicinal plants of Thailand IX. 11-Hydroxy-10-methoxy-dracaenone and 7,11-dihydroxy-10-methoxy-dracaenone, J. Nat. Prod., in press.
13. D. Meksuriyen, L.-J. Lin, G.A. Cordell, S. Mukhopadhyay, and S.K. Banerjee, Studies of colchicine and its photoisomers, β- and γ-lumicolchicines, J. Nat. Prod., in press.
14. M. Suffness and G.A. Cordell, Antitumor Alkaloids, in: A. Brossi (Ed.), The Alkaloids, Vol. XXV, Academic Press, New York, N.Y., 1985, p. 178 and p. 188.
15. R. Stadler, T.M. Kutchan, S. Loeffler, N. Nagakura, B. Cassels and M.H. Zenk, Revision of the early steps of reticuline biosynthesis, Tetrahedron Lett., 28 (1987) 1251-1254.
16. D.W. Hughes and D.B. MacLean, in: R.G.A. Rodrigo (Ed.), The Alkaloids, Vol. XVIII, Academic Press, New York, N.Y., 1981, p. 217-262.
17. M. Shamma and D.M. Hindenlang, Carbon-13 NMR Shift Assignments of Amines and Alkaloids, Plenum Press, New York, N.Y., 1979.
18. T.A. Broadbent and E.G. Paul, Heterocycles, 20 (1983) 863-980.
19. V. Simanek, in: A. Brossi (Ed.),The Alkaloids, Vol. 26, Academic Press, New York, N.Y., 1985, p. 185-240.
20. B.D. Krane, M.O. Fagbule, M. Shamma and B. Gozler, The benzophenanthridine alkaloids, J. Nat. Prod., 47 (1984) 1-43.
21. G. Blaskó, G.A. Cordell, S. Bhamarapravati, and C.W.W. Beecher, Carbon-13 NMR assignments of berberine and sanguinarine, Heterocycles, submitted for publication.
22. V. Preininger, L. Hruban, V. Simanek and F. Santavy, Colln. Czech. Chem. Commun., 35 (1970) 124-135.
23. R.A. Labriola, A.M. Kuck and J. Comin, Anales Asoc. Quim. Argentina, 54 (1966) 29-36.
24. C. Konno, H.-Z. Xue, Z.-Z. Lu, D.D. Soejarto, G.A. Cordell, H.H.S. Fong and W. Hodgson, 3β-(3,4-Dihydroxy-cinnamoyl)-erythrodiol and 3β-(4-hydroxy-cinnamoyl)-erythrodiol from Larrea tridentata, Phytochemistry, In press.
25. C. Konno, H.-Z. Xue, Z.-Z. Lu, B.-X. Ma, C.A.J. Erdelmeier, G.A. Cordell, D.D.Soejarto, D.P. Waller, and H.H.S. Fong, New 1-aryl tetralin

lignans from Larrea tridentata, J. Nat. Prod., Submitted for publication.

26. C. Konno, H.-Z. Xue, Z.-Z. Lu, B.-X. Ma, C.A.J. Erdelmeier, G.A. Cordell, D.D. Soejarto, D.P. Waller, and H.H.S. Fong, H.H.S. Unpublished results, presented at the 28th Annual Meeting of the American Society of Pharmacognosy, University of Rhode Island, Kingston, RI, July 1987.
27. R.I. Geran, N.H. Greenberg, M.M. McDonald, A.M. Schumacher and B.J. Abbott, Protocols for screening chemical agents and natural products against animal tumors and other biological systems, Cancer Chemother. Rep., 3(2) (1972) 1-93.
28. M. Arisawa, J.M. Pezzuto, C. Bevelle and G.A.Cordell, Plant anticancer agents XXXI. N-Demethylation of fagaronine, J. Nat. Prod., 47 (1984) 453-458.
29. M.O. Hamburger, G.A. Cordell, P. Tantivatana and N. Ruangrungsi, Traditional medicinal plants of Thailand VIII. Isoflavonoids of Dalbergia candenatensis, J. Nat. Prod., 50, (1987) 696-699.
30. All of the nOe's observed for **4a** and **5** were negative and comparatively weak. NOE difference experiments with the simple isoflavan vestitol **7** under identical conditions showed positive enhancement. The negative sign observed for **4** and **5** may be explained by the slow tumbling rate of the molecule at room temperature due to the high viscosity of DMSO-d_6. NOE experiments were not performed at higher temperature, because of the eventual degradation of **5** and the faster tautomerization of **4a**.
31. K. Kurosawa, W.D. Ollis, B.T. Redman, I.O. Sutherland, O.R. Gottlieb and H.M. Alves, The absolute configurations of the animal metabolite, equol, three naturally occurring isoflavans, and one natural isoflavanquinone, J. Chem. Soc., Chem. Commun., (1968) 1265-1267.
32. J. Reisch, I. Novak, K. Szendrei and E. Minker, Savinin (hibalactone) ein lignan-derivat aus den Wurzeln von Ruta graveolens, Die Pharmazie 22 (1967) 220.
33. J. Reisch, K. Szendrei, I. Novak, and E. Minker, Die Lignane der Wurzeln von Ruta graveolens: Helioxanthin, Die Pharmazie, 25 (1970) 435-436.
34. N. Ozaki, S. Hasegawa, and Y. Hirose, Terpenoids from the seed of Chamaecyparis obtusa, Phytochemistry, 22 (1983) 1771-1773.
35. J.E. Batterbee, R.S. Burden, L. Crombie, and D.A. Whiting, Chemistry and synthesis of the lignan (-)-cubebin, J. Chem. Soc., Sec. C, (1969) 2470-2477.
36. M. Kuhn and A. von Wartburg, Podophyllum-Lignane: Structur und Absolutkonfiguration von Podorhizol-β-D-glucoside (=Lignan F), Helv. Chim. Acta, 50 (1967) 1546-2477.
37. C.-T. Che, M.S. Ahmed, S.S. Kang, D.P. Waller, A.S. Bingel, A. Martin, P. Rajamehendran, N. Bunypraphatsara, G.A.Cordell D.D. Soejarto, R.O.B. Wijesekera and H.H.S. Fong, Studies on Aristolochia III. Isolation and biological evaluation of constituents from Aristolochia indica roots for potential fertility-regulating activity, J. Nat. Prod., 47 (1984) 331-341.
38. H. Inouye, S. Ueda, K. Inoue, H. Nayeshiro and N. Moritome, Biosynthesis of Prenylnaphthoquinones in cultured cells of Streptocarpus dunnii and Tabebuia argentea, Plant Tissue Culture, (1982) 375-376.
39. H. Wagner, B. Kreher and O. Seligmann, unpublished results.
40. R. Benassi, U. Folli, A. Mucci, L. Schnetti and F. Taddei, Long-range ^{13}C-^{1}H spin-spin coupling constants in the conformational analysis of formyl derivatives of furan and thiophene, Magn. Reson. Chem. 25 (1987) 804-810.

N.M.R. Solutions to Problems of Connectivity in the Structural Elucidation of Natural Products

DAVID S. RYCROFT
Department of Chemistry, University of Glasgow,
Glasgow, G12 8QQ, Scotland, U.K.

1 INTRODUCTION

Structural elucidation in general involves determining an extended sequence of bond connectivities, and as someone who is not specifically an *organic* chemist I am very interested in methods which enable one to build up a structure from scratch in a logical manner as one then requires a minimum of biogenetic knowledge and chemical intuition. Indeed we have been quite successful in solving structures where initially we had no idea what class of compound was involved. Scapaniapyrone A (1) is an example where the combined use of the proton-coupled ^{13}C n.m.r. spectrum and 2D long-range δ_C/δ_H correlation led inexorably to the novel naphthyl pyrone structure,[1] without having first to propose a working structure and then test it to see if it fit. However the connectivity problem that I wish to discuss in this chapter is the

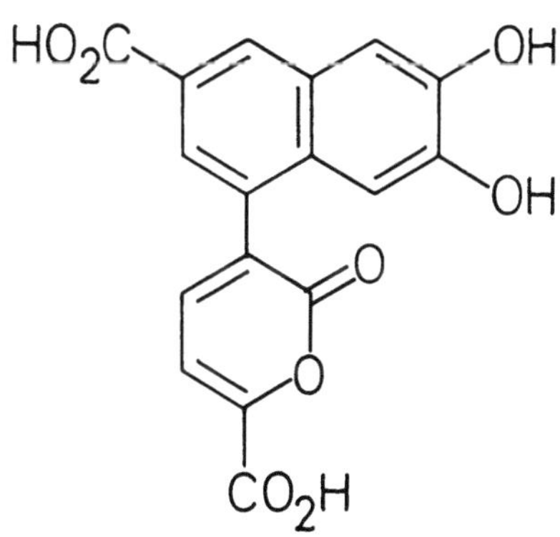

1

more common situation where one has to work out how to fit together several pieces whose structures individually have been relatively easy to elucidate. To illustrate the various types of problem that can arise I wish to describe the structural elucidation of oligosaccharides from several sources and of some compounds isolated from liverworts (Hepaticae).

2 OLIGOSACCHARIDES

2.1 Sequencing of oligorhamnosides from *Cleistopholis glauca*

The Annonaceae have been found to be a good source of interesting, biologically-active compounds (*e.g.* acetogenins) and we have already been given further examples in this symposium. Along with collaborators in Cameroon we decided to investigate *Cleistopholis glauca*.[2] No acetogenins were found but six compounds were isolated which turned out to oligosaccharides. On acetylation five of these gave the same peracetate **2** and this simplifies the presentation of the structural elucidation somewhat.

The ^{1}H NMR spectrum of **2** showed signals from acetates, secondary methyl groups, an alkyl chain, and sugar protons. The first problem to solve was the nature of the alkyl chain, which was too long to give resolved signals for every carbon atom. Acid hydrolysis fortuitously gave the alkyl rhamnopyranoside **3** whose FAB MS showed that the hydrocarbon chain was dodecanyl ($[M + Na]^+$ 355, with cleavage at the glycosidic C-O bond, m/z 147 and 185). There was also a small amount of the *bis*-homologue, $[M + Na]^+$ 383. COSY at 360 MHz enabled us readily to tabulate (Table 1) the chemical

TABLE 1
^{1}H n.m.r. chemical shifts (and couplings) of **2** (300 MHz, $CDCl_3$)

	1-H (d)	2-H (dd)	3-H (dd)
	4.66 (1.4)	5.21 (3.4, 1.6)	5.24 (9.4, 3.4)
a	4.94 (1.9)	5.01 (3.1, 2.0)	3.98 (9.8, 3.2)
b	4.85 (1.7)	4.97 (3.4, 1.7)	3.92 (9.9, 3.4)
c	4.85 (1.7)	5.05 (3.5, 1.9)	5.15 (10.2, 3.3)

	4-H (t)	5-H (dq)	6-H_3 (d)
	3.64 (9.3)	3.80 (9.3, 6.2)	1.34 (6.2)
a	5.07 (9.8)	3.89 (9.8, 6.3)	1.22 (6.4)
b	5.06 (9.7)	3.71 (9.7, 6.3)	1.16 (6.3)
c	5.02 (10.0)	3.81 (9.6, 6.3)	1.17 (6.3)

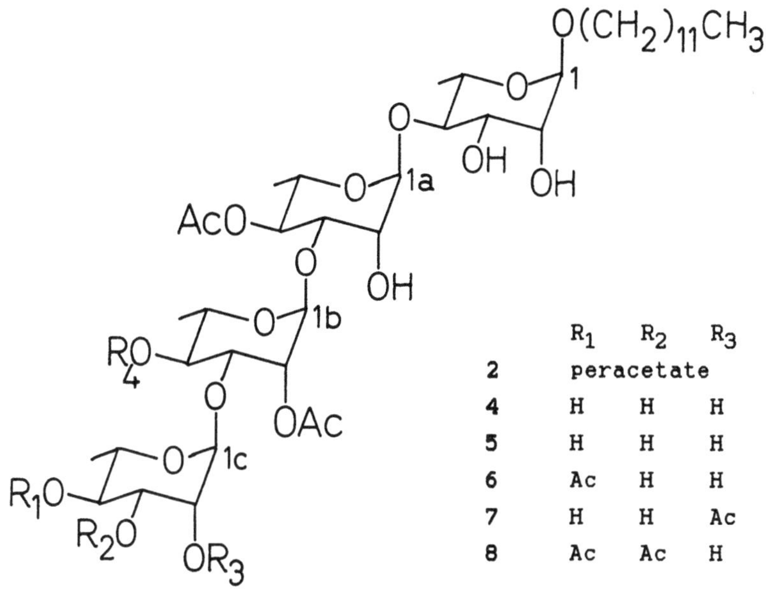

	R_1	R_2	R_3	R_4
2	peracetate			
4	H	H	H	H
5	H	H	H	Ac
6	Ac	H	H	Ac
7	H	H	Ac	Ac
8	Ac	Ac	H	Ac

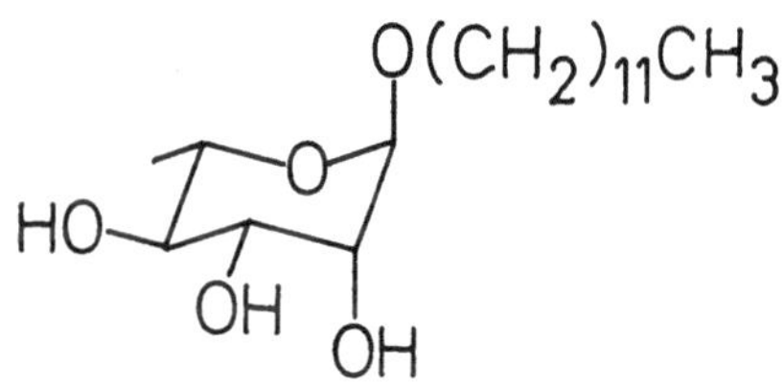

3

shifts and coupling constants of four sugar units whose splitting patterns showed that they were rhamnose. Hydrolysis of **2** with TFA gave the free sugar and the optical rotation showed that it was L(+)-rhamnose. FAB MS of **2** showed $[M + H]^+$ 1149 and also peaks for mono-, di-, tri-, and tetra-saccharide fragments (m/z 273, 503, 733, and 963). The chemical shifts showed that of the non-terminal units one was 1→4 and two were 1→3 linked. In order to determine the sequence of these units we wanted to look for spin-spin coupling across the glycosidic links.

The addition of delays before and after the mixing pulse in the COSY sequence gives a relative enhancement to correlations arising from small coupling constants.[3] This technique is called "delayed COSY" (or "small J COSY"); it should be noted that the term "long-range COSY" is inappropriate for homonuclear proton experiments because use of it can encourage people to overlook the fact that correlations may arise from small *vicinal* coupling constants. An example of the application of this method is a study of saponins from alfalfa,[4] and similar experiments were performed on the present oligosaccharides at the University of Reims.

Fig. 1 is an expansion of the delayed COSY spectrum (with delays of 250 ms before and after the mixing pulse) of **2** which shows correlations of the three anomeric protons involved in links to other sugars. Starting at the terminal unit we see cross-peaks for 1c-H to a 3-proton, which enables us to identify the b set of protons, and hence cross-peaks from 1b to 3a and from 1a to 4 (as well as cross-peaks from couplings within individual sugar rings). These results give us the sequence of the sugar units in **2**.

Before considering the anomeric configurations in **2** I wish to make a slight detour and look at the general question of determining anomeric configurations using rhamnose tetra-acetate as a model system. Acetylation of rhamnose gave a 2:1 mixture of anomers which we examined without separation. The 1-H signals are very different (Fig. 2), the major (α) anomer showing a coupling of 0.6 Hz to 5-H. This observation did not help us as the spectra of **2** did not show this sort of resolution. We also note that the values of J[(1-H)-(2-H)] are not usefully different (1.8 and 1.2 Hz in the α and β anomers respectively).

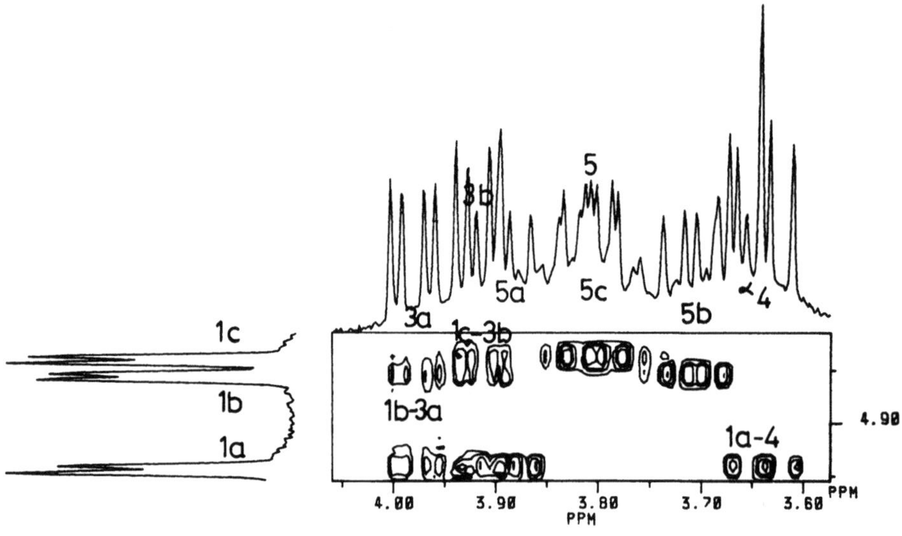

Fig. 1. Delayed COSY experiment of **2** (300 MHz, $CDCl_3$).

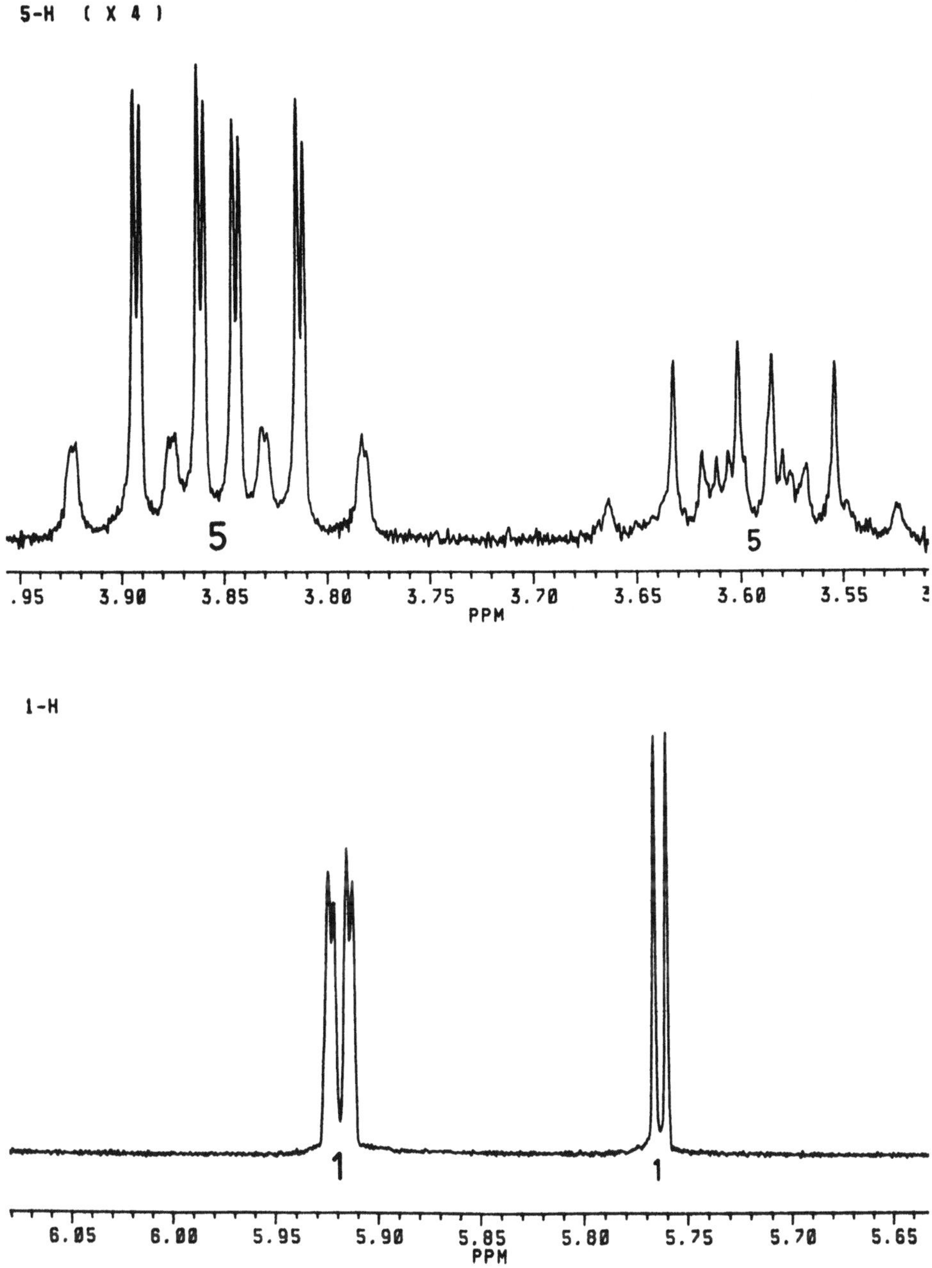

Fig. 2. ^{1}H n.m.r. spectrum of α- (major) and β- (minor) rhamnose tetra-acetate (200 MHz, $CDCl_3$).

The chemical shifts of C-3 and C-5 should be sensitive to the anomeric configuration because the change from a *γ-trans* to a *γ-gauche* interaction on going from β to α should cause considerable shielding. This necessitates an assignment of the carbon spectrum. A routine 2D direct C/H chemical shift correlation covering all the CH signals is adequate in most cases, but an expansion of the region of C-2, 3, 4, and 5 (Fig. 3) shows that the digital resolution (at 4.7 T) is inadequate for our purposes. Fig. 4 shows the same region with the same size of data matrices but with smaller spectral widths and hence better digital resolution. The

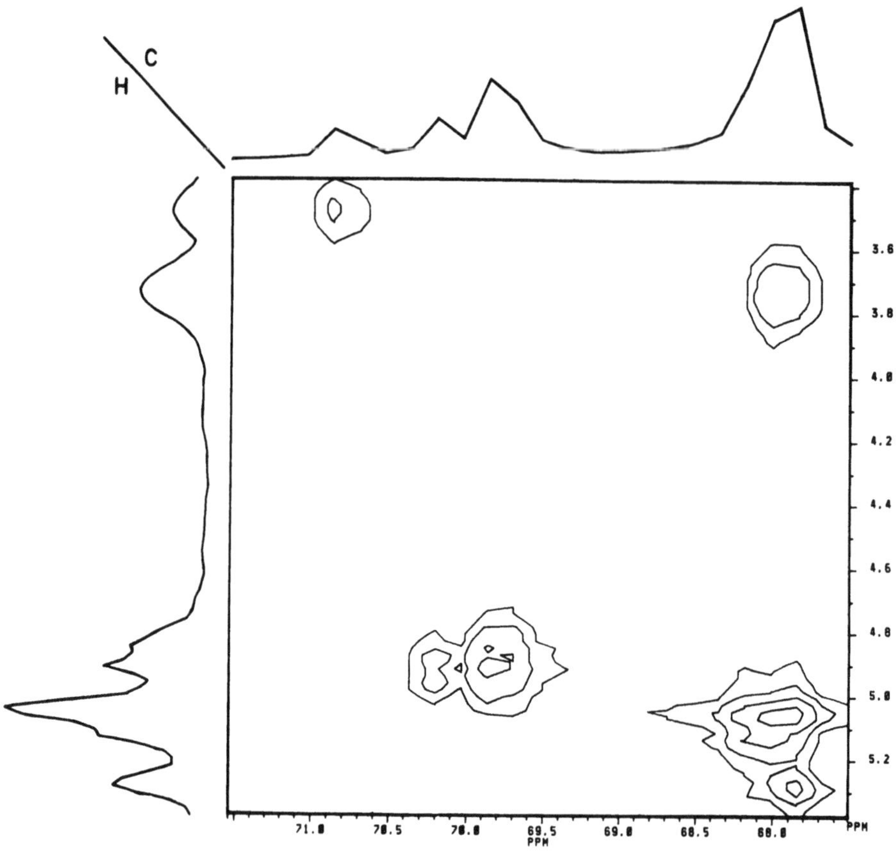

Fig. 3. δ_C/δ_H correlation of rhamnose tetra-acetate (4.7 T, $CDCl_3$ solution). Projections are drawn along the axes.

contours now show useful features, and we can immediately see a shift difference of 3 p.p.m. between the C-5 signals. Further expansion in Figure 5 to see the C-2, 3, and 4 region shows that whereas the two C-2 signals (narrow dd) and the two C-4 signals (t) are very similar in shift there is a shift difference of 2 p.p.m. between the two C-3 signals (wide dd). There are two assignments of the carbon spectrum of rhamnose tetra-acetate in the literature, both based on analogy arguments, and they differ in the assignments of one pair of signals.[5,6] This 2D spectrum gives the assignments unequivocally and shows that the assignment in reference 6 is correct.

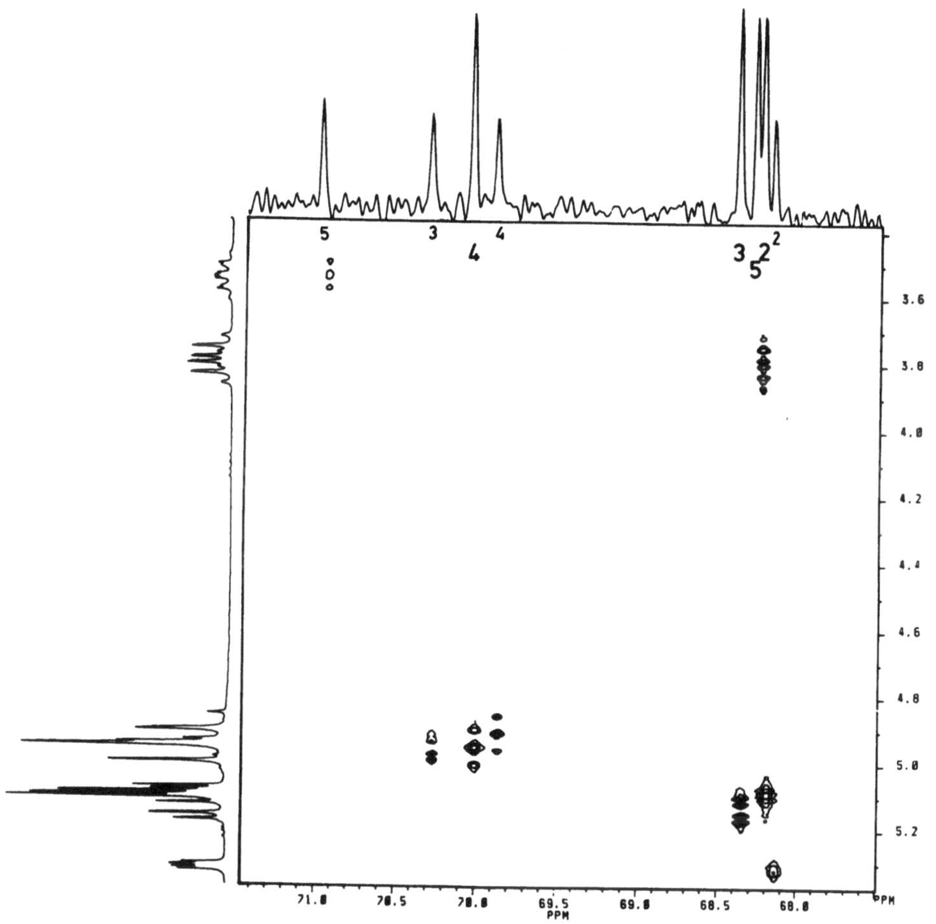

Fig. 4. δ_C/δ_H correlation of rhamnose tetra-acetate with better digital resolution than Fig. 3. 1D spectra are drawn along the axes.

The data in Fig. 5 illustrate very nicely a less well known feature of 2D C/H shift correlation. The ^{1}H cross-sections of C-2 and C-3 (Fig. 6) show first order multiplets for protons which are strongly coupled in the 1D ^{1}H spectrum. This is because the 2D spectrum is dealing with the ^{13}C satellites in the ^{1}H spectrum (one carbon at a time) and whereas H-2 is strongly coupled to H-3, the ^{13}C satellites of H-2 are not strongly coupled to H-3 and *vice versa*. As a corollary of this one must also be aware that it is possible for the 2D experiment to give second order multiplets when the 1D spectrum is first order.

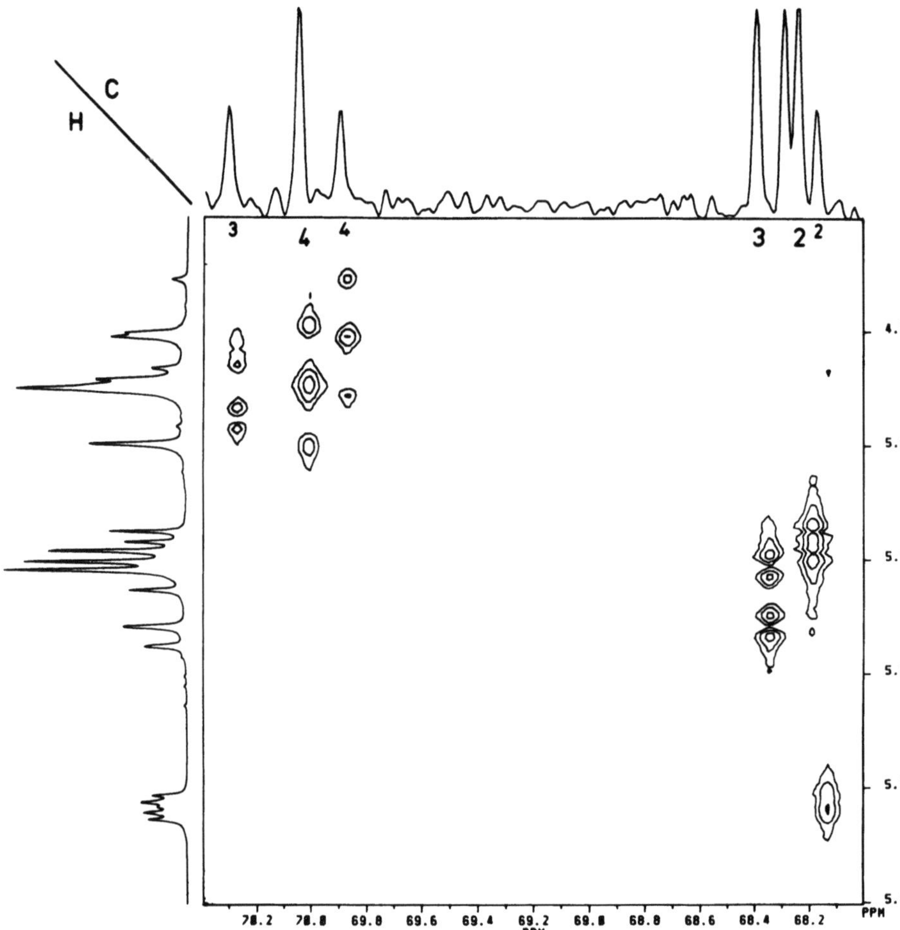

Fig. 5. Expansion of Fig. 4.

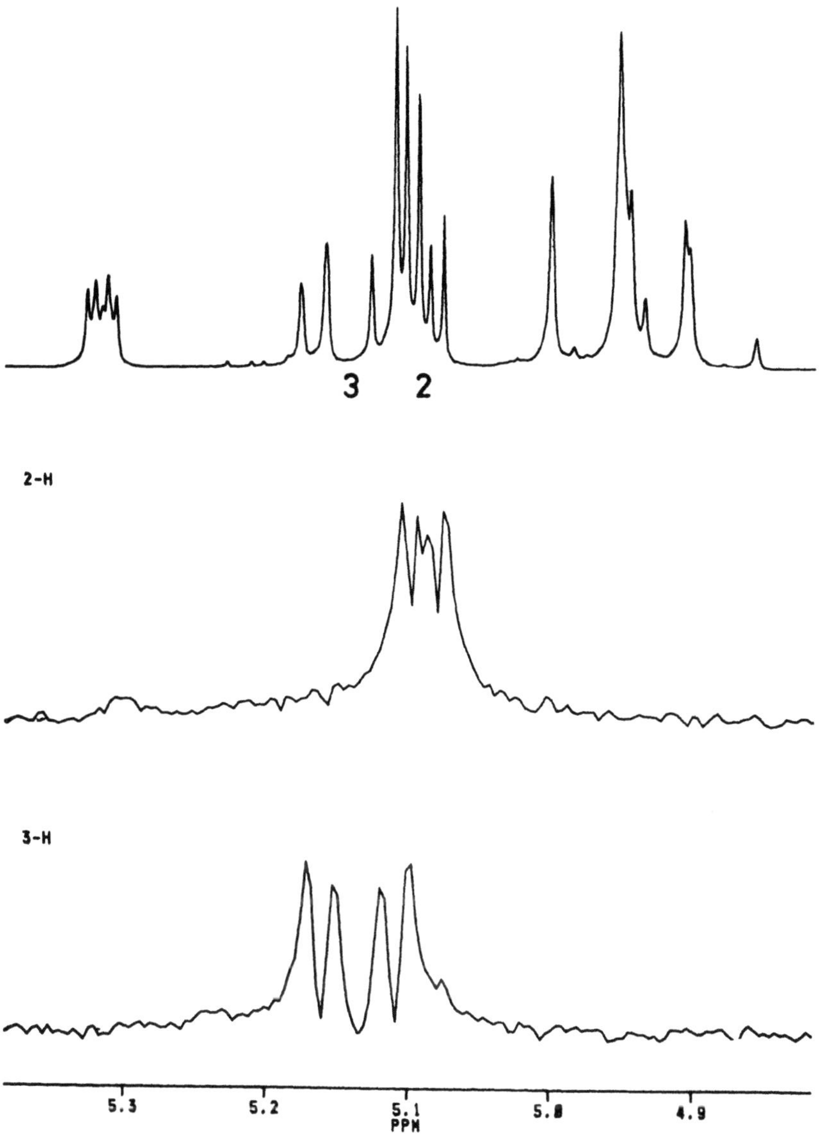

Fig. 6. Proton cross-sections through C-3 and C-2 in Fig. 5 showing first order multiplets for 3-H (bottom) and 2-H (middle) with the second order multiplets in the 1D ^{1}H spectrum shown (top) for comparison.

To return to the configuration question, the shift differences observed are not large and the method would involve assigning the more congested spectra of the tetrasaccharides. In any case the ^{13}C method of choice for configurational assignment would be to measure $^{1}J(CH)$ of the anomeric carbons.[7] A ^{1}H method, which is more sensitive, is to use NOE measurements. Fig. 7 shows these for rhamnose tetra-acetate. The ß-, but not the α-, anomer shows sizeable NOEs at 3-H (5%) and 5-H (7%) when 1-H is irradiated. It can also be seen that there is an NOE (6%) at 2-H of both anomers. There is also an NOE at 5-H (0.6%) when 1-H of the α-anomer is irradiated, but it is very small. This observation gives me the

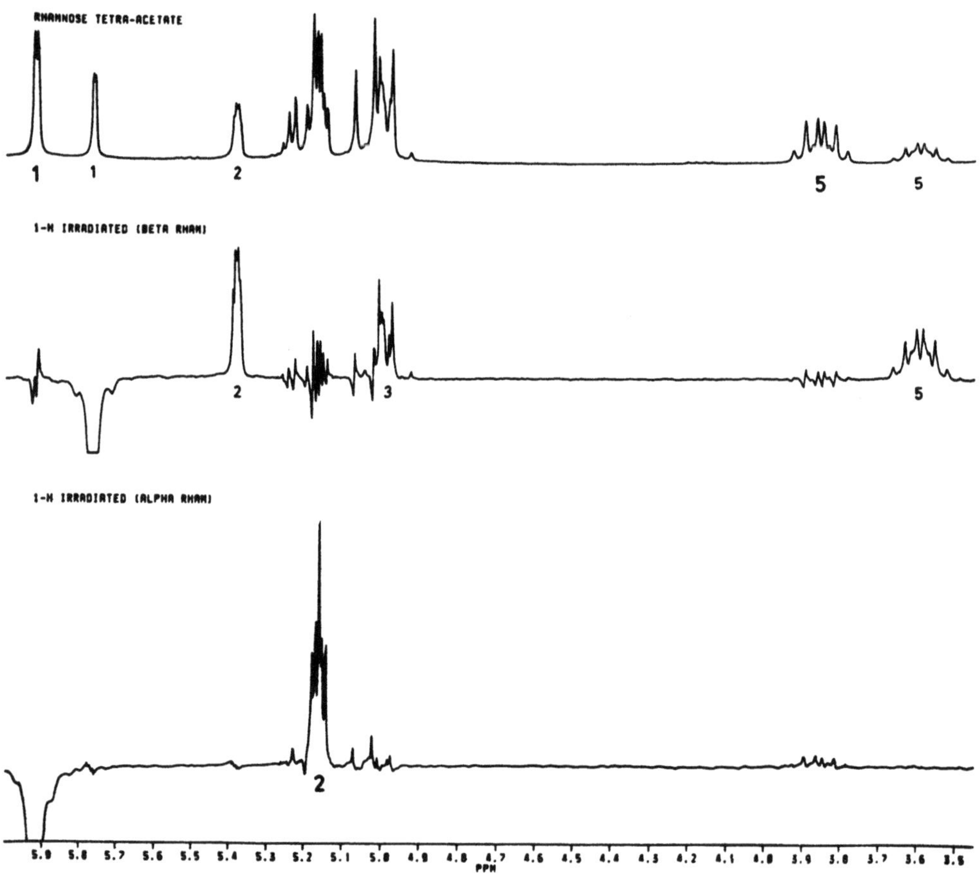

Fig. 7. NOE difference spectra of rhamnose tetra-acetates with irradiation of 1-H of the α- (bottom) and ß-anomer (middle) compared with (top) the control spectrum (200 MHz, $CDCl_3$ solution).

opportunity to make a general point about the troublesome topic of the reporting of NOEs in the literature. In the present case we could validly make independent statements that on irradiation of 1-H we observe an NOE at 5-H for the α-anomer and for the ß-anomer, and it is obvious that we cannot conclude that 1-H and 5-H are in a *syn*-1,3-diaxial arrangement in both cases. The measurability of NOEs depends *inter alia* on the S/N of the spectra and **it is common to observe NOEs which have no significance** in the sense that the mere observation of an NOE does not distinguish whether or not two protons are "close" to each other. It is the relative sizes of NOEs that have to be considered when stereochemical deductions are made and hence it is important that NOEs are measured and reported quantitatively.

Returning now to the *Cleistopholis* compounds, NOE difference experiments where the 1-Hs are irradiated are shown in Fig. 8(i)-(iv). Discrimination between incompletely resolved multiplets, *e.g.* 1b- and 1c-H, is possible by restricting irradiation to only one line of each doublet. Here we see NOEs across the glycosidic links (5%) which confirm the delayed COSY results and also NOEs of the 2-Hs (3%) but not of 3-H or 5-H. Hence we conclude that the glycosidic links are axial (α in the L sugar).

Fig. 8 also illustrates another feature of difference spectroscopy. The out-of-phase signal at the centre of the control spectrum comes from the residual $CHCl_3$, folded in from outside the spectral window. The difference artifact from this signal is worst in the spectra at the centre of Fig. 8. The experiments were performed in the usual way, by doing a series of on-resonance irradiations (in this case 8) followed by one off-resonance control irradiation, and repeating the whole cycle several times for signal averaging. It is apparent that the further away in time the spectrum is from the control, the worse the difference artifact. In fact Fig. 8(viii) is better than 8(i), consistent with the fact that (viii) always precedes the control (ix) whereas (i) follows (ix) every time except the first cycle, when it is most remote from (ix). Instrumental instability is the cause of these difference artifacts and the lesson to be learnt here is that the cycle of irradiations should kept as short as possible and repeated many times, not just for S/N improvement but also to average the instability and reduce the size of the difference artifacts.

So far the structure **2** of the peracetate of the *Cleistopholis* compounds has been deduced. The original compounds are partially

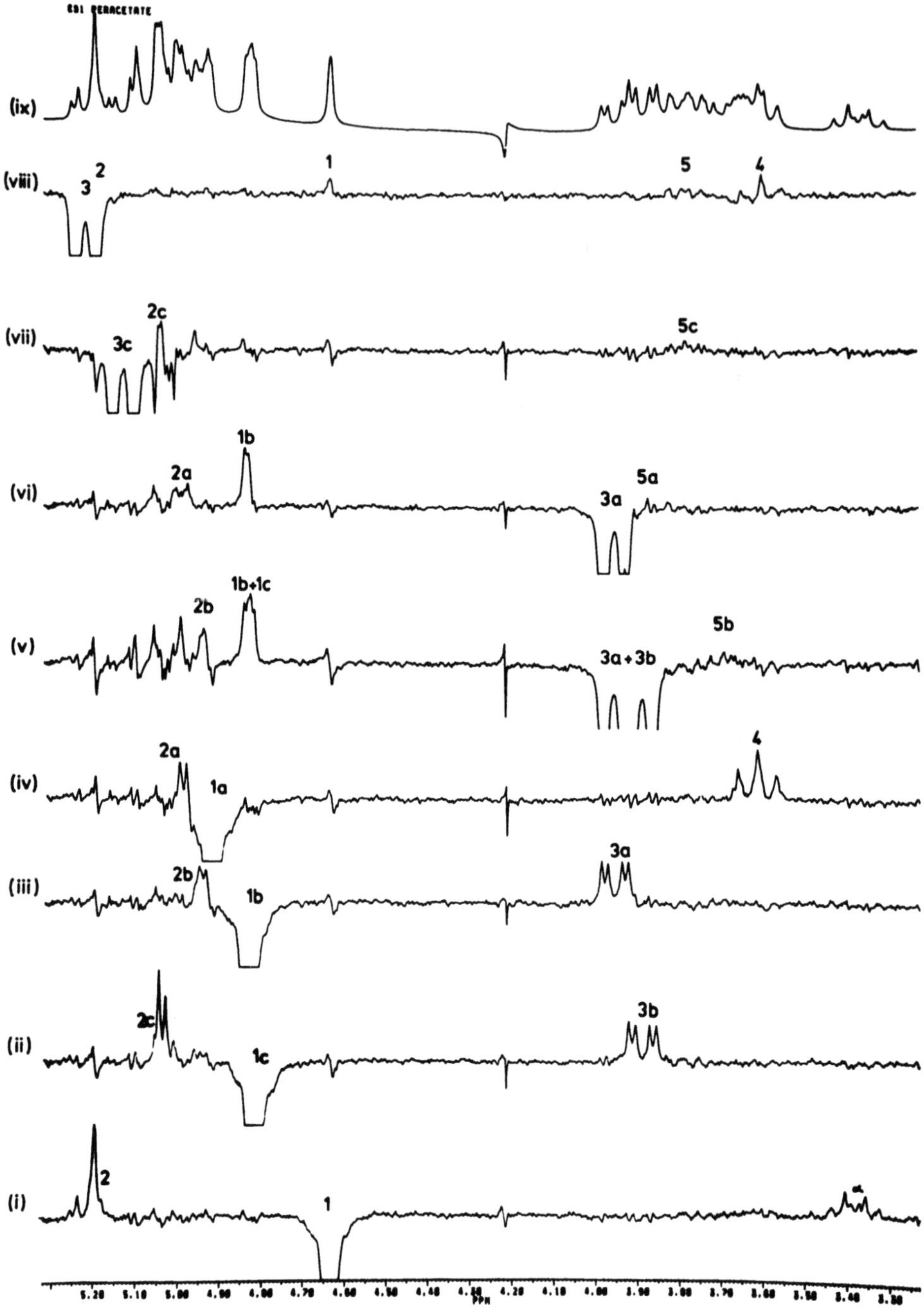

Fig. 8. (i)-(viii). NOE difference spectra of **2** compared with (ix) the control spectrum (200 MHz, $CDCl_3$ solution).

acetylated tetrarhamnosides and to determine the structure and sequence of them it was necessary to perform the COSY and delayed COSY experiments with each of them. In this respect solving the structure of the peracetate has not helped to solve the structures of the original compounds. By way of example, Fig. 9 shows an expansion of the delayed COSY spectrum of the tetra-acetate **7**. As the assignments within the sugar rings are known from the COSY spectrum, the ring b set of protons is labelled by finding the 3-H which has a correlation with the 1-H of the terminal ring c. Hence the ambiguity in this case as to whether there is a free hydroxyl group on ring a or on ring b is resolved. By these means we arrived at the structures of all five compounds **4**, **5**, **6**, **7**, and **8**.

The sixth compound **9** was a trirhamnoside with four acetates and again with the dodecanyl ether, $[M + Na]^+ = 815$, and we deduced the structure without needing the delayed COSY experiment. FAB MS showed that three acetates were on the terminal ring (m/z 273), and one on the central ring (m/z 461, = 273 + 188). The proton shifts showed that the acetylated positions were C-2, -3, and -4 (twice).

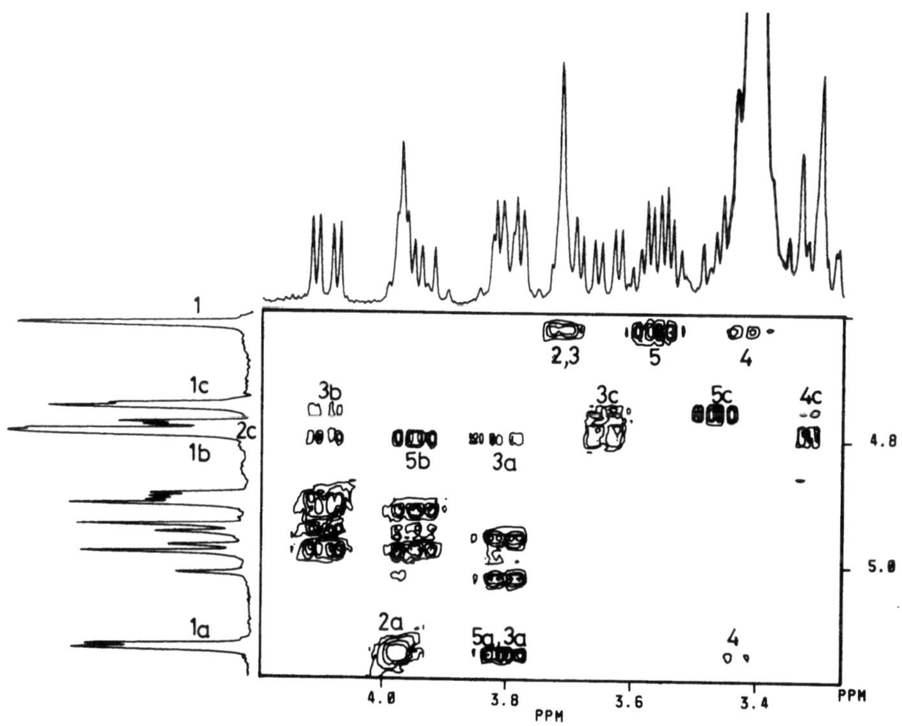

Fig. 9. Delayed COSY of **7** (300 MHz, $CDCl_3$).

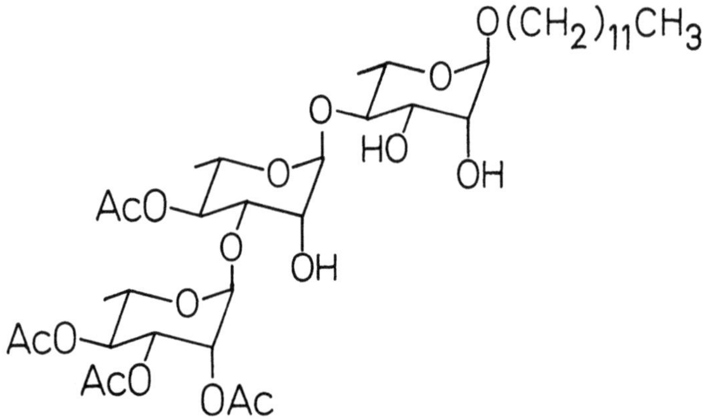

9

Acetylation gave a peracetate where the newly acetylated positions were C-3 and -2 (twice). So by process of elimination we conclude that the glycosidic links were 1→4 and 1→3, and that the structure is **9**.

2.2 A chromone glycoside from *Schumanniophyton magnificum*

A compound was isolated from a member of the Rubiaceae, the Cameroon species *Schumanniophyton magnificum*.[8] The ^{1}H n.m.r. spectrum of the acetate showed an olefinic proton and a methyl group, two *meta*-coupled protons, and seven acetates. The rest of the spectrum consisted of 13 sugar protons which turned out to come from a glucose and an apiose unit. The working structure is shown as **10**. The unconventional representation of the apiose with the L

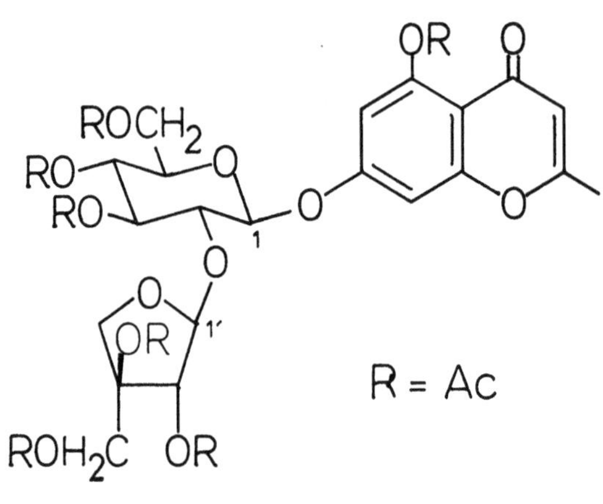

10

configuration is intended to highlight the fact that, as far as I am aware, the D configuration usually drawn has been proved only once, in apiin,[9] and of course where the presence of apiose has been demonstrated by hydrolysis the configurational integrity of C-3 is lost. We have not yet resolved this point or the question of the anomeric configuration of the apiose for **10**. Meanwhile I want to describe what we can learn from the delayed COSY spectrum of **10**.

As well as correlations from coupling of 1'-H to one 4'-H and of 2'-H to the other 4'-H, correlations can be seen from coupling across the glycosidic link, *viz.* from 1'-H to 2-H. There are cross-peaks resulting from coupling of 1-H (of the glucose) to both aromatic protons and for the moment it is assumed that these are both *ortho*, although I am not aware that it has ever been shown that it is not possible to see the effects of coupling to a proton *para* to the glycosidic oxygen in this sort of system, even though the protons would then be 7 bonds apart (it is worth remembering that in toluene ^4J, ^5J, and ^6J of the methyl protons to the aromatic protons are all of the same order of magnitude, *viz.* -0.7, 0.4, and -0.6 Hz respectively;[10] on the other hand in anisole the methoxyl protons couple significantly only to the *ortho* protons, mainly due to a through-space mechanism[11]).

Perhaps the most surprising cross-peaks seen in the delayed COSY spectrum of **10** arise from coupling of sugar protons to acetate protons (in three cases). Cross-peaks involving intense, sharp peaks always require to be checked carefully and in this case we were satisfied that they were not artifacts resulting from mutual enhancement of horizontal and vertical t_1 ridges, or symmetrization artifacts resulting from symmetrically related points of strong t_1 ridges. Although we do not gain any additional information about **10** from these observations because all the esters are acetates, in compounds where this is not so it might prove to be a useful method of assigning positions of ester attachment. Indeed if we had known this earlier we would have used it to look at the sugar esters discussed in the next section.

2.3 <u>Attachment of esters in sugars from *Salpiglossis sinuata*</u>

Compounds **11** and **12** were isolated from *Salpiglossis sinuata*, a member of the Solanaceae native to Chile.[12] Quite recently some similar compounds have been reported from *Solanum berthaultii* (also Solanaceae) where they occur as sticky exudates produced when the leaves are disturbed by foraging insects.[13] The ^{1}H n.m.r. spectrum of **11** initially showed only one compound, but this soon epimerised and Fig. 10 shows how a COSY-45 spectrum can be used to unravel such a mixture. The connectivity of protons 1 to 5 is drawn separately for the two epimers on opposite sides of the diagonal.

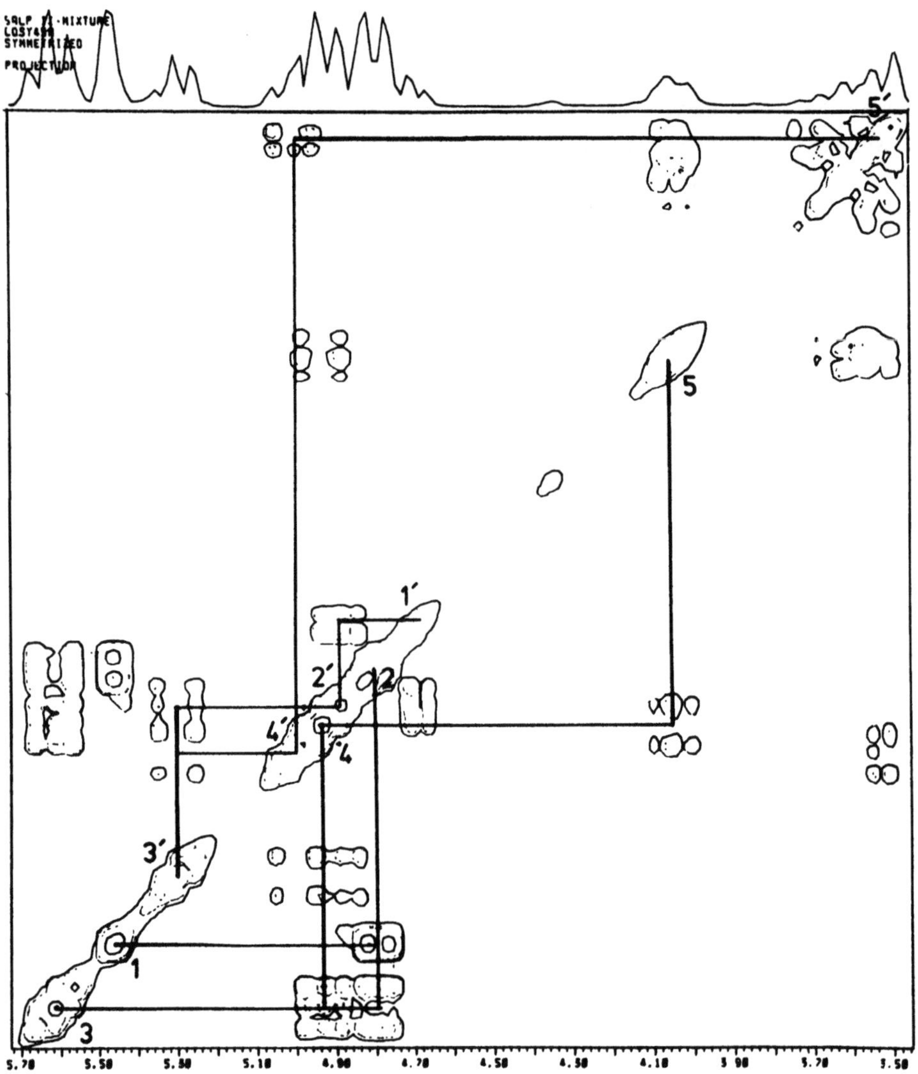

Fig. 10. COSY-45 spectrum of **11** after epimerisation. The coupling paths shown are for the α- and β-isomers below and above the diagonal respectively (200 MHz, $CDCl_3$ solution).

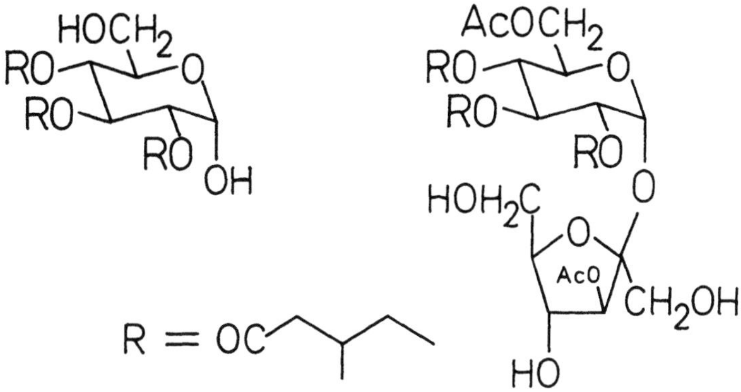

11 12

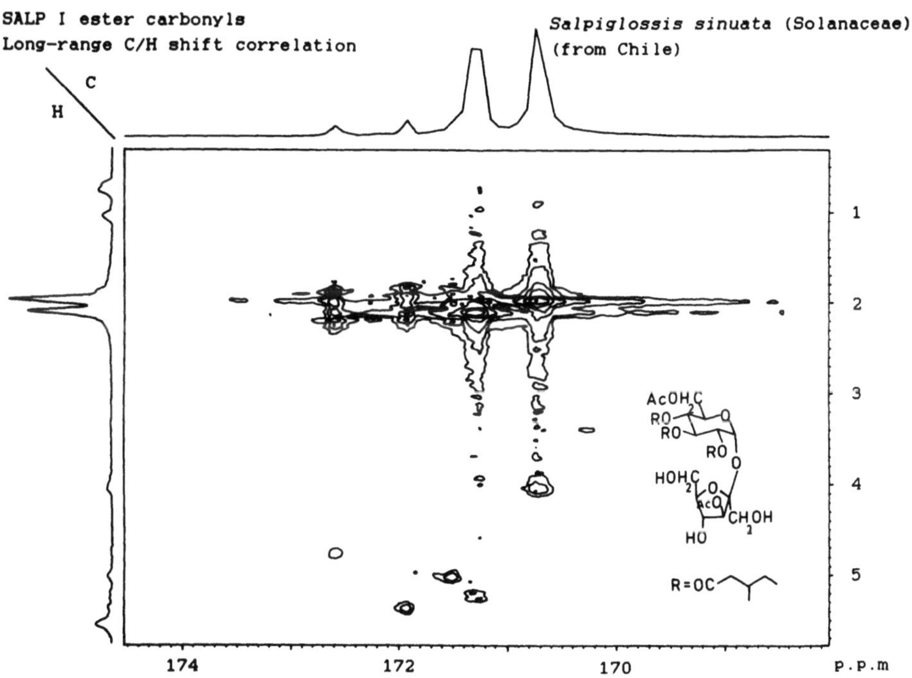

Fig. 11. Long-range δ_C/δ_H correlation spectrum of **12** (4.7 T, $CDCl_3$ solution). Projections are drawn along the axes.

The more complex compound, **12**, was a sucrose derivative with both acetates and 3-methylvalerates, and the basis we adopted to position the esters was to assign, *via* the long-range couplings, the carbonyl carbon signals to particular sugar and ester protons by means of 2D long-range δ_C/δ_H correlation (in this case the basic sequence as carbonyls do not have directly attached protons).[14] The contour plot in Fig. 11 shows large peaks for the acetate methyl protons and, at the same carbon shifts (δ_C 170.7 and 171.3), peaks for 6-H_2 of the glucose and 2-H of the fructose unit (identified by prior assignment of the ^1H spectrum). The other three carbonyls (δ_C 171.5, 171.9, and 172.6) belong to 3-methylvalerates and show correlations to 4-, 3-, and 2-H, respectively, of the glucose unit. Hence structure **12** was deduced.

3 COMPOUNDS FROM LIVERWORTS (HEPATICAE)

This section briefly describes some results we have obtained from working with liverworts, primitive plants classified along with mosses (Musci) as Bryophyta. Liverworts are usually intolerant of drought and many thrive in the climate of the west of Scotland. The leaves of most liverworts contain oil-bodies, and the terpenoids found in them are often enantiomeric to those occurring in higher plants. Terpenoids with unusual carbon skeletons have been identified, e.g. tamariscol, **13**, for which we used 2D INADEQUATE to trace out the carbon connectivity.[15] Novel aromatic compounds have also been isolated, including **1**, scapaniapyrone A,[1] which was mentioned in section 1.

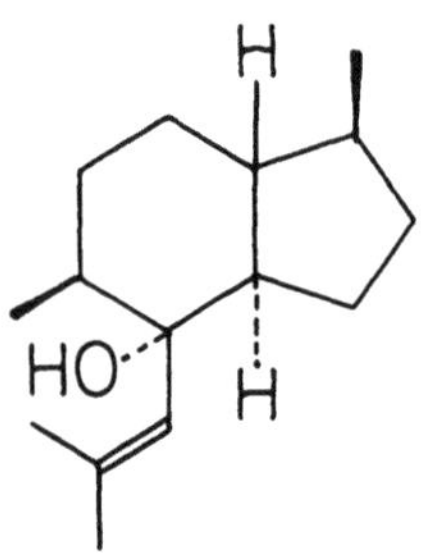

13

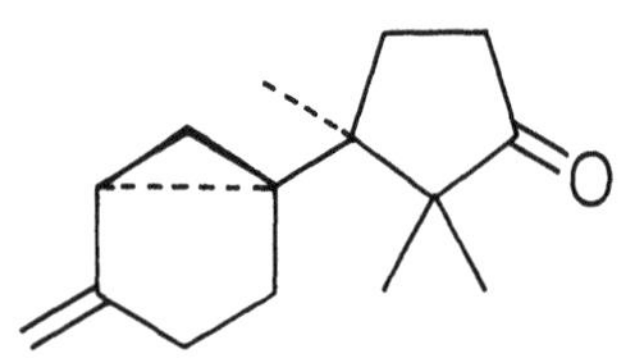

14

3.1 Pakyonol, a cyclic bisbibenzyl diether from *Mannia fragrans*

Mannia fragrans is a thalloidal liverwort which when growing in central Europe or Mongolia can sometimes be detected by smell when it is several metres away. The compound responsible was isolated by Dr. Siegfried Huneck before 1975 and is grimaldone, **14**. The structure came from X-ray studies, and it is worth noting in passing that the failure to solve the structure by using n.m.r. arose because we failed to recognise that a 2 Hz coupling in the ^1H spectrum arose from a 4- rather than a 3-bond interaction.[16] When Dr. Huneck collected the liverwort in Korea there was no smell and the only compound he could isolate was pakyonol, **15** (structure in Fig. 12), which was shown to be a cyclic bisbibenzyl diether.[17] It was relatively easy to identify the four aromatic rings, and then assign the bibenzyls from small *J* H/H correlations of methylene and *ortho* aromatic protons (confirmed by NOE difference measurements). The remaining problem was the arrangement of the rings as ABDC or ABCD. At the time we chose to use NOE difference experiments to resolve this question (although with the benefit of our present knowledge I would now try delayed COSY with long delays). We have to see NOEs across the ether links which means distinguishing the effects of irradiating 3'-H and 10'-H. {3'-H} gives a bigger NOE at 6-H than {10'-H} and *vice versa* for 11,13-H_2 (Fig. 12). I used multiple selective irradiation[18] of all the lines of each multiplet when performing these experiments and inevitably affected both signals but differentially. Better selectivity would be obtained by irradiating just one line of each multiplet or at least by not irradiating the overlapping lines [*cf*. Fig. 8 (ii) and (iii)]. This is the first compound of this type with this substitution pattern (both ether links *para* in one of the bibenzyls). Related compounds are already known from the Hepaticae and have been reported to be biologically active.[19]

3.2 Bibenzyl derivatives from *Plagiochila spinulosa*

The work described so far has involved extensive international collaboration, but I would like to conclude by returning to Scotland and discussing some compounds isolated[20] from a leafy liverwort *Plagiochila spinulosa*, a plant described as having an Atlantic distribution (*i.e.* liking a damp climate) and collected by myself near Glasgow. One of the new compounds isolated, **16**, is a close relative of lunularic acid, **17**, a growth inhibitor which seems to be ubiquitous in liverworts.

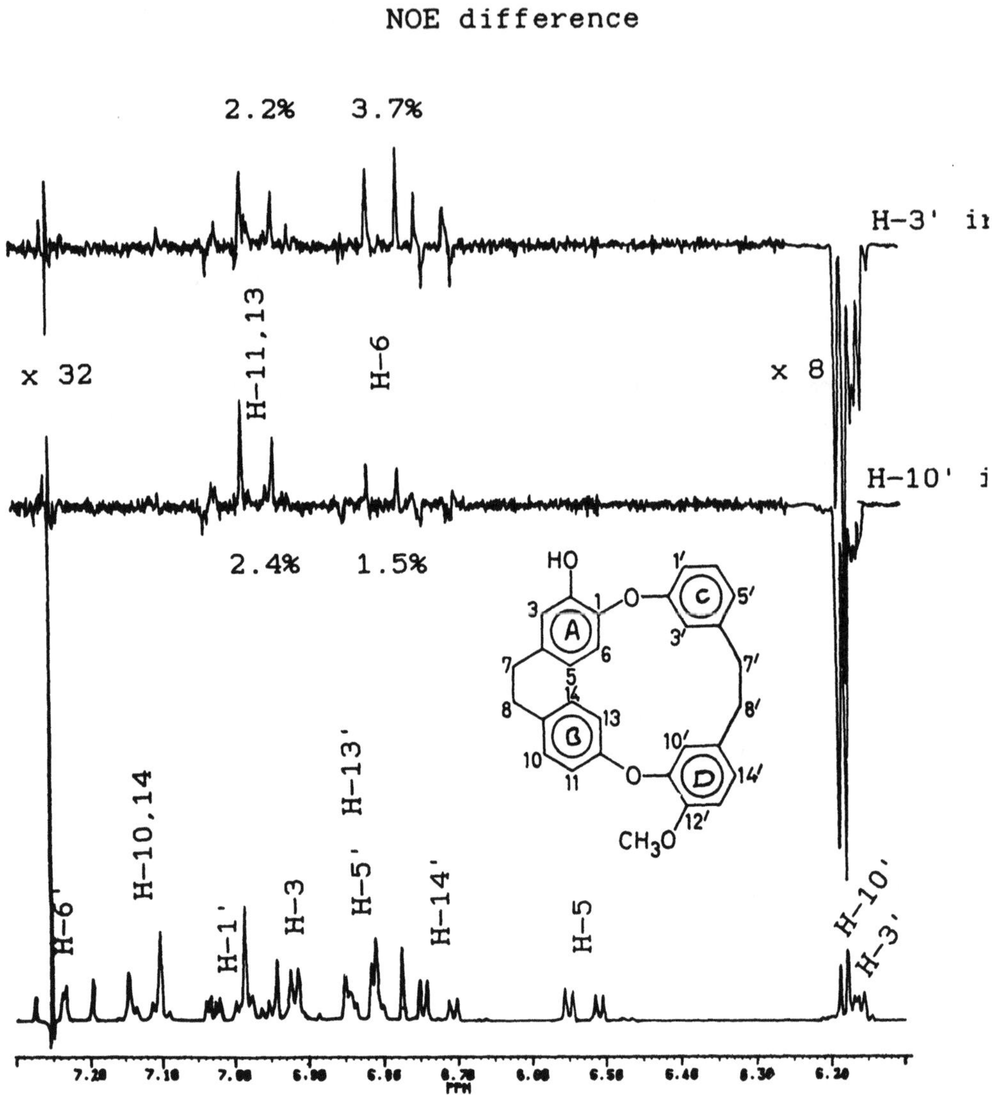

Fig. 12. N.m.r. spectrum (200 MHz, $CDCl_3$ solution) of the aromatic protons of pakyonol, **15** (structure shown), with NOE difference spectra for irradiation of 3'-H (top) and 10'-H (middle). The percentages indicated refer to the enhancements actually observed and do not take account of the different degrees of saturation in the two multiple selective irradiation experiments.

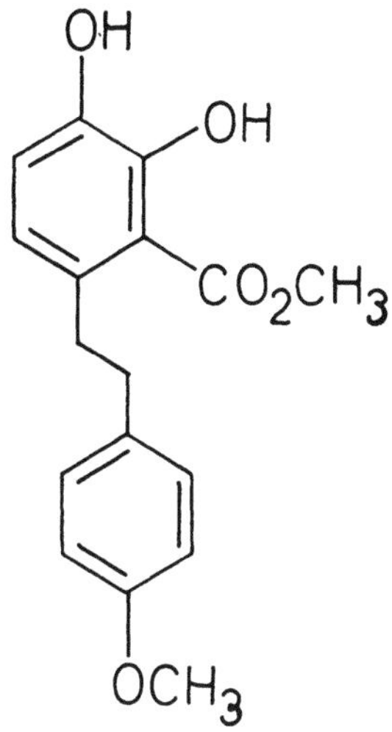

16

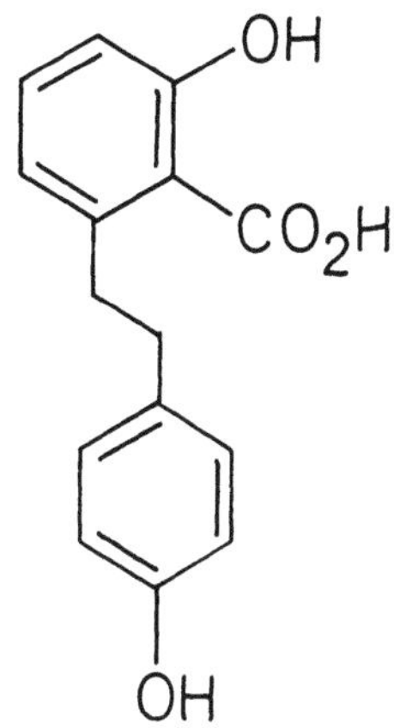

17

Another compound, **18**, remained unsolved for quite some time. The ^1H and ^{13}C n.m.r. spectra showed aromatic and methoxyl signals similar to those of **16**, but in addition there were terpenoid-like signals which were mostly always rather broad. On the basis of some limited n.m.r. experiments we proposed working part-structures (which ultimately turned out to be correct as far as they went) of a bibenzyl joined to a fusicoccane diterpenoid through two ether

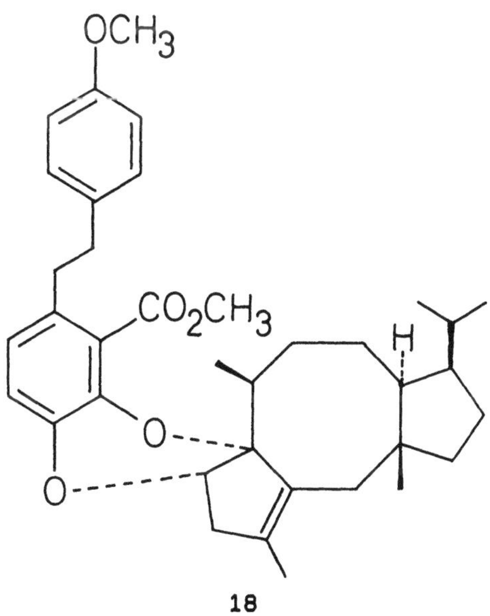

18

linkages, but could not prove them, mainly because the second-order character and broadness of the ^{1}H spectrum resulted in 2D n.m.r. experiments giving unsatisfactory results. Persistence eventually paid off and the third person working on the extraction isolated enough of the compound that 2D INADEQUATE should have been possible. However while waiting in the n.m.r. room the compound crystallised, and solutions gave inadequate 2D INADEQUATE spectra. Fortunately the crystals were suitable for an X-ray structure determination, which showed that the compound was **18**. This is an interesting curiosity as it is an example of typical liverwort structures, *viz.* terpenoids and bibenzyls, occurring in the same molecule. The fusicoccane diterpenoid skeleton is uncommon, but we have found it previously[21] in the liverwort *Anastrepta orcadensis* and more recently it has been reported[22] in the fusicoplagins from a Japanese *Plagiochila* species. By analogy to the last compounds we have named **18** spinuloplagin.

4 ACKNOWLEDGEMENTS

I should like to thank Professor Atta-ur-Rahman and his colleagues for their hospitality during the symposium and my subsequent stay at the H.E.J. Research Institute of Chemistry, the British Council (Karachi) for a travel grant, and the following people for their contributions to various aspects of the work described here: Dr. J.D. Connolly, Dr. A.A. Freer, Dr. L.J. Harrison, Dr. Patience Ifeadike, Dr. J. Singh, and Mr. N. Woods (Glasgow, Scotland), Dr. S. Huneck (Halle, GDR), Dr. R. Mues (Saarbrücken, FRG), Dr. Cecilia Labbé (Santiago, Chile), Dr. J.F. Ayafor and Mr. P. Tané (Yaoundé, Cameroon), Dr. G. Massiot and Dr. Catherine Lavaud (Reims, France), and Dr. I.H. Sadler and Dr. D. Reed (S.E.R.C. High Field N.M.R. Service, Edinburgh, Scotland).

REFERENCES

1 J.D. Connolly, S. Huneck, R. Mues, and D.S. Rycroft, in preparation.
2 P. Tané, J.F. Ayafor, B.L. Sondengam, C. Lavaud, G. Massiot, J.D. Connolly, D.S. Rycroft, and N. Woods, *Tetrahedron Letters*, in press.
3 A. Bax and R. Freeman, *J. Magn. Resonance*, **44** (1981) 542.
4 G. Massiot, C. Lavaud, D. Guillaume, L. Le Men-Olivier, and G. Van Binst, *J. Chem. Soc., Chem. Commun.* (1986) 1485.
5 C. Laffite, A.M. Nguyen Phuoc Du, F. Winternitz, R. Wylde, and F. Pratviel-Sosa, *Carbohydrate Res.*, **67** (1978) 105.
6 V. Pozsgay and A. Neszmélyi, *Carbohydrate Res.*, **80** (1980) 196.
7 *E.g.* L.D. Hall and G.A. Morris, *Carbohydrate Res.*, **82** (1980) 175.
8 P. Tané, J.F. Ayafor, B.L. Sondengam, J.D. Connolly, and D.S. Rycroft, in preparation.
9 R.K. Hulyalkar, J.K.N. Jones, and M.B. Perry, *Canad. J. Chem.*, **43** (1965) 2085.
10 M.P. Williamson, R.J. Kostelnik, S.M. Castellano, *J. Chem. Phys.*, **49** (1968) 2218.
11 T. Schaefer, T.A. Wildman, and J. Peeling, *J. Magn. Resonance*, **56** (1984) 144.
12 C. Labbé, J.D. Connolly, P. Ifeadike, D.S. Rycroft, and N. Woods, in preparation.
13 R.R. King, R.P. Singh, and L.A. Calhoun, *Carbohydrate Res.*, **166** (1987) 113.
14 J.D. Connolly, C.O. Fakunle, and D.S. Rycroft, *J. Chem. Research (S)* (1984) 368.
15 J.D. Connolly, L.J. Harrison, and D.S. Rycroft, *Tetrahedron Letters*, **25** (1984) 1401.
16 S. Huneck, J.D. Connolly, A.A. Freer, and D.S. Rycroft, *Phytochemistry*, in press.
17 S. Huneck, J.D. Connolly, D.S. Rycroft, and N. Woods, *J. Chem. Research (S)* (1988) 78.

18 A. Pfaltz, B. Jaun, A. Fassler, A. Eschenmoser, R. Jaenchen, H.H. Gilles, G. Diekert, and R.K. Thauer, *Helv. Chim. Acta,* **65** (1982) 828; M. Kinns and J.K.M. Sanders, *J. Magn. Resonance,* **56** (1984) 518.

19 Y. Asakawa, *Progr. Chem. Org. Nat. Prod.,* **42** (1982) 1; Y. Asakawa, M. Toyota, Z. Taira, T. Takemoto, and M. Kido, *J. Org. Chem.,* **48** (1983) 2164; M. Tori, M. Toyota, L.J. Harrison, K. Takikawa, and Y. Asakawa, *Tetrahedron Letters,* **26** (1985) 4735.

20 J.D. Connolly, A.A. Freer, L.J. Harrison, P. Ifeadike, D.S. Rycroft, and J. Singh, in preparation.

21 S. Huneck, G. Baxter, A.F. Cameron, J.D. Connolly, and D.S. Rycroft, *Tetrahedron Letters,* **24** (1983) 3787.

22 T. Hashimoto. M. Tori, Z. Taira, and Y. Asakawa, *Tetrahedron Letters,* **26** (1985) 6473.

Newer Applications of Circular Dichroism in Natural Products Chemistry

R.C.CAMBIE,** P.C.HO,* K.NETZKE,* W.SCHOENFELDER,* F.SNATZKE,* G.SNATZKE*, and J.SCHULTE* (Affiliations see page 87)

1 GENERAL REMARKS

Optical activity of a substance leads at least to three spectroscopic phenomena which are, however, related: Optical Rotatory Dispersion (ORD), i.e. the rotation of the plane of linearly polarized light by traversing the substance, Circular Dichroism, expressed as $\Delta\varepsilon \equiv \varepsilon_L - \varepsilon_R$ (if the relative molar mass of the molecules is known, or - in case of biopolymers - the mean residual relative molar mass), and Molar Ellipticity, $[\theta] = 3300 \times \Delta\varepsilon$. A necessary condition for optical activity of a substance is the Chirality of its molecules, i.e. the property of being not superposable onto its mirror image. This allows still the presence of (proper) axes of rotation C_p, but not improper ones (S_p, including $S_1 \equiv$ mirror plane, and $S_2 \equiv$ centre of inversion). Theoretically, the area under a CD-band is given by the scalar (or dot) product of the electric times the magnetic transition moments associated with the transition and this value is proportional to the Rotational strength

$$R_{0i} = 22.97 \times 10^{-40} \int (\Delta\varepsilon/\lambda) d\lambda = \mu_{0i} \times m_{i0} \times \cos(\varphi) \qquad (1)$$

(φ is the angle between the two transition moments, index 0 refers to the ground, and i to the excited state). Thus the CD is positive if is acute (including parallel moments), and negative if this angle is oblique (including antiparallel moments). R_{0i} has been defined in analogy to the Dipol Strength D_{0i}, given by

$$D_{0i} = 91.88 \times 10^{-38} \int (\varepsilon/\lambda) d\lambda = \mu_{0i}^2 \qquad (2)$$

which describes quantitatively the area under an UV-absorption band. $\vec{\mu}_{0i}$ describes the Translation of Charge within the electron cloud during the excitation from the ground state Ψ_0 into the electronically excited state Ψ_i (or from the k-th MO into the l-th), $\vec{m}_{i0}$ the associated Rotation of Charge. In order to obtain an estimate of these one can apply the following recipes:

$\vec{\mu}_{0i}$	$\vec{m}_{i0}$
a) Identify the MOs Ψ_k and Ψ_l between which the electron "jumps" during excitation.	
b) Multiply these formally, using the rule that the product of MO-lobes of identical/opposite signs is positive/negative.	b) Note, whether by transforming Ψ_k into Ψ_l any rotation of charge (besides around 180°) is involved.
c) Invert the signs and determine the centre of gravity of the positive and negative "transition charges".	c) If yes, then apply the right-hand-rule to the rotation around the smaller of the two possible angles.
d) If by this procedure a dipole results it characterizes $\vec{\mu}_{0i}$ in direction (+ to -) and magnitude.	d) The thumb gives now the direction of $\vec{m}_{i0}$, the magnitude of which is greatest if this angle of rotation is 90°.

All chiral molecules can for our purposes be classified into one of four categories, viz. **A**) that for which the chromophore itself is chiral like hexahelicene, a twisted biphenyl, or a non-coplanar styrene (molecules with "inherently chiral chromophore"), **B**) such in which the chromophore is achiral, but incorporated into a chiral skeleton ("inherently achiral chromophore in a chiral environment"). One can further subdivide this class into **B1**), where the chromophore is built into a chiral ring (e.g. tetralins, tetrahydroisoquinolines) and **B2**), where this is not the case (e.g. benzohomoadamantenes). If no such ring is present then the torsional angle between the benzene ring and the attached substituent may replace it. The fourth class **C**) comprises such chiral molecules, which contain two isolated chromophores, and whose associated (strong) electric transition moment vectors are chirally arranged to each other (<u>Exciton Coupling</u>). The recipies given before can be directly applied only to case **A**), but with some additional steps can often be adapted for the classes **B**). A special rule is valid for the exciton case. For a more detailed discussion and worked out examples cf. e.g. (refs. 1-3).

If we discuss only $\pi\rightarrow\pi^*$ - transitions of benzene derivatives then most rearrangements within the electron cloud can be characterized by charge movements parallel to the ring plane, the resulting transition dipole $\vec{\mu}_{0i}$ lies thus within this plane, and according to the right-hand-rule the built up magnetic transition moment $\vec{m}_{i0}$ will be perpendicular to it. The angle between these is then 90°, and therefore $R_{0i} \equiv 0$. The chiral "environment" (i.e. the chiral rest of the molecule, but it may also be the chiral molecules of an optically active solvent) perturbs the MOs in such a way that they will be slightly mixed with each other, and the missing transition moments are "stolen" from another transition, thus

$$R_{0i} = \mu_{0i}(x,y) \times m_{st}(x,y) + \mu_{st}(z) \times m_{i0}(z). \qquad (3)$$

Since these two terms may have opposite signs, and (at least for the α-band) are often small the prediction of the CD-sign for a given absolute configuration needs careful individual discussion in each case.

2 HELICITY RULES

Fig. 1 shows a simple but typical case, viz. that of 6,7-dimethoxy-1,2,3,4,4a,9,9a,10-octahydroanthracene (**1**), whose molecules are chiral, but they still have a C_2-axis. The first two Cotton effects are negative for the given absolute configuration and are relatively small; the strong ß,ß'-bands are split and have opposite CD-signs. This behaviour is expected, since only the latter transitions are electrically dipole allowed already in benzene. For this type of chormophore (with C_2-symmetry) the rule holds, that the first CD-band (order always from long to short wavelengths) is negative for the helicity shown in Fig. 1. This type of Helicity Rule is applicable when the benzene ring is built into another chiral ring and no further perturbing substituents are close to this chromophore. It must be emphasized, that the directions of both $\vec{\mu}$ and $\vec{m}$ are determined by the substitution pattern of the benzene chromophore, so for the same helicity also positive Cotton effects may be observed with other groups or other positions of substitution on the aromatic ring.

3 SECTOR RULES

Chirally arranged groups will also contribute to the CD if they are not too distant from the chromophore, and their influence acts through bonds and also through space. The first interaction needs an antiperiplanar train of bonds (like in long-range coupling in NMR), and is observed only for near groups, for the second type of contributions Sector Rules are applicable. In general these conform to the local symmetry of the substituted chromophore, thus for a tetralin it must at least be a quadrant rule, but one may also expect on the other extreme a rule with altogether 16 sectors (cf. fig. 2)! Also for a sector rule the signs of contributions will, however, again depend on the substitution pattern. For the four most important sectors the signs given in fig. 2 were proposed mainly on the basis of CD-curves of tetrahydroisoquinolines (refs. 4,5). We could recently prove this unequivocally (ref. 6) by the synthesis of **2** with known absolute configuration: the skeleton has a plane of symmetry, chirality is introduced only by the 2 gem-dimethyls, which lie both in the same sector. The CD within the α-band is indeed positive (fig. 2) as predicted by the previously proposed sector rule.

The important influence of the substitution pattern upon the sign of the α-band CD within 3 series of homochirally analogous benzene derivatives shows fig. 3. The data are in part from the literature (ref. 7) and in part our own;

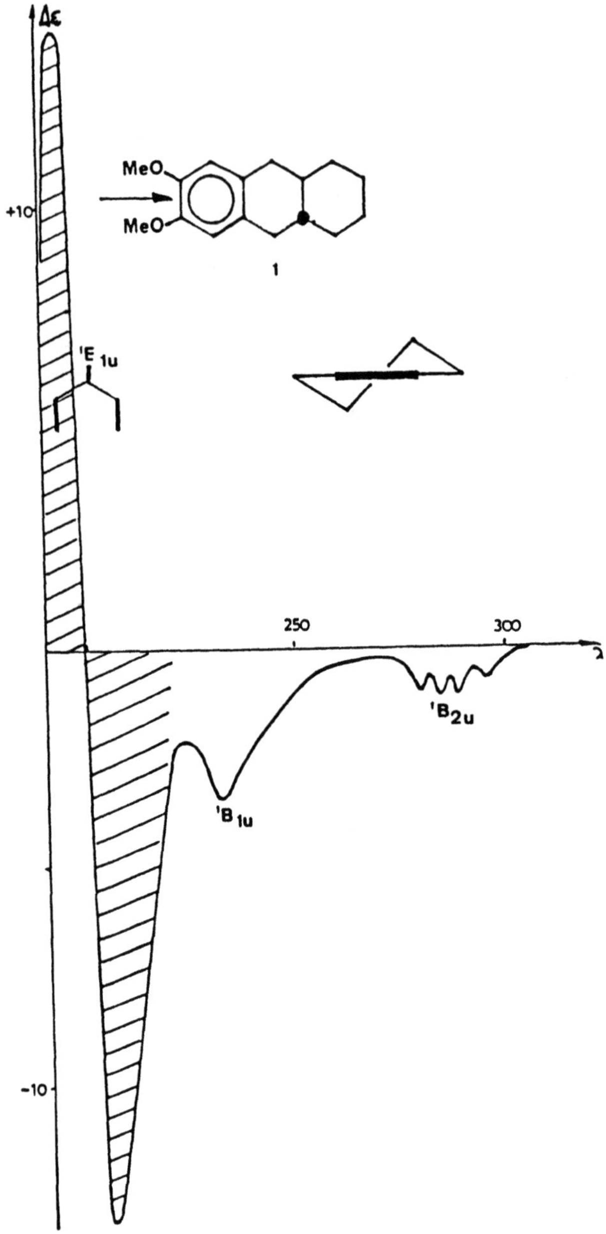

Fig. 1:Typical CD-curve for a relatively simple benzene derivative: same signs of the first two CD-bands, split CD (appearance like in CD-couplet) within the ß,ß'-absorption band.

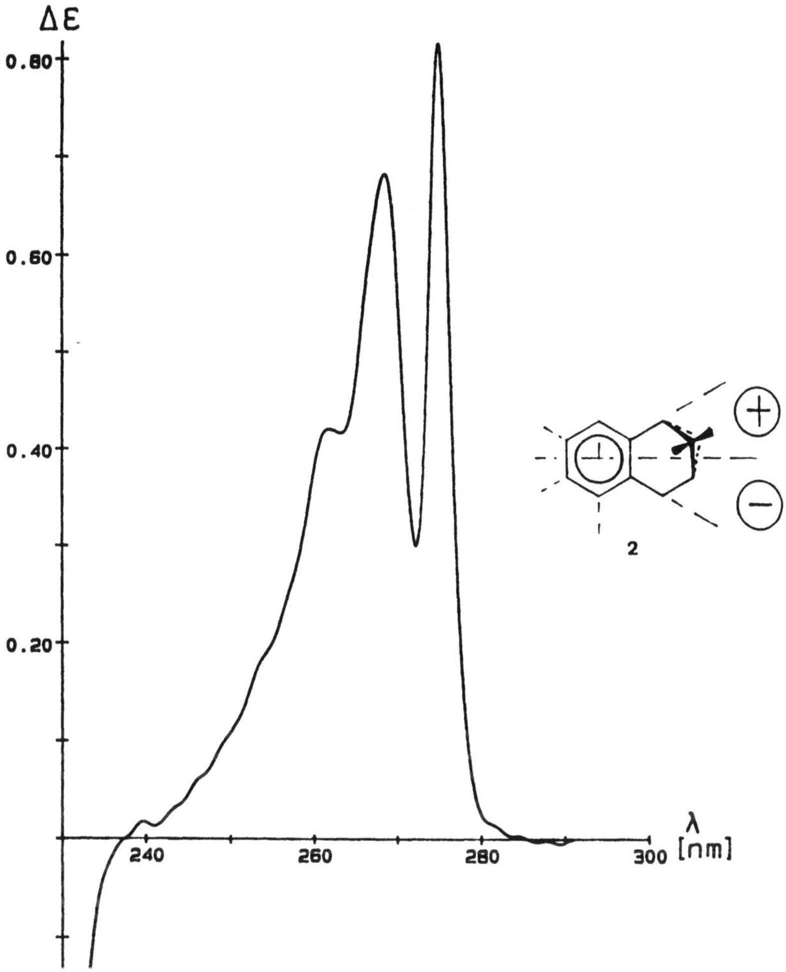

Fig. 2:CD (within the α-band) of benzonorpinene.

for tyrosine we observed even a dependence of the sign on the solvent used for the measurement. Another example of "warning" may be cited (ref. 8): the CD-spectra of the two closely related and homochirally analogous alkaloids (+)-methyl anhalonidine (**3**, positive) and (-)-anhalonine (**4**, negative). As X-ray diffraction showed (ref. 9) the piperideine rings of these two alkaloids have opposite helicities also in the crystals.

Another case of change of the CD-sign due to a small change of substituent may be cited from our same review (ref. 8): the CD within the α-band of estradiol (**5a**) or its methyl ether **5b** is negative, that of its acetate **5c** is positi-

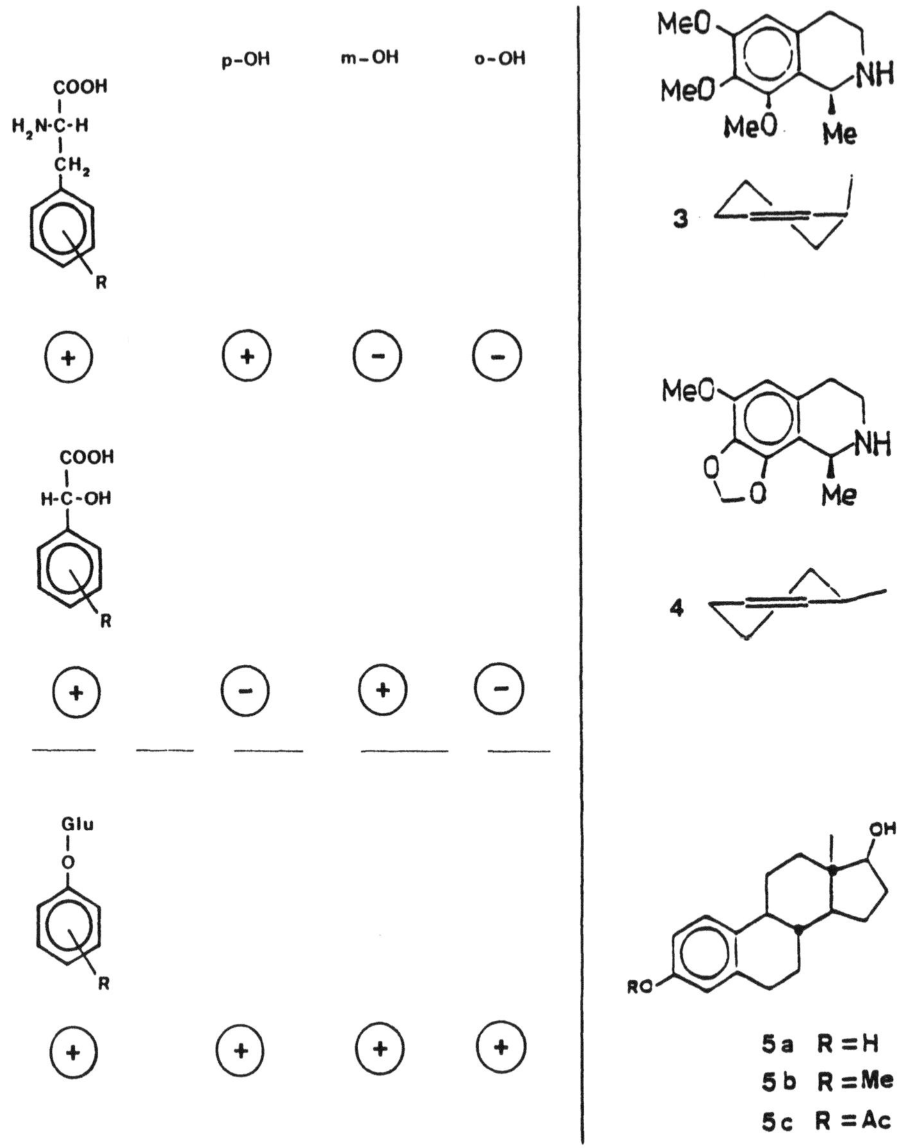

Fig. 3:Signs of the first Cotton effect within three related series of aromatic compounds.

ve. These results can be understood (but yet not always unequivocally predicted) by making use of the so-called "spectroscopic moments" introduced by Platt (ref. 10), and elaborated later in more detail by Petruska (ref. 11). These and other examples may be found in refs 12-15. Here we would like to give more recent ones, all having one additional benzene ring attached to the 2,3-position of ring A of cholest-2-ene, and differing in the further substitution (fig. 4, ref. 16).

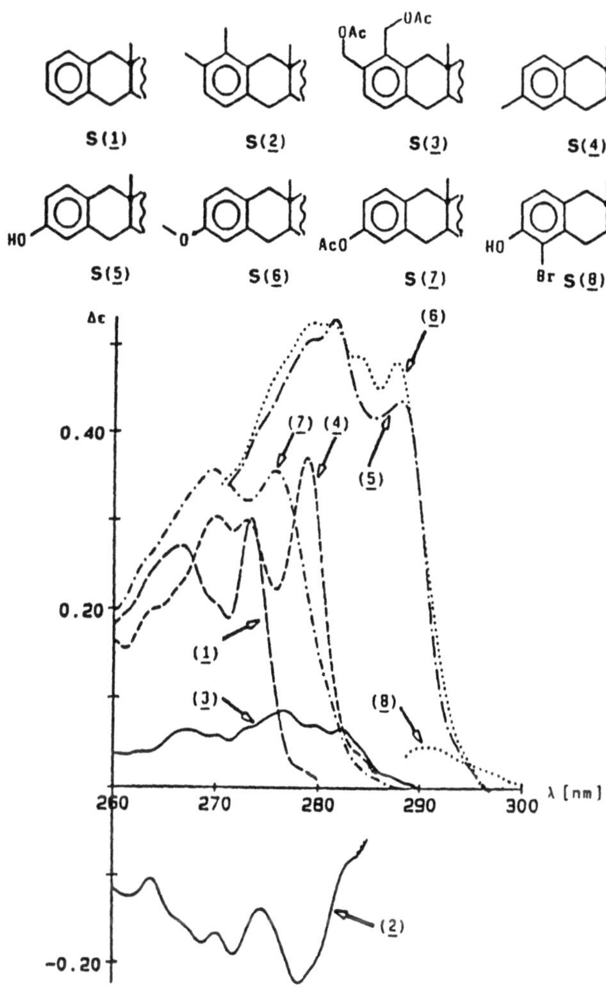

Fig. 4:CD-spectra (α-band only) of 8 differently substituted benzo[2,3]cholest-2-enes.

Another difficulty for the interpretation poses the case of CD-spectra in which the Cotton effect changes sign within one single band system. Usually such behaviour is explained by assuming the presence of (at least) two conformers in equilibrium, but this must by no means be the only reason. As has been proven experimentally in partially oriented matter (ref. 17) conjugated enones may show different signs for different progressions, i.e., for different orientation of $\vec{\mu}$. That similar results may be expected for benzene derivatives shows fig. 5. The ß,ß'-band system is electrically dipole-allowed in x- and y-direction, but magnetically forbidden. In the same figure the electron flow during excitation is also indicated. In z-direction within the chiral ring attached to

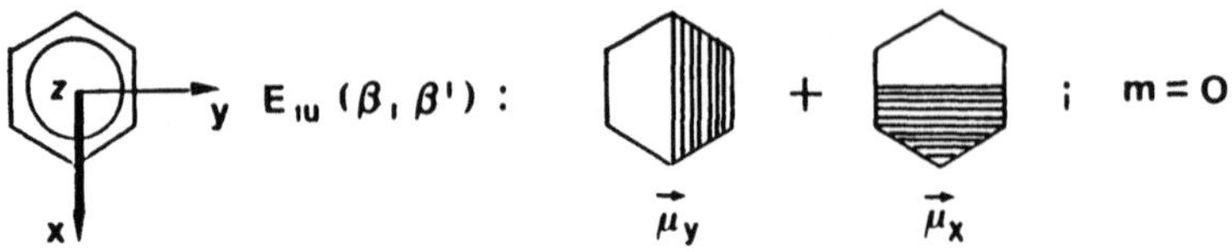

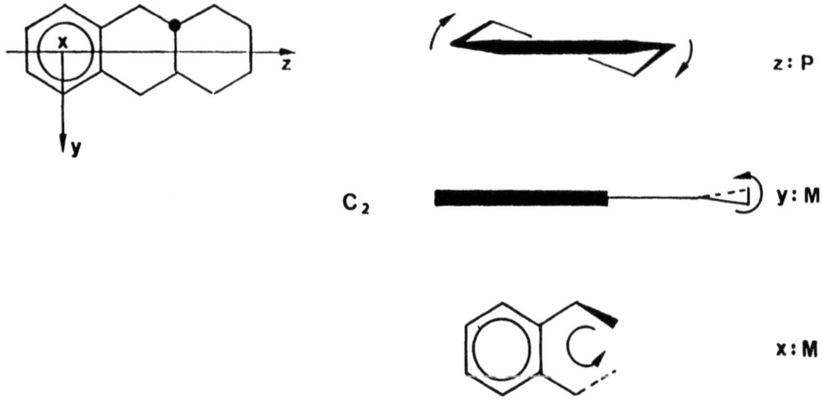

Fig. 5:Helicities (P- or M-) for the projections along the three co-ordinate axis of a decahydroanthracene.

the benzene ring a righthanded helical movement of electrons is induced during any transition within the α-band, whereas along the y- and the x-axis such electron flux follows clearly an M-helix, if the absolute configuration drawn in fig. 5 is assumed. Occurrence of different helicities within one and the same molecule due to different directions of projections may be better known from chemistry of complexes: when projected along the two-fold axis C_2 the helicity has opposite sense to that along the threefold axis C_3 (ref. 18). Since the respective progressions of lines differ for different orientations within the molecule, the bisignate behaviour (ref. 19) of many CD-band systems within one single benzene transition becomes understandable. In fig. 6 two extremely pronounced finestructure series within the α-band are shown for 2 related compounds **6** (ref. 20) and **7** (ref. 21) (see later).

Another way of looking into this problem of bisignated CD-curves is the following: the α-band of benzene with D_{6h}-symmetry is electrically as well as magnetically forbidden, the two branches (ß,ß') of the degenerate E_{1u}-transition are magnetically forbidden, but electrically allowed (and polarized in x- and y-direction). Addition to the two rings as in fig. 7 reduces symmetry to

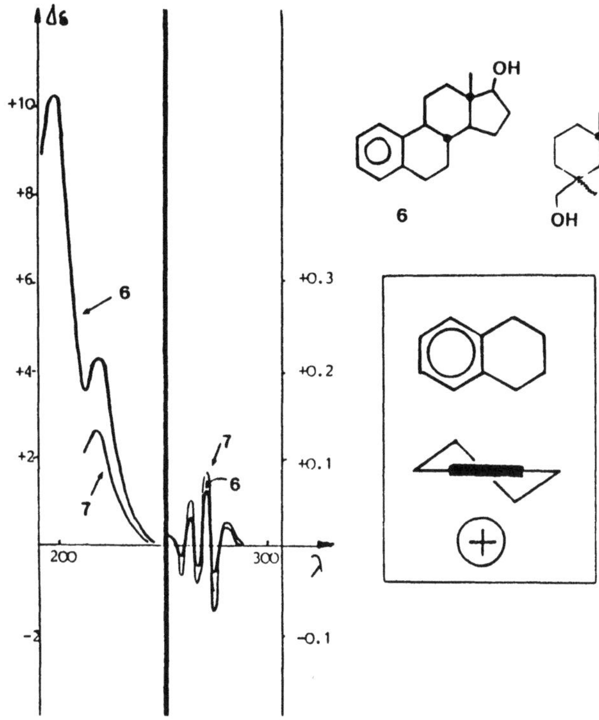

Fig. 6: CD-curves of **6** and **7** with unusually pronounced bisignate finestructure.

C_2, thus the B_{2u} - transition becomes an A-type ($\vec{\mu}$ and $\vec{m}$ both being allowed), the degenerate E_{1u} - transition is split into a B- ($\mu_y \neq 0$, $m_y \neq 0$) and an A-transition ($\mu_z \neq 0$, $m_z \neq 0$). Since excited one-electron configurations of same symmetry mix with each other, this will also be so with the two A-transitions, but by this the finestructure is barely influenced. In order for the B-state to mix in, one needs one quantum of the non-totally symmetric vibration of symmetry b, thus the 0-0 - lines of these two partial band systems will not coincide, and their CD-signs must not be the same. In this latter case the "bisignate" picture of the CD within the α-band may be observed.

If chiral arrangements of benzylic bonds as chosen in figs. 5 and 6 lead to positive CD-contributions, then the opposite must give a negative one: the benzylic substituent $X \neq H$ in quasiequatorial conformation of fig. 7 has indeed such a strong negative contribution, that the originally positive CD ($X = H$) is more than overcompensated by the OH (fig. 8). On the contrary, an axial OH in that position, lying nearly perpendicularly to the plane of the benzene ring, is not expected to influence the CD, and fig. 8 shows that this is indeed the case (compounds **8a**, **8b**, and **8c** of fig. 8). Even these delicate influences of the stereochemistry of a benzylic OH are thus correctly predicted in a semiempirical way.

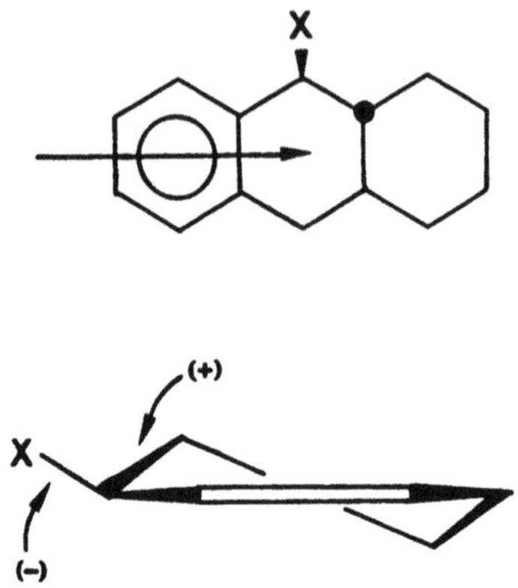

Fig. 7:Sign of contribution of a quasiequatorial benzylic X to the CD of a decahydroanthracene, in comparison to that of the chiral ring. A quasiaxial substituent is predicted to have nearly no influence, since it lies (nearly) in a nodal plane.

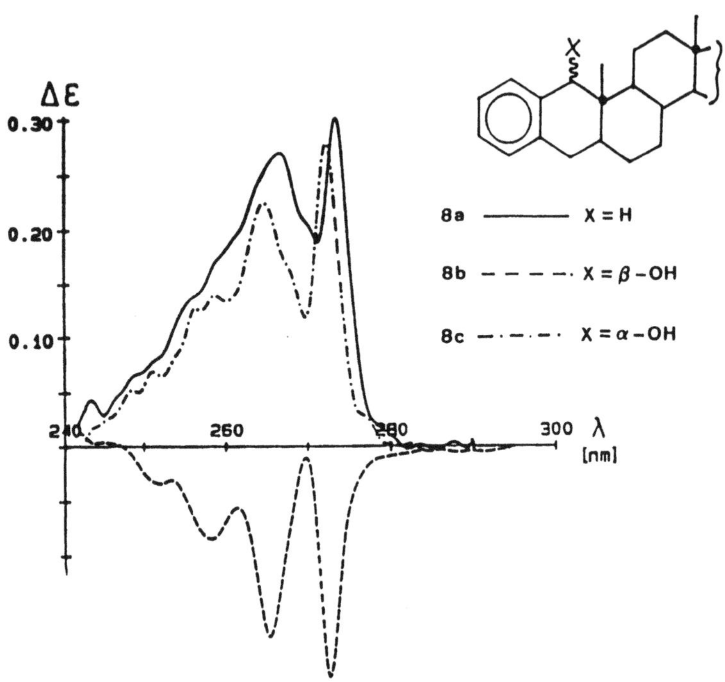

Fig. 8:CD-curves (α-band only) of 3 benzo[2,3]cholest-2-enes (**8a,b,c**) with different substituents in benzylic position.

4 SPECTROSCOPIC MOMENTS AND CD OF POLYSUBSTITUTED BENZENE DERIVATIVES

As mentioned, the electric transition moment of a polysubstituted benzene derivative can be estimated with the help of so-called "spectroscopic moments"; for the -band this is symbolized by a vector perpendicular to the bond towards the substituent in question; their magnitudes are tabulated, and for most substituents they point alternately clockwise and anticlockwise when going around the benzene ring (ref. 10,11). Is the spectroscopic moment "q", however, negative, then the corresponding arrow has bo be inverted. The translation of electron charge during the excitation is proportional to that arrow, but it takes, of course, place <u>within</u> the π-system of the ring, and not outside. The position as well as the magnitude of the 0-0 - line of a UV-band can very accurately be predicted by simple formulae, and we have tried to use these for the characterization of CD-spectra, too. Fig. 9 shows the CD of two p-disubstituted benzene derivatives; in the first (**9**) the two (chiral) substituents are identical, and the corresponding Cotton effect is negative (and bisignate). Is one of these groups replaced by the achiral but otherwise very similar ethyl group (**10**) then the CD-values are appr. half of those of the chirally disubstituted benzene, which seems reasonable, since the substitution pattern remained the same, but only "half the chirality" is present. A similar result is obtained by replacement of two of the three chiral substituents of **11** by the achiral neopentyl groups: the CD of **12** dropped to appr. 1/3 of that of **11**, but position and finestructure of the band remains mostly unchanged (fig.10).

Most interesting is now the comparison of the CD-spectra of the three analogues **13** through **15**: the q-value of the chiral group R* is appr. -4.5, that of the neopentyl group has the same absolute value, but is positive, and thus we do expect a compensation instead of addition of the two transition moments, the overall electric transition moment vector is practically zero, and so should then be the CD. This is indeed nicely born out by comparison of the Cotton effects: that of **15** is barely detectable in the noise (fig. 11), whereas that of **13** is small, but clearly seen in the spectrum. The q-value of the methyl group is also positive, and its absolute value is larger than that of the chiral group R*, viz. +5. For **14** a sign inversion is thus expected, if the magnetic transition moment is more or less independant of the substitution within the achiral substituent, and such has been found experimentally. The inversion of sign of the first Cotton effect by such a negligible change in the substitution pattern is remarkable, and it is very gratifying to note that the simple description of the benzene absorption spectra with the help of the q-values does also work appropriately for the CD-spectra.

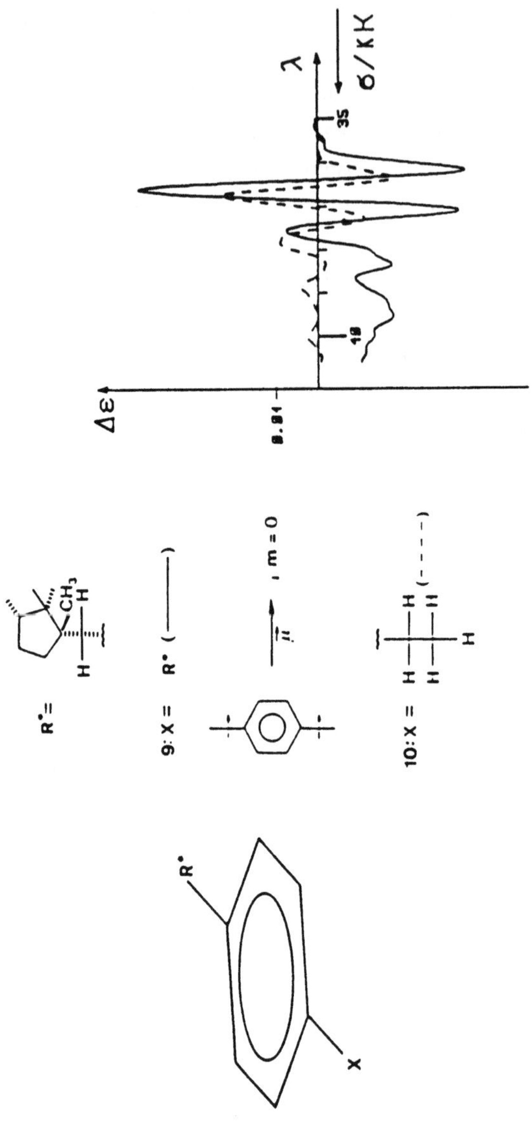

Fig. 9:CD-curves (α-band only) of **9** (two identical chiral p-substituents) and of **10** (one chiral, one achiral one).

5 APPLICATION TO SOME DITERPENE DERIVATIVES WITH OCTAHYDROPHENANTHRENE STRUCTURE

In a system like **1** (cf. fig. 1) the M-helicity of the chiral ring attached to the chromophore leads to a negative CD within the α-band, and correspondingly for P-helicity a positive CD is observed. In fig. 4 several examples for the influence of the substitution pattern onto the CD of a similar system has been shown, but this must not necessarily also hold for octahydrophenanthrenes,

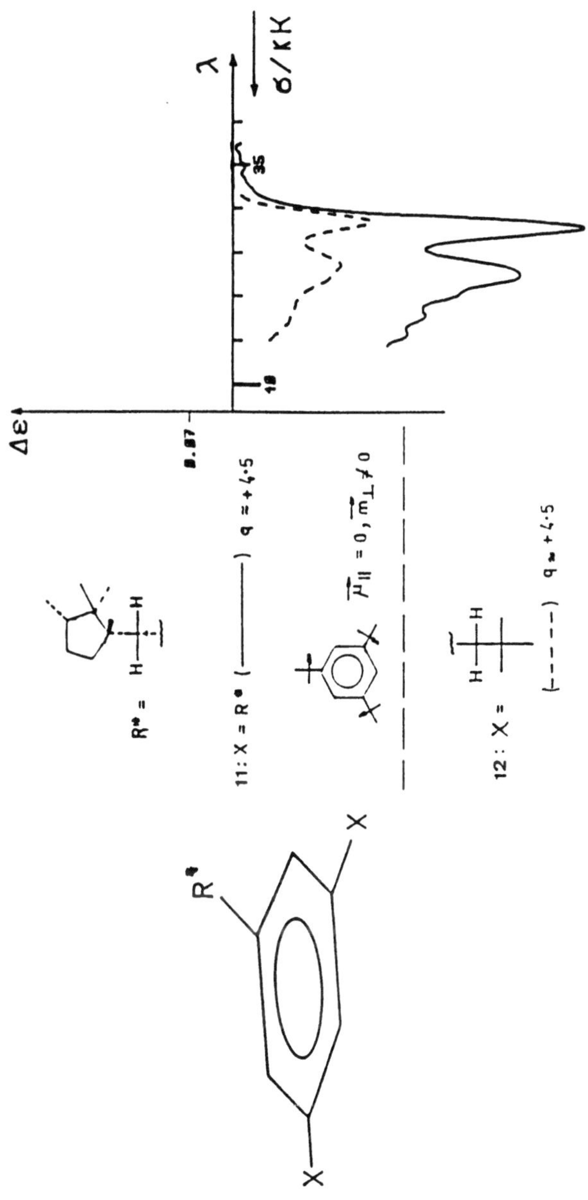

Fig. 10: CD-curves (α-band only) of the homochirally trisubstituted benzene derivatives **11** and of analogue **12** with only one such chiral group and two achiral ones.

since for them the two benzylic positions are differently substituted - one is secondary, the other tertiary. Fig. 6. proves that this is the case for not further substituted benzene chromophores, and in this special case the CD is bisignate with nearly equal positive and negative contricutions within the individual finestructure lines. The same figure furthermore shows that the difference

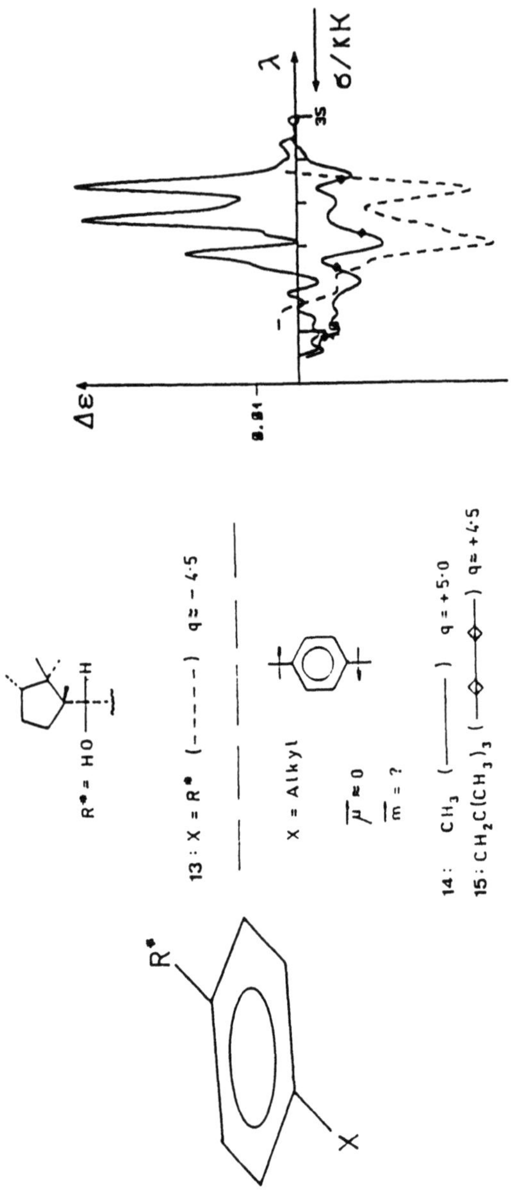

Fig. 11:CD-curve (α-band only) of 3 related p-disubstituted benzene derivatives **13** through **15**.

between secondary/tertiary and secondary/quarternary C-atom combinations does not result in a great difference of the CD-spectra. One is thus tempted to extrapolate that the two methyl ethers **5b** and **17** should show similar CDs, and this is indeed the case, although the magnitudes differ somewhat. Even introduc-

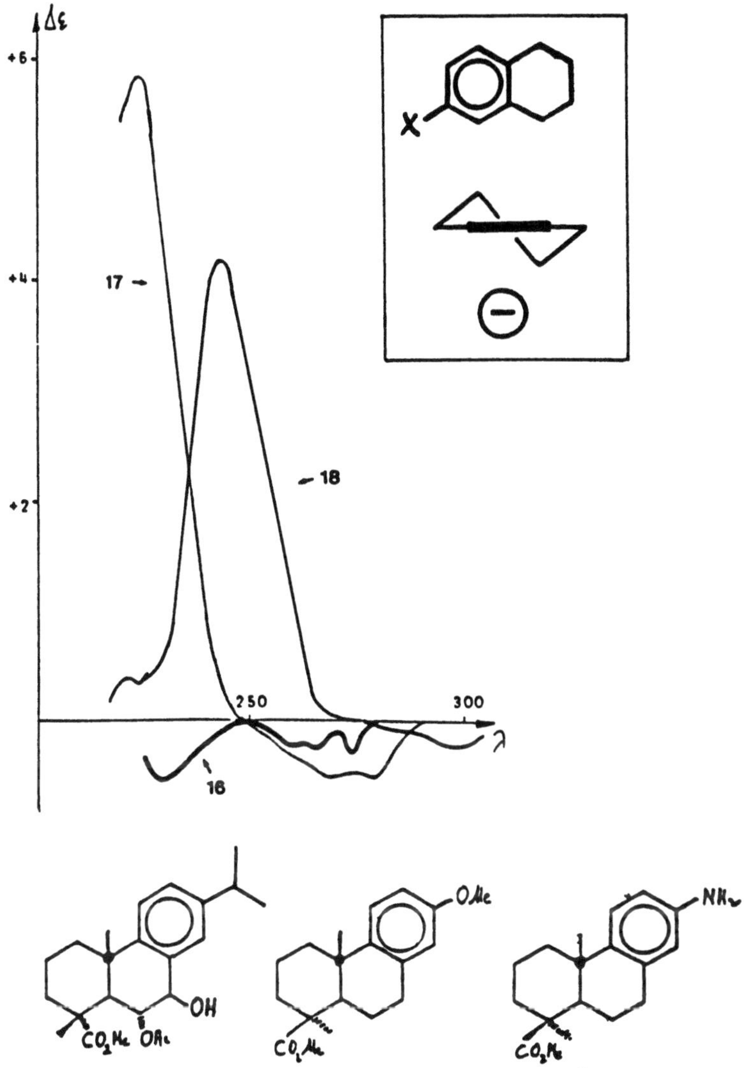

16 17 18

Fig. 12:CD-curves of 3 diterpene derivatives (**16**, **17**, **18**) with only one additional substituent on the benzene chromophore.

tion of a group like NH_2 with such a large q-value as 24 does not change this (negative) sign of the 1. Cotton effect (**18**), nor does so, of course, the isopropyl group of **16** (fig. 12). Although the situation is not fully symmetric when one switches over to the isomeric substitution type (**19** through **24**) the CD within the α-band retains its (negative) sign besides in one single case (**24**). It is, however, interesting to note that for **19** and **20** with additional RO-substitution in the benzylic position this Cotton effect is unisignate (fig. 13),

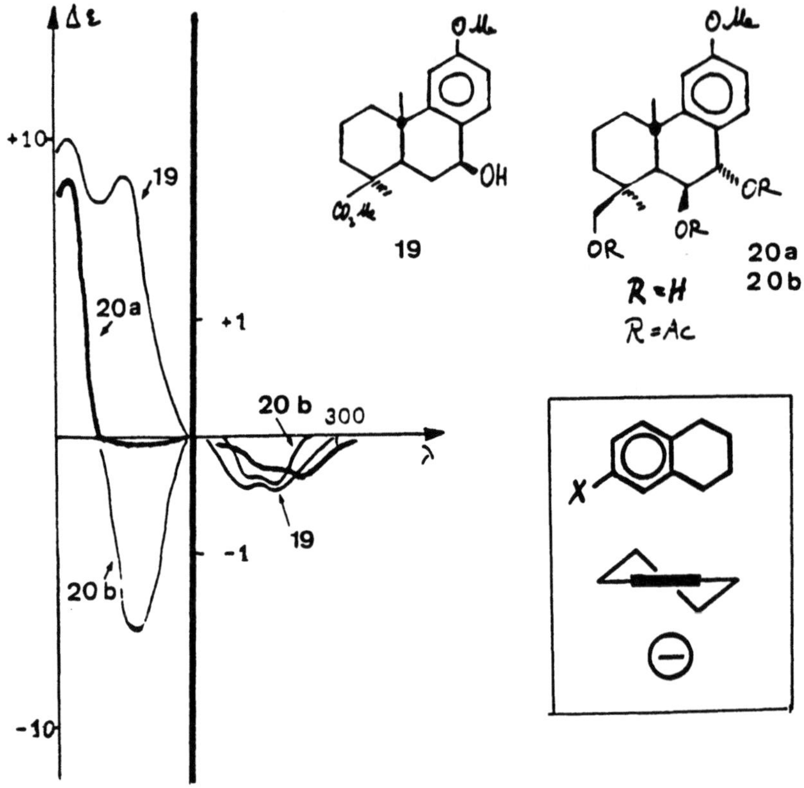

Fig. 13:CD-curves of the methoxy-decahydrophenanthrenes **19** and **20**, containing benzylic RO-groups.

whereas without this (**21** through **23**) again a bisignate Cotton effect is observed (fig. 14). As mentioned before, these CDs have to be called "negative", although the main area under the curve is rather positive (for **21** and **23**). We had not enough material to check whether the positive sign for **24** has to do with a possible H-bridge, which is, of course, absent in all the other compounds.

With the symmetric substitution type as found in **25** through **30** (figs. 15,16) a positive CD is more probable, and this is found so experimentally. **31** (fig. 16) seems to be an exception (bisignate "negative" CD), but one should keep in mind that for CH_2OH, which is the closest analogue found in the tables (refs. 10,11), the q-value is negative, thus the direction of the corresponding vector is inverted!

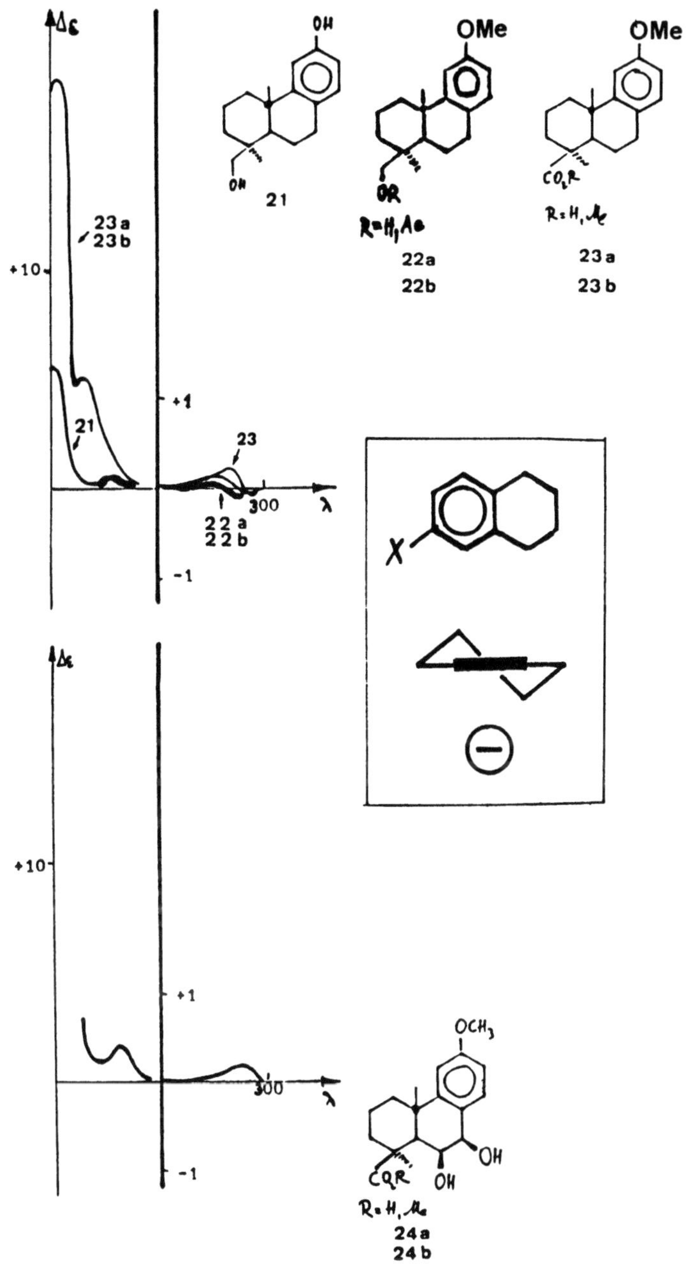

Fig. 14:Bisignate CD-curves within the α-band of 3 related oxygensubstituted decahydrophenanthrenes, and CD of a related glycol allowing internal hydrogen bridging.

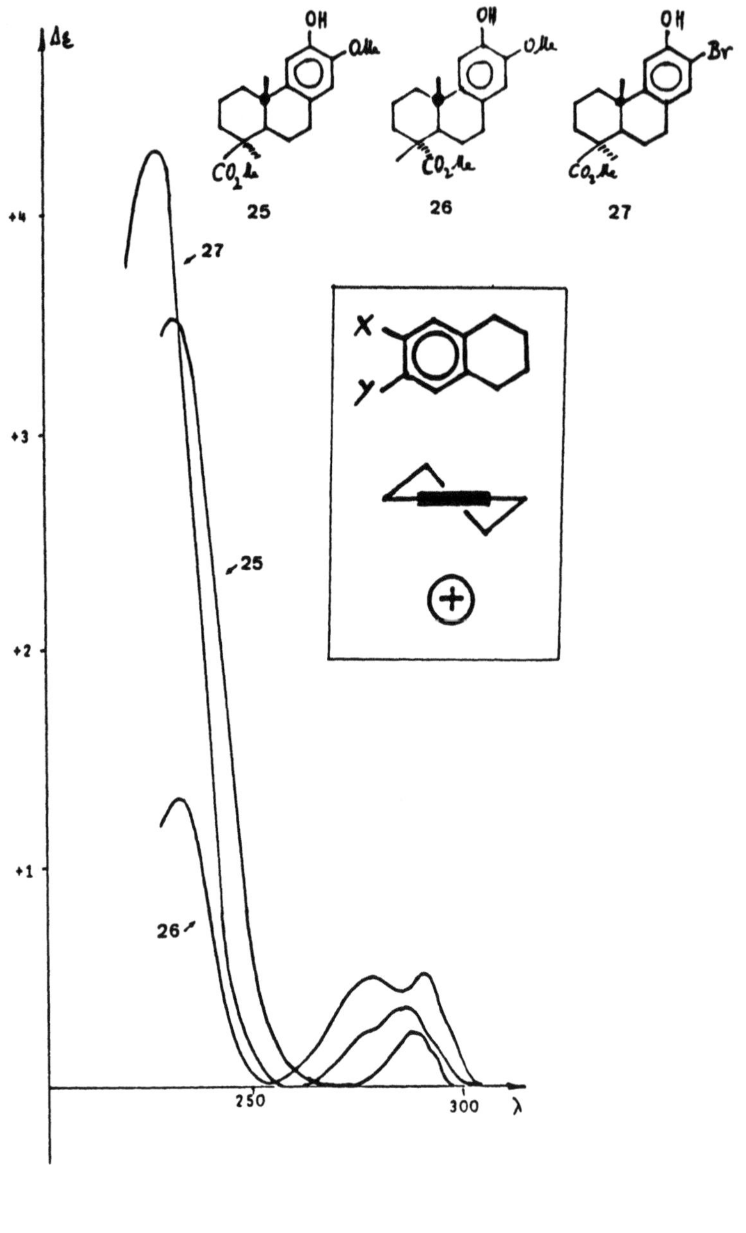

Fig. 15:CD of 3 related symmetrically substituted decahydrophenanthrenes.

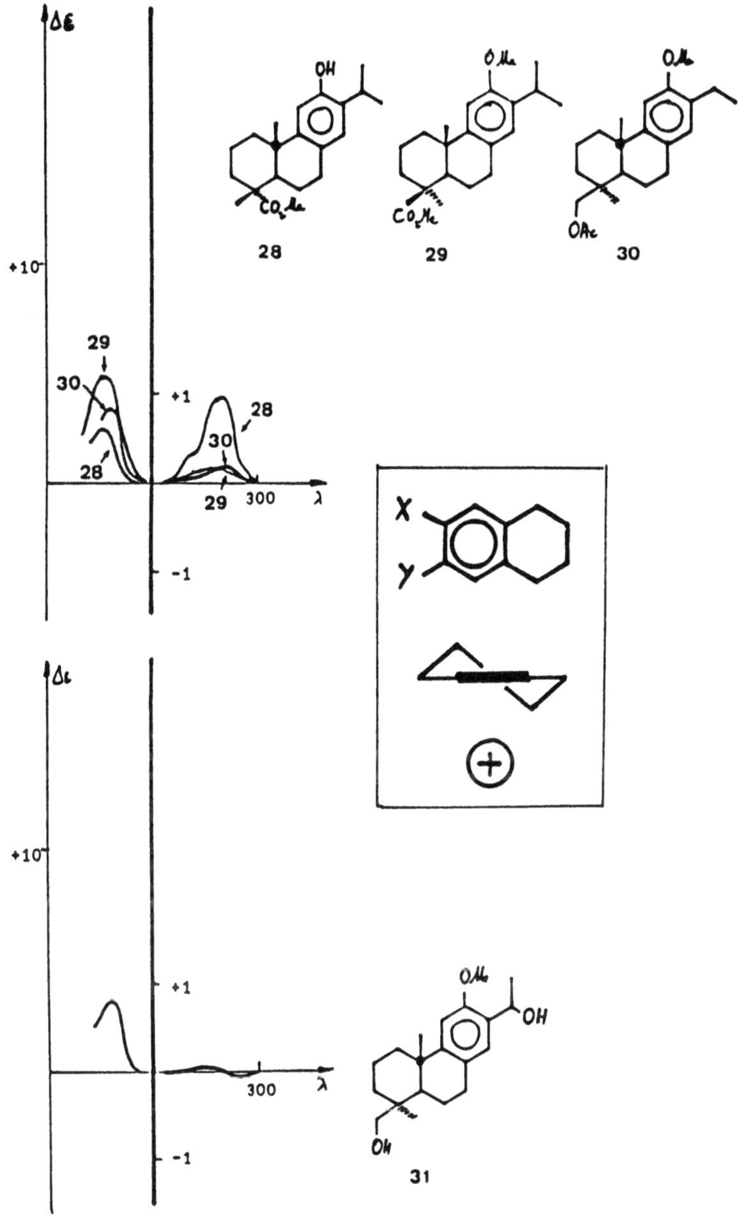

Fig. 16:CD curves of 4 decahydrophenanthrenes with same substitution pattern as in fig. 15. Top: both substituents with positive q-values. Bottom: q-value of one substituent (hydroxyethyl) negative.

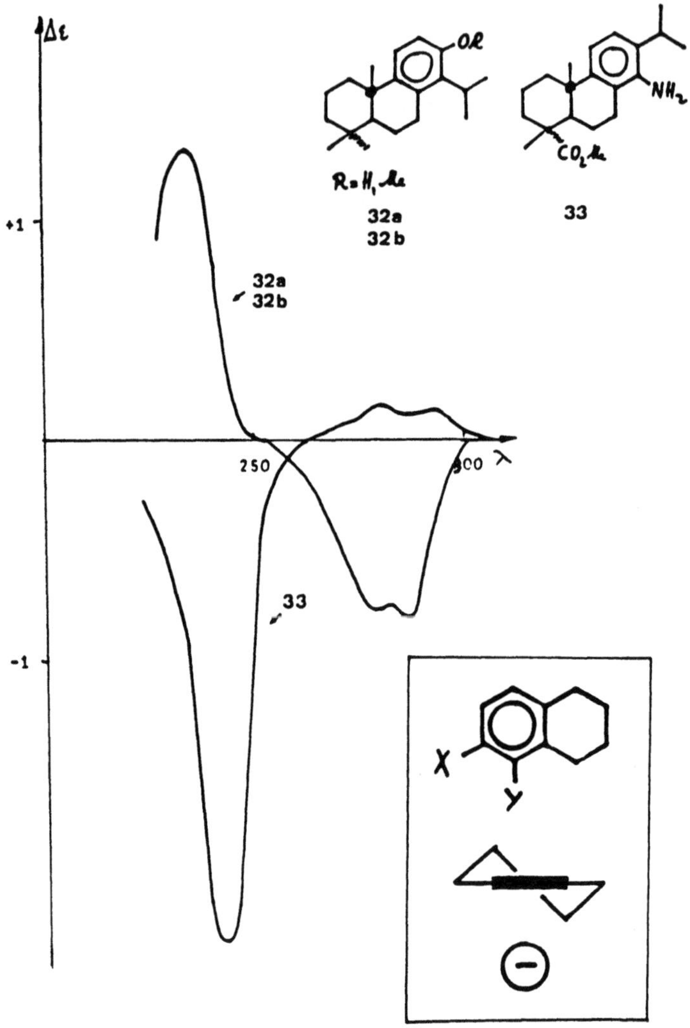

Fig. 17:N- vs. C-substitution on benzene chromophore (for details cf. text).

Finally, for the substitution type of **32** (fig. 17) a negative Cotton effect is also expected, and this is found so. The example of **33** shows, however, that one has also to consider the magnitudes of these q-values: the amino group of **33** is such a strong perturber, that it by far overrides the influence of the also present isopropyl group. The direction of the transition moment in **33** for NH_2 is (appr.) antiparallel to that of the attached alicyclic ring, thus practically no rotation of the sum vector is introduced as compared with the unsubstituted parent compound **7**. A moderately positive CD should then arise from this special substitution type, and this is again in agreement with the experiment.

ACKNOWLEDGEMENT - G.S. thanks Dr. H. Klein for many programmes and valuable technical help, Mr. U. Wagner for skilful measurements, Mrs. E. Sauerbier and Mr. P. Wolff for technical help, many collegues for providing me with interesting substances, HOECHST AG, Deutsche Forschungsgemeinschaft and Fonds der Chemischen Industrie for financial support.

REFERENCES

1 G.Snatzke, Angew. Chem. **91** (1979) 380.
2 G.Snatzke, Chemie in uns. Zeit **15** (1981) 78; - **16** (1982) 160.
3 N.Harada und K.Nakanishi, CIRCULAR DICHROIC SPECTROSCOPY - Exciton Coupling in Organic Stereochemistry - University Science Books, Mill Valley, 1983.
4 K.Kuriyama, T.Iwata, K.Moriyama, K.Kotera, Y.Hamada, R.Mitsui and K.Takeda, J.Chem.Soc. (B) (1967) 46.
5 G.Snatzke and P.C.Ho, Tetrahedron **27** (1971) 3645.
6 J.Schulte and G.Snatzke, in preparation.
7 T.M.Hooker and J.A.Schellman, Biopolymers **9** (1970) 1319.
8 G.Snatzke, Pure & Appl.Chem. **51** (1979) 769.
9 A.Brossi, J.F. Blount, J.O'Brien and S.Teitel, J.Am.Chem.Soc. **93** (1971) 6248.
10 J.R.Platt, J.Chem.Phys. **17** (1949) 484.
11 J.Petruska, J.Chem.Phys. **34** (1961) 1120.
12 G.Snatzke, M.Kajtár and F.Snatzke, XXIIIrd Interntl. Congr. Pure & Appl. Chem., Vol. **7** (1971) 117, Butterworths, London.
13 G.Snatzke, M.Kajtár and F.Werner-Zamojska, Tetrahedron **28** (1972) 281.
14 G.Snatzke, M.Kajtár and F.Snatzke, in F.Ciardelli and P.Salvadori (Eds.), Fundamental Aspects and Recent Developments in ORD and CD, Heyden & Son Ltd., London, 148 (1973).
15 G.Snatzke, Angew.Chem. **80** (1968) 15.
16 J.Schulte and G.Snatzke, FEBS, 2nd Internatl. Conference on C.D.,Budapest, 15.-18.8.1987.
17 H.-G.Kuball, J.Altschuh, R.Kulbach and A.Schönhofer, Helv.Chim.Acta **61** (1978) 571.
18 S.F.Mason, in F.Ciardelli and P.Salvadori (Eds.), Fundamental Aspects and Recent Developments in ORD and CD, Heyden & Son Ltd., London, 196 (1973).
19 W.Klyne and D.N.Kirk, Tetrahedron Let. (1973) 1483.
20 M.Legrand, personal communication.
21 W.Schoenfelder and G.Snatzke, Israel J.Chem. **20** (1980) 142.

Affiliations

** Department of Chemistry, University of Auckland, Private Bag, Auckland, New Zealand

* Lehrstuhl für Strukturchemie, Universität Bochum, Postfach 10 21 48, 4630 Bochum 1, FRG

Novel Carbohydrate Transformations Discovered en Route to Natural Products

B. FRASER-REID, R. MOOTOO, V. DATE, and S. HANDA

Department of Chemistry, Paul M. Gross Chemical Laboratory, Duke University, Durham, North Carolina 27706, USA

ABSTRACT

Progress in the synthesis of the ansa chain of streptovaricin A is discussed. The target molecule is an 11-carbon array containing 9 contiguous chiral centers, 2 of which are tertiary alcohols, and a third is a methoxycarbonyl group. The synthesis begins with levoglucosan (1,6-anhydro-β-D-glucopyranose), which by a process of pyranosidic homologation, is elaborated into a dipyranoside, and thence a tripyranoside which carries eight of the nine chiral centers. The ninth is installed by an acyclic stereoselective process.

There are two digressions. The first describes a convergent strategy to multiple contiguous chiral centers which involves aldol condensation of two sugars. The second reports a procedure for chemoselective activation of the anomeric oxygen.

Complex natural products present excellent opportunities for organic chemists to exercise their creative skills, and also for testing the limits of current synthetic methodology. In the course of these adventures, unexpected reactions are almost certain to occur, and although these are initially unwanted, it is from such reactions that new science arises. The very fact that they were "unexpected" means that the potential exists for the advancement of the discipline, even more so than might have been accomplished by the initial undertaking. In this lecture, I hope to discuss such a situation--our progress toward a complex natural product, and the benefits of some "unexpected" reactions.

1.1 INTRODUCTION

Following Masamune's pioneering synthesis of methimycin (ref. 1), other macrolides became less daunting synthetic targets, and since most macrolides are polychiral and heavily oxygenated, the development of compatible synthetic strategies became of subsidiary interest. The visual relationships between sugars and macrolides are obvious since both are polyoxygenated, and the possibility of a biosynthetic relationship had been advanced in an early scholarly treatise by Celmer (ref. 2). While it is true that the idea of a macrolide synthesis (erythronolide) from a sugar was first advanced by Miljkovic and co-workers (ref. 3), the earliest and most substantial advances were made by the Hanessian group (ref. 4). Since then, sugar-based approaches to macrolides have been numerous (refs. 5-8) and there have been spectacular successes.

An additional advantage of sugar-based syntheses was that the products arising therefrom would (most likely!) be optically active. However, our early interest in this avenue to natural products centered around stereoselectivity--not chirality (ref. 9). Chirality is a gift of nature and cannot be created--although it can be destroyed! On the other hand, stereoselectivity can be created by the ingenuity of the investigator. Indeed, this objective is a major component of the enterprise of synthetic organic chemists.

We were intrigued by the fact that reactions of carbohydrates were frequently more stereoselective than in comparable carbocyclic compounds. The reasons for this circumstance have long interested us (ref. 10) since our objective was to utilize the attributes of sugar to achieve the highest stereocontrol.

However, the carbohydrate literature shows that stereocontrol is usually excellent only at "on-template" sites (e.g., C2, C3, and C4) (refs. 5-8). Stereocontrol at "off-template" sites (e.g., C6) is less predictable, and C1 and C5 hold the template together and are therefore not available. Thus, there are only three reliable sites for synthetic manipulation (i.e., C2, C3, and C4), and the limitations imposed by this realization can be appreciated when one contemplates a sugar-based approach to the ansa chain of streptovaricin A, **4** (Scheme 2).

1.1.1 The Problem

Scheme 1

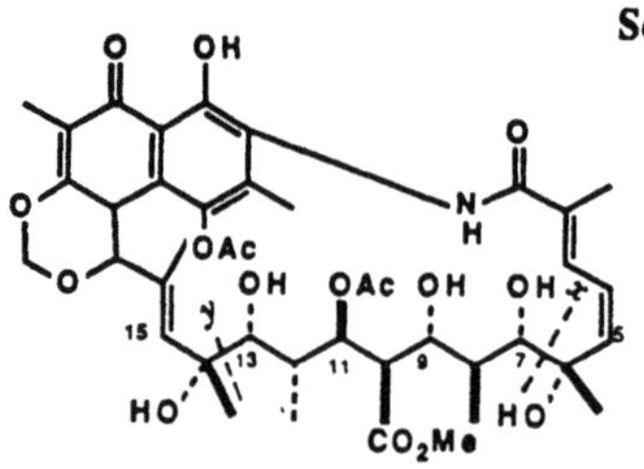

1
STREPTOVARICIN A

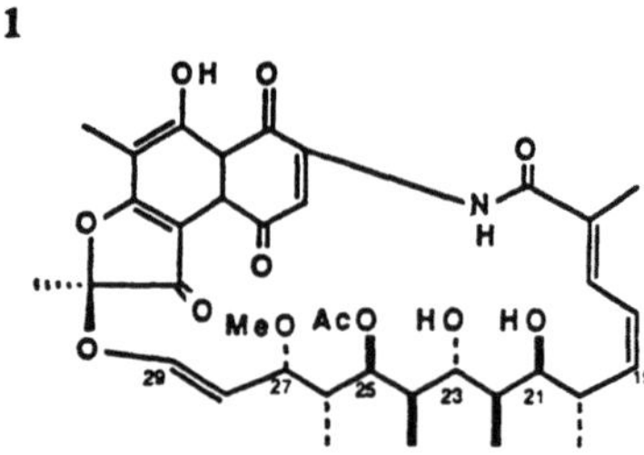

2
RIFAMYCIN S

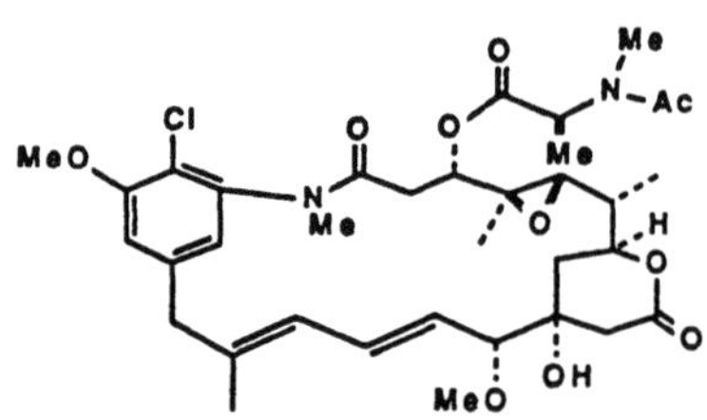

3
MAYTANSINE

Streptovaricin and rifamycin belong to the ansamycin group of antibiotics, which also includes maytansine **3**. The ansa chains (shown in bold in **1**, **2**, and **3**) are polychiral, but that of streptovaricin A is the most complex. Not only is the number of contiguous chiral centers (9) the greatest, but their "quality" makes it the most daunting. Thus, there are two tertiary alcohols, and an epimerizable methoxycarbonyl center flanked by adjacent oxygens which are prone to β-elimination. Indeed, the ansa chain of streptovaricin A, **4**, is the most complex assembly of contiguous chiral centers in the literature, and unlike the counterparts in **2** and **3**, it has thus far defied synthesis.

Our approach to this challenging synthetic target was designed to test the attributes of sugars for (a) stereoselective manipulation, and (b) ready structure proof by NMR techniques. The synthesis was intended to be self-consistent without having to be authenticated by comparison of the synthetic product with material obtained by degradation of the natural product--as is the usual practice (refs. 11-13). The latter prospect was particularly significant since, unlike the situation in rifamycin **2** (ref. 11), the ansa chain of streptovaricin cannot be obtained intact because cleavage occurs at the sites marked *x* and *y* in **1** (ref. 14).

1.1.2 Retrosynthesis

The nine chiral centers of **4** (Scheme 2) are contiguous and hence, there are no logical sites for disconnection. Thus, the synthetic union of sugar-derived

Scheme 2

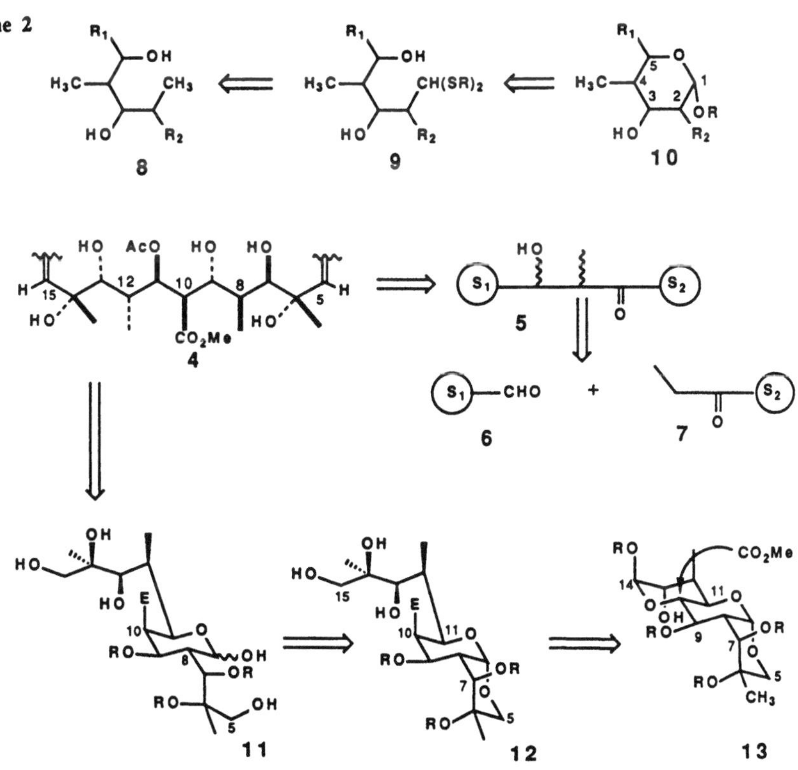

E = COOMe

subunits (e.g., by the aldol condensation of **6** and **7**) would generate a zone of uncertain stereochemistry at the point of union in **5** and authentication of the reaction products would be problematic. Efforts invested in the crafting of the subunits **6** and **7** could therefore be compromised or eroded by the uncertainties related to **5**.

If carbohydrate-based transformations are indeed as stereoselective as we had initially claimed (ref. 5), it should be possible to avoid such stereochemical uncertainties. Thus, our objective is to synthesize **4** in a manner that is not only highly stereocontrolled, but also does not require verification by external methods.

Our retrosynthetic plans have been detailed elsewhere (ref. 15), but the salient features arise from recognition that there is a repeating pattern of vicinal CH_3 and OH groups in the ansa chain, **4**, as a consequence of its biosynthesis from propionate residues. These features can be summarized by the fragment **8**, which could be derived from a pyranoside, **10**, *via* the dithioacetal **9**, as pioneered by Emil Fischer (ref. 16), and Raney nickel reduction, as pioneered by Wolfrom (ref. 17).

The basic methodology is therefore well-grounded in the carbohydrate domain, and the further retrosynthetic considerations continue this trend. Thus, the C8-CH_3 can be used to hold two rings together by formation of an internal acetal, **12**. Sacrifice of the C10 and C14 stereocenters allows three rings to be assembled in **13**.

We therefore go from a mono- to di- to tripyranoside (**11** ---> **12** ---> **13**) in a process which we have defined as "pyranosidic homologation" (ref. 15). The tripyranoside **13** can accommodate eight of the nine chiral centers of **4** on the three pyranoside rings.

1.1.3 The Tripyranoside 13

The first plateau was the preparation of the tripyranoside **13**. A number of approaches were tried and the details of these have been published (refs. 18-20). The salient features of the preferred strategy are summarized in Scheme 3. The starting material was levoglucosan, **14** (obtainable by pyrolysis of starch (ref. 21), which is readily converted into the epoxide **15** (ref. 22). Reaction of the derived branched-chain compound **16** with acidic catalysts effected acetal exchange leading to both **17** and **18**. On the basis of the anomeric effect (ref. 23), the axial anomer, **17**, is expected to be favored, as was indeed found to be the case.

Compound **17** is reminiscent of a *cis*-fused decalin, and reactions are therefore expected to occur preferably from the *exo* (convex) face. Indeed, epoxidation of **17** gave **19** exclusively, and acid catalyzed hydrolysis proceeded in the expected *trans*-diaxial manner. A W-coupling, ascribable to the H5 and H7 protons shown in **20**, verified the C7 stereochemistry and, by corollary, the adjacent C6 center. One of the tertiary alcohols of **4** was therefore secure in **20**.

Scheme 3

(For details see references 18 and 19)

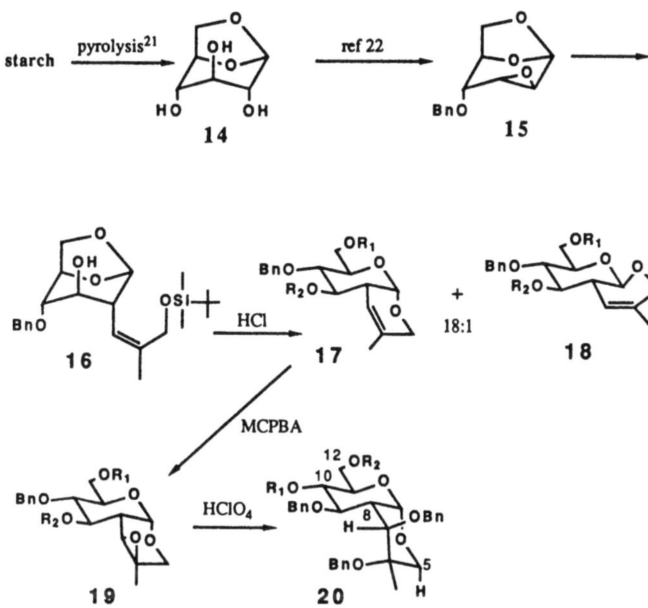

With the dipyranoside **20** in hand, we were ready to add the third ring. The most efficient route involved intermediacy of the acetal **21**, prepared by olefination of the corresponding aldehyde with the Bestmann/Wittig reagent (ref. 24) shown in Scheme 4. Although compound **21** was obtained as a *cis/trans* mixture, our studies showed that reaction of both geometric isomers with an alcohol in the presence of pyridinium *p*-toluene-sulfonate (ref. 25) afforded the unsaturated glycoside, **23**. The specific role of pyridinium *p*-toluenesulfonate in this reaction seemed to imply that the reaction proceeded *via* conjugate addition of pyridine to the enalonium ion **22**, as outlined in Scheme 4.

Notably, compound **23** was obtained as the axial anomer--again, because of the anomeric effect (ref. 23), and again, this stereochemistry was necessary to induce hydroxylation of the double bond from the opposite face. The diol **24** was converted into compound **25** by a procedure developed specifically for this requirement, and thence into the desired tripyranoside **13** (ref. 27).

The three chiral centers at C9, C10, and C11 of **13** were transferred from levoglucosan **14**. The other six were introduced with such stereocontrol that, with the exception of **18**, epimers were not detected. In addition, the ^{1}H NMR spectrum of **13** was so well resolved that the stereochemistry of the entire molecule could be assigned unambiguously at 200 MHz.

Scheme 4

(For details see reference 26)

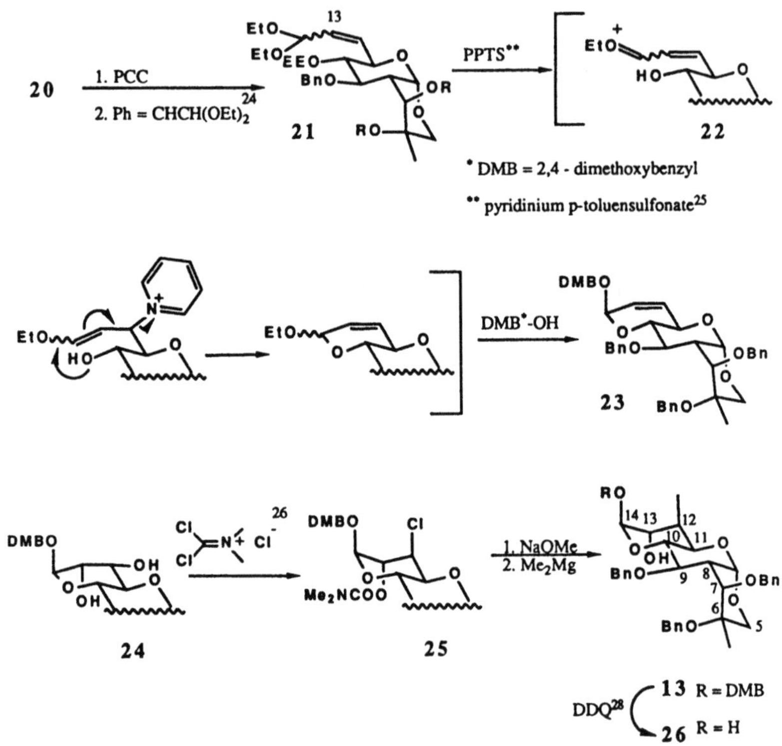

These two observations meant that the twin advantages of pyranosidic homologation, namely, stereocontrol and structure proof which were declared as our initial objective had been fulfilled.

1.1.4 The C10 and C14 Centers

Re-examination of the retrosynthetic plan (Scheme 2) shows that with the tripyranoside **13** in hand, the C10 and C14 stereocenters now had be addressed. It is evident that before either task could be undertaken, it would be necessary to open the "upper" pyranoside ring. It was in anticipation of this requirement that we had used the dimethoxybenzyl alcohol for the glycosidation leading to **23**. Oikawa and co-workers have shown that this protecting group can be removed chemospecifically by oxidation with DDQ (ref. 28), and the hemiacetal **26** was obtained thereby.

It was necessary to replace the C10 oxygen of **26** with inversion of configuration; however, all attempts at SN2-type displacements were completely unavailing (ref. 29). We therefore had to resort to a oxidation/Wittig/hydrogenation strategy (Scheme 5a) which led to a rich mixture of the isomers **29** and

Scheme 5

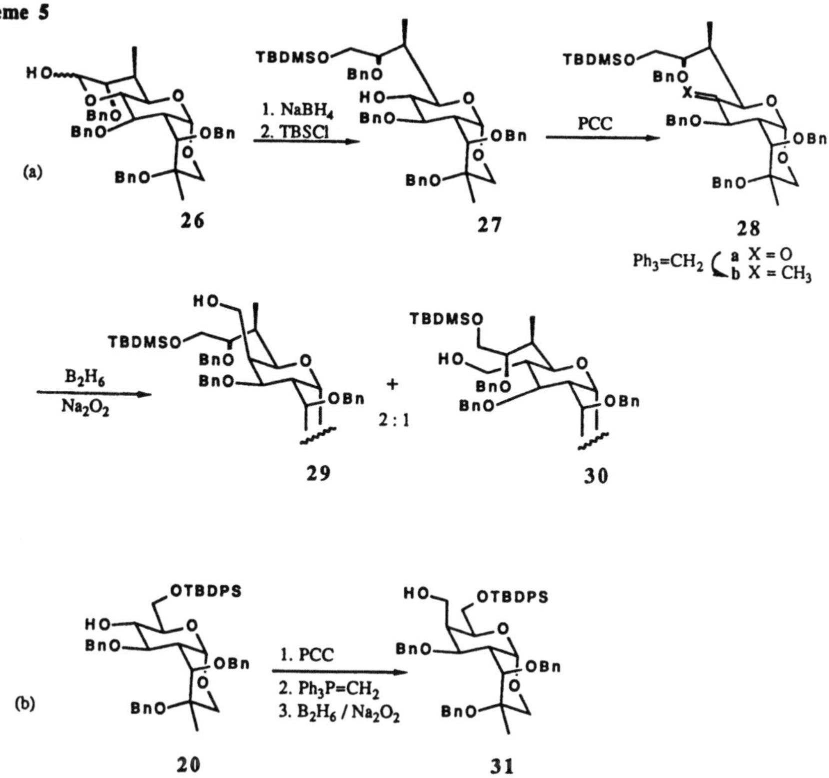

30. This result was disappointing, but in an occurrence, which is presently unexplainable, it was discovered that if these operations were carried out on the dipyranoside **20** (Scheme 5b) (i.e., before undertaking elaboration of the C12 and C13 stereocenters), the process was completely stereoselective, giving **31** as the only primary alcohol isolated (ref. 29).

Scheme 6
(For details see reference 20)

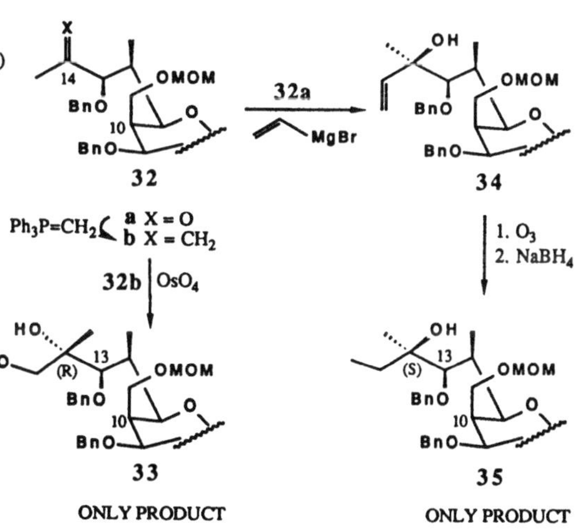

Located at an "off-template" site, we expected the C14 center to be more difficult to establish and verify. We were going to have to rely on precedents of acyclic stereoselection, and based on Kishi's empirical observations (ref. 30), hydroxylation of olefin **32b** should have given the R epimer **33**, preferentially. On the other hand, Cram chelate addition (ref. 31), as elaborated by Still (ref. 32), to the α-benzyloxy ketone **32a**, should give the allyl alcohol **34** from which the S-epimer **35** could be derived. Indeed, both processes were entirely stereoselective, and we were able to prepare the two possible products, each as a single isomer.

However, at this stage, it should be emphasized that the C14 configurations shown for **34** and **35** are based on empirical precedents only (refs. 30-32), and on these grounds, **35** is the isomer we need. Concrete structure proof remains to be undertaken.

1.1.5 <u>A Shorter, but Less Stereoselective Route</u>

In light of the results relating to the C10 and C14 centers, we can now claim to have obtained eight of the nine contiguous chiral centers of **4**, and with the exception of C14, we are absolutely sure of the configuration of each. With this strategy in hand, we have been able to explore a shorter route.

We noted above (Scheme 5b) that the bipyranoside **20** gave **31** as a single isomer. Processing of the latter led to the methyl ketone **36a** and thence olefin **36b**. As indicated in Scheme 7, one or other of the epimers of **37** could be obtained as the major product, depending on which hydroborating agent was used. The structure of **37** was assigned as shown because subsequent processing led to the allylic alcohol benzyl ether **32b**, which was identical to the product obtained in Scheme 6.

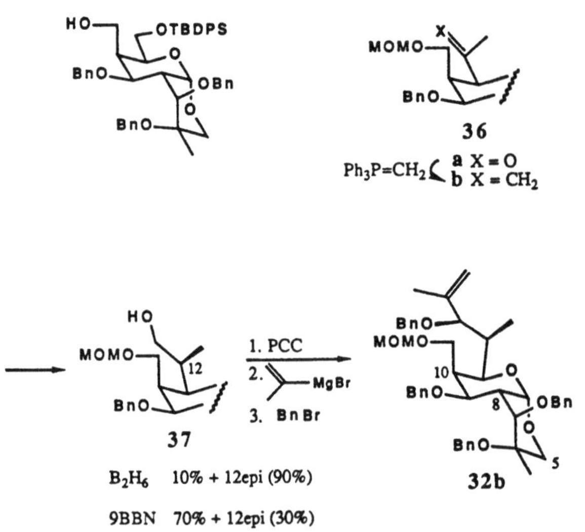

Scheme 7

The route summarized in Scheme 7 is therefore much shorter, although not as stereoselective, nor as easily verified as that shown in Scheme 4.

2.1 TWO DIGRESSIONS

The subject of "shorter" routes, provides an opening for discussion of other work currently underway in our laboratories. A convergent strategy for systems such as **4**, which involve the combination of prefabricated subunits, would be ideal. However, in the Introduction, we commented that such strategies (e.g., the aldol condensation of **6** and **7**, Scheme 2) might give unpredictable mixtures of **5** whose stereochemistry would be difficult to prove. In instances where such convergent reactions have been carried out (refs. 11, 12), the subunits (S_1 and S_2 in Scheme 2) were prepared from sugars. However, the principles of sugar chemistry played no role in the actual reaction event in which the two subunits were combined. Thus, the mixture **5** would have the same composition, regardless of the source of its precursors **6** and **7**.

2.1.1 Aldol Routes to Multiple Contiguous Chiral Centers

We have been exploring aldol condensations of sugars where their reaction course is directed by stereoelectronic factors which are unique to the carbohydrate domain.

Klemer and Rodemeyer (ref. 33) showed that the reaction of the dibenzylidinated mannopyranoside **38** with butyllithium led to ketone **40a** (Scheme 8). The enolate **39** was postulated as an intermediate and this has been established subsequently in our laboratory (ref. 34) and Chapleur's (ref. 35) by *in situ* alkylation to give **40c**. We established, indirectly, that alkylation gave axial adduct **40b**, with subsequent epimerization to **40c**, by means of a second alkylation of the latter, which led to the stereoisomer **40d**, exclusively. Thus, the alkylation of enolate **39** occurs from the β-face (ref. 36).

In short, the kinetic products of reaction at C2 are axial in these alkylation (electrophilic) reactions, just as with the well-precedented examples of nucleophilic reactions at C2 (ref. 37).

The surprising stability of enolate **39** towards β-elimination of the axial glycosidic OCH_3 is obviously related to its carbohydrate nature, since such an axial substituent on a carbocyclic ring would not have been expected to survive. We attribute this stability to the *exo* anomeric effect (ref. 38) which can be described in molecular orbital terminology by the $n\sigma^*$ overlap depicted in **41** (Scheme 8). This interaction confers some double bond character which strengthens the glycosyl-O-bond (Scheme 8).

We have shown that this phenomenon can also be exploited for aldol condensations (ref. 39). The kinetic (axial) products from the reactions, **42** and **43**, show much less tendency to epimerize than the alkylation counterparts, **40b**.

However, even more gratifying was the observation that the simple S and R sugar aldehydes, gave single epimers at C7 of **42** and **43**, respectively. These glyceraldehyde derivatives have been found to undergo Felkin-type additions (ref. 40), and the results in Scheme 8 show that the elements of stereocontrol in each of the reaction partners, operate jointly. As a result, only one of the four possible diastereomers is formed from each of the aldol condensations in Scheme 8.

Scheme 8

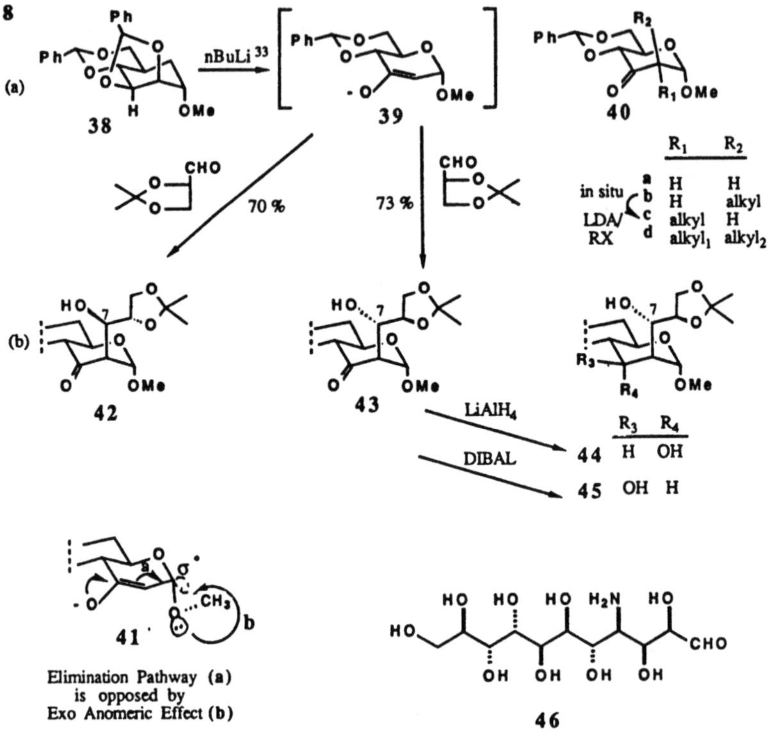

This fortunate observation could be extended even further. Thus, it is well established that reduction of sp^2 centers at C3 of pyranosides proceeds from the β-face (ref. 41), and it was therefore not surprising that reaction of **43** with lithium aluminum hydride afforded the axial C3-OH **44**, predominantly. However, by a fortuitous observation, it was found that reduction with diisobutylaluminum hydride led to the equatorial counterpart, **45**. Rationalizations for different reaction courses in these reductions are still being sought; however, the results in Scheme 8b show that phenomenological events, which are unique to the carbohydrate systems, can be exploited to elicit complete stereocontrol in the assembly of multiple contiguous chiral centers in convergent syntheses involving aldol condensations.

The relevance of this strategy for the elaboration of higher carbon sugars, such as hikizimycin **46**, is obvious.

The aldol strategy in Scheme 8b emanated from our interest in the alkylation of **39**, a reaction which was explored in connection with our sugar-based approach to the polyquinanes (ref. 36). Thus, our application of carbohydrates for natural product synthesis has returned to benefit the carbohydrate domain in view of the new potential that it has opened for the synthesis of higher carbon sugars (e.g., **46**). In the next section of this lecture, we will present yet another area of carbohydrate chemistry which has benefitted from our forays into natural product synthesis.

2.1.2 Chemospecific Activation of the Glycosidic Oxygen

In the course of examining strategies for creating the C-14 chiral center in Scheme 6, we had tried to obtain the bromohydrin of alkene **32b** by reaction with N-bromosuccinimide in an aqueous medium. Instead, the product was the

Scheme 9

(For details see refernce 42)

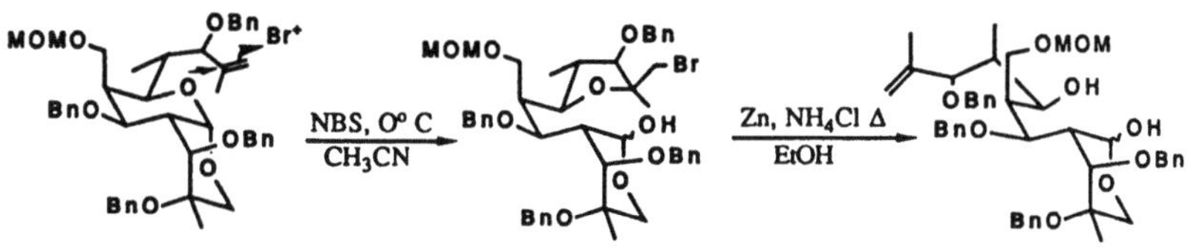

bromomethyl tetrahydrofuran **47** (Scheme 9), and subsequent reduction with zinc afforded alkene **48**. We had in effect "hydrolyzed" the dipyranoside by this tandem oxidation/reduction process (ref. 42). This was of interest to us because early attempts to effect hydrolysis of the internal acetal **32b** ---> **48** by means of protic or Lewis acid protocols resulted in destruction of the molecule.

The mechanistic rationalization is presented in Scheme 10 where the essential component of **32** is captured in **I**. A cascade of "onium" ions **II** --->

Scheme 10

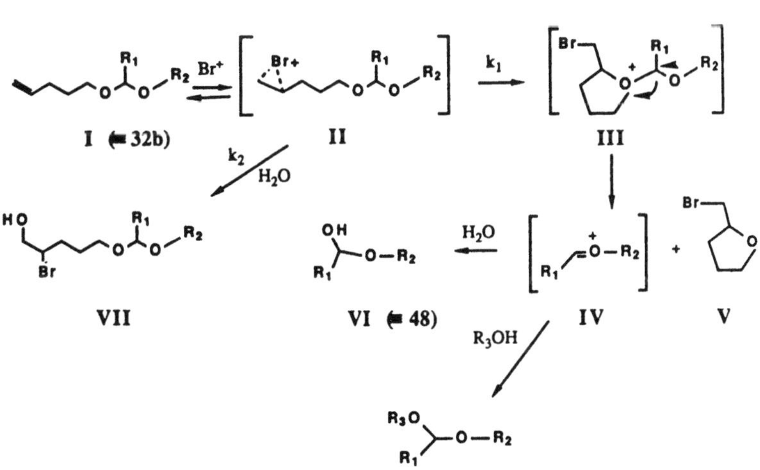

III ---> **IV** results in the formation of **VI** (≡ **47**). As a result of this analysis, an even more intriguing possibility can be envisaged which provides a procedure for general "hydrolysis" of acetals without using acid. This is particularly relevant to sugar chemistry since it implied that n-pentenyl glycosides could be "hydrolyzed" without the use of acids.

The particular advantages of such a procedure had been tested with substrates **49** (a-f) (Scheme 11). Thus, compound **49a** has three acetal groupings, but only the n-pentenyl glycoside is removed in the reaction with NBS. Other acid labile groups (e.g., silyl in **49b**) are also unaffected. Base labile groups (e.g., **49c**) also survive. Furthermore, the brominolysis conditions are so mild that other oxidizable protecting groups, such as benzyl (**49d**), and even dimethoxybenzyl (**49e**) also survive.

Bromohydrin formation (which had been sought in the original reaction of **32b**) is not a problem, and this observation implied that cyclic bromonium ions, such as **II** (Scheme 10), react preferentially with the acetal oxygen to give **III**. By corollary, this means that the intramolecular reaction rate (e.g., k_1) is faster than the bimolecular reaction k_2, and evidence for this comes from our observations that bromohydrin formation increases with the concentration of water.

Since formation of the cyclic bromonium ion is known to be reversible (ref. 43), other double bonds elsewhere in the molecule, which are incapable of forming an intermediate comparable to **III**, should be unable to compete with the acetal oxygen.

This postulate was tested by means of the allyl protected derivative **49f** (Scheme 11) which underwent deglycosidation to give **50f** in 72% yield. Therefore, this result suggests that the allyl groups undergo reversible bromination, but since they cannot form suitable cyclic oxonium ions, the equilibration favors formation of the oxonium ion, **III**.

From this study, it is seen that the n-pentenyl residue is a unique protecting group for the anomeric center permitting specific and chemoselective deglycosidation.

Scheme 11

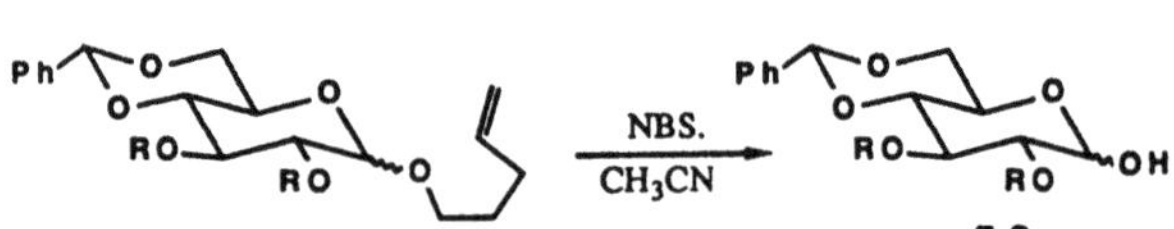

49 **50**

a R = EE

b R = $tBuMe_2Si$

c R = Ac

d R = $C_6H_5CH_2$

e R = p-$OMeC_6H_4CH_2$

f R = 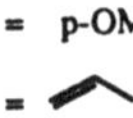

Scrutiny of Scheme 10 suggests that it should be possible to capture the intermediate **IV** with an alcohol. Such a reaction would lead to **VIII**, and the overall process **I** ---> **VIII** would be an acetal exchange--or in the case of sugars, a glycoside exchange. This has proved to be possible.

Thus, the n-pentenyl glycoside **51** (Scheme 12a) reacted in the presence of methanol to give the methyl glycoside **52**. An even more exciting development is shown in Scheme 12b where the "alcohol" is diacetone glucose. The reaction **51** ---> **53** allows the direct formation of the disaccharide from the activated glycoside.

The potential of this procedure for the syntheses of oligosaccharides is obvious. However, various questions must be answered (some dealing with mechanistic details, stereocontrol of the glycoside exchange, etc.) before the merits of the reaction can be appreciated.

Scheme 12

GLYCOSIDE EXCHANGE

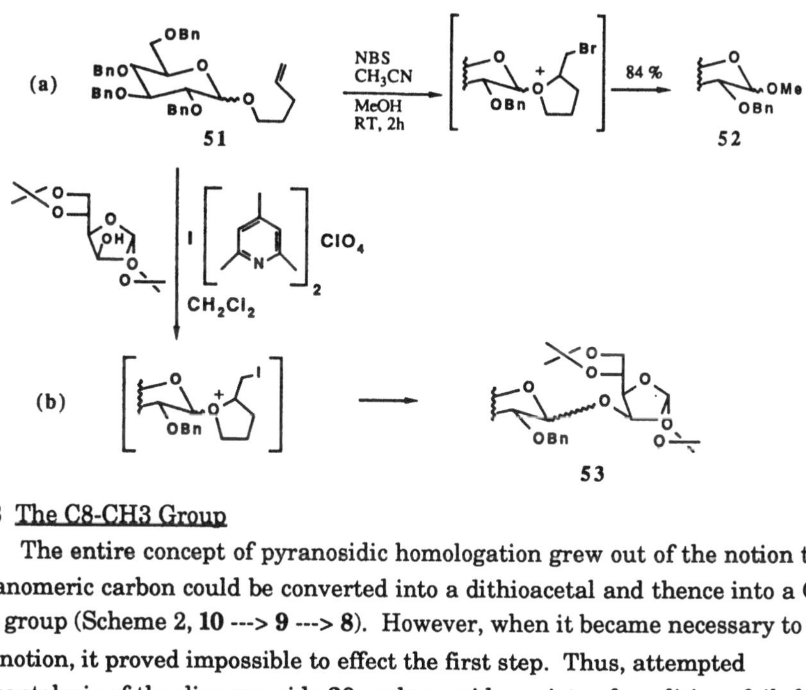

2.1.3 <u>The C8-CH3 Group</u>

The entire concept of pyranosidic homologation grew out of the notion that the anomeric carbon could be converted into a dithioacetal and thence into a C-CH_3 group (Scheme 2, **10** ---> **9** ---> **8**). However, when it became necessary to test this notion, it proved impossible to effect the first step. Thus, attempted mercaptolysis of the dipyranoside **20** under a wide variety of conditions failed to give the desired dithioacetal **55** (Scheme 13). That this was due to the acid lability of the system was shown by treating **20** with boron trifluoride etherate which caused ring contraction to **54**. Acetolysis then gave **56c**; however, mercaptolysis stopped at the thioglycoside **56b**.

Fortunately, this obstacle was removed when the oxidation/reductive hydrolysis (Scheme 9) was discovered. The dipyranoside could now be opened (**32b**

---> **48**). Could the anomeric carbon of the resulting mannopyranose, **48**, be reduced to CH_3?

Scheme 13

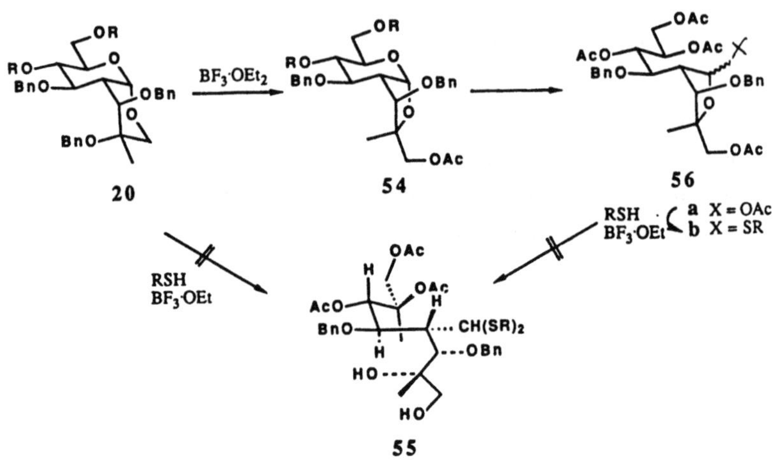

For an answer to these problems, we examined the model compound **58**, which was prepared from the dipyranoside **20** *via* the iodide **57** (Scheme 14). However, mercaptolysis also proved to be ineffective, leading only to the thioglycoside **59**. It became apparent that compounds such as **56** and **58** resist opening by conventional acid catalyzed protocols, and hence, new chemistry had to be developed.

In this connection, we had been impressed with some of Weinreb's science (ref. 43) (Scheme 15) which showed (1) that lactones (e.g., **IX**) could be opened with $Me(MeO)NAlMe_2$ (**Xa**), and that the resulting amide, **Xa**, showed no tendency for recyclization, even in the case of γ-lactones. (2) Reduction of amides (e.g., **Xa**) stop cleanly at the aldehyde stage **XI**.

In exploring this procedure, we found that the N,N-dimethyl analogue, **Xb**, was as efficient as the hydroxylamine equivalent (**Xa**) used by Weinreb (ref. 43). Thus, the hemiacetal **58** was processed to give lactone, **60**, which was then opened to the N,N-dimethylamide **61**. Reduction of the latter to aldehyde **62** was effected by using the Mukaiyama reagent (ref. 45).

For the deoxygenation of the aldehyde, a modern protocol involving intermediacy of the *p*-toluenesulfonyl hydrazone, as outlined below, was

$$RCHO \longrightarrow RCH{=}NNHTs \xrightarrow{NaBH_4} RCH_3 \qquad (2.1)$$

Scheme 14

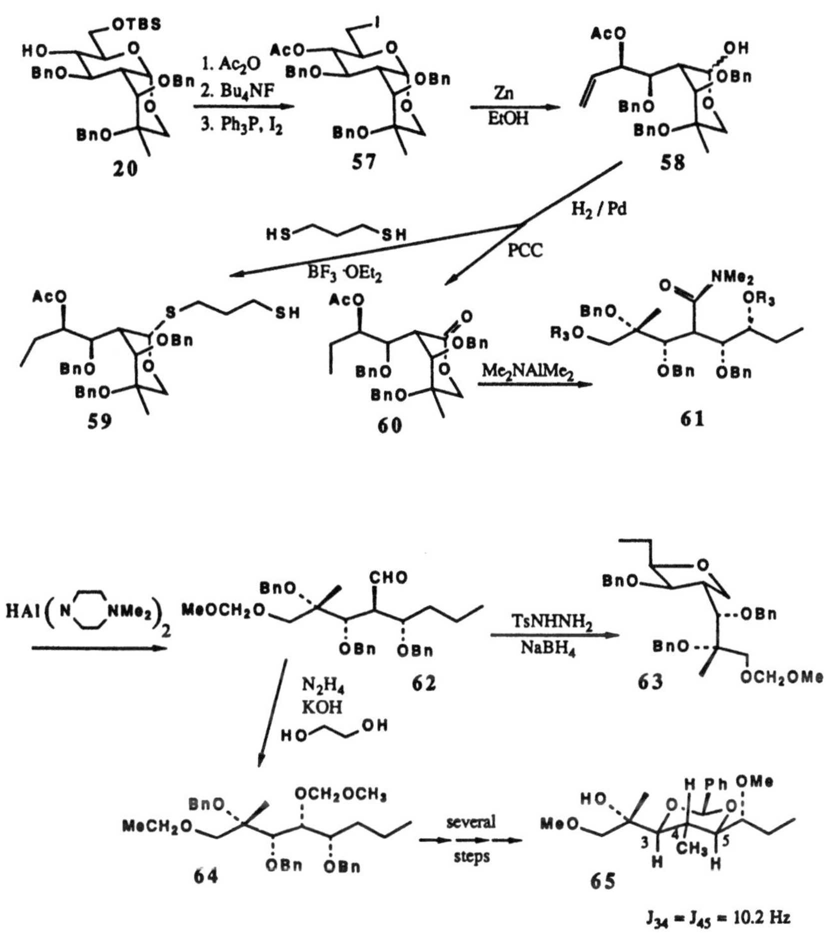

Scheme 15 [44]

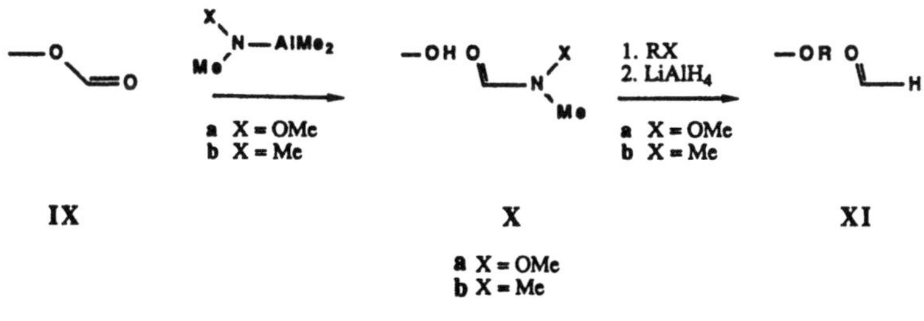

attempted. However, application of this to aldehyde **62** resulted in the formation of furan **63**. The overriding tendency for the ring formation in this system had again defeated our efforts.

It transpired that the Wolff-Kishner reaction could be executed without event, and **64** was obtained by the traditional methodology.

However, the intermediates **60**, **61**, and **62**, were potentially epimerizable. It was therefore necessary to establish unambiguously that the C8-CH_3 in **64** did, in fact, have the desired orientation. To this end, the molecule was processed for preparation of the benzylidene derivative **65**. The coupling constants (Scheme 14) established unambiguously that the CH_3 was equatorial and that in fact, compound **64** had the proper orientation. Thus, the Weinreb protocol may be applied without risk of epimerization.

With the feasibility studies for the C14-OH (Scheme 6) and the CH-CH_3 (Scheme 14) completed, we are now position to undertake the final task of pulling all the various model studies together. This undoubtedly will be a major undertaking, since this work has so far demonstrated the perils of dealing with molecules with multiple contiguous chiral centers. Remote groups can participate in reactions--even when protected, but such unexpected events, although initially frustrating, can turn out to be of enormous benefit. Thus, the oxidative hydrolysis of glycosides (Scheme 9) has the potential to provide new chemistry of practical, as well as theoretical significance. This potential is currently being explored in our laboratories for the synthesis of molecules which are completely unrelated to the ansa chain **4**, whose conquest began this entire story.

ACKNOWLEDGEMENT

We are grateful to the National Science Foundation (CHE 8703916) and the National Institutes of Health (GM 37380) for financial support.

REFERENCES

1 S. Masamune, C. U. Kim, K. W. Wilson, G. O. Spesard, P. E. Georghiou, G. S. Bates, J. Am. Chem. Soc., 97 (1975) 3512; S. Masamune, H. Yamamota, S. Kamamata, A. Fukuzawa, J. Am. Chem. Soc., 97 (1975), 3513.
2 W. D. Celmer, Pure Appl. Chem., 28 (1971), 413.
3 M. Miljkovic, M. Gligorijevic, T. Satoh, D. Miljkovic, J. Org. Chem., 39 1974) 1379.
4 S. Hanessian, G. Rancourt, Can. J. Chem., 55 (1977) 1111; S. Hanessian, G. Rancourt, Y. Guindon, Can. J. Chem., 56 (1978) 1843.
5 B. Fraser-Reid, R. C. Anderson, Prog. Chem. Org. Nat. Prod., 39 (1980) 1.
6 S. Hanessian, Total Synthesis of Natural Products: The Chiron Approach, Pergamon Press, New York, 1983.
7 T. D. Inch, Tetrahedron, 40 (1984) 3161.
8 For a recent review see: J. Patterson, M. M. Mansuri, Tetrahedron, 41 (1985), 3569.
9 R. C. Anderson, B. Fraser-Reid, J. Am. Chem. Soc., 97 (1975) 3870.

10 B. Fraser-Reid, R. Tsang, D. Lowe, Chemica Scripta, 25 (1985) 177.
11 M. Nakata, H. Takao, Y. Ikeyama, T. Sakai, K. Tatsuta, M. Kinoshita, Bull. Chem. Soc., Jpn, 54 (1981) 1749.
12 S. Hanessian, J.-R. Pougny, I. K. Boessenkool, J. Am. Chem. Soc., 104 (1982) 6164.
13 H. Nagaoka, W. Rutsch, G. Schmid, H. Iio, M. R. Johnson, Y. Kishi, J. Am. Chem. Soc., 102 (1980) 7962; H. Iio, H. Nagaoka, Y. Kishi, Y., J. Am. Chem. Soc., 102 (1980) 7965.
14 K. L. Rinehart, Jr., L. S. Shield, L. S., in: Progress in the Chemistry of Organic Natural Products, W. Herz, H. Grisebach, G. W. Kirby, (Eds.), Vol. 33, Springer-Verlag, New York, 1976, pp. 231-307.
15 B. Fraser-Reid, L. Magdzinski, B. Molino, J. Am. Chem. Soc., 106 (1984) 731.
16 E. Fischer, Chem. Ber., 27 (1894) 673.
17 M. L. Wolfrom, R. U. Lemieux, S. M. Olin, J. Am. Chem. Soc., 71 (1949) 2780.
18 B. Fraser-Reid, L. Magdzinski, B. F. Molino, D. R. Mootoo, J. Org. Chem., 52 (1987) 4495.
19 B. Fraser-Reid, B. F. Molino, L. Magdzinski, D. R. Mootoo, J. Org. Chem., 52 (1987) 4505.
20 D. R. Mootoo, B. Fraser-Reid, B., J. Org. Chem., 52 (1987) 4511.
21 Levoglucosan is obtained in 30-40% by the vacuum pyrolysis of starch: R. B. Ward, Methods Carbohydr. Chem., 2 (1963) 394.
22 M. Cerny, J. Stanek, Jr., Adv. Carbohydr. Chem. Biochem., 34 (1977) 24.
23 (a) R. U. Lemieux, N. J. Chu, N. J., Abstracts of Papers, Am. Chem. Soc., 33 (1958) 31N; (b) A. J. Kirby, The Anomeric Effect and Related Stereoelectronic Effects at Oxygen, Springer-Verlag, New York, 1983.
24 H. J. Bestmann, K. Roth, M. Ettlinger, Chem. Ber., 115 (1982) 161; H. J. Bestmann, K. Roth, M. Ettlinger, M., Angew. Chem. Int. Ed. Engl., 18 (1979) 687; S. Trippett, D. M. Walker, J. Chem. Soc., (1961) 1266.
25 M. Miyashita, A. Yoshikoshi, P. A. Grieco, J. Org. Chem., 42 (1977) 3772.
26 U. Sunay, D. Mootoo, B. Molino, B. Fraser-Reid, Tetrahedron Lett., 27 (1986) 4697.
27 B. F. Molino, L. Magdzinski, B. Fraser-Reid, Tetrahedron Lett., 34 (1983) 5819.
28 Y. Oikawa, T. Yoshioka, O. Yonemitsu, Tetrahedron Lett., 23 (1982) 885.
29 B. F. Molino, J. Cusmano, D. R. Mootoo, R. Faghih, B. Fraser-Reid, J. Carbohydr. Chem., 6 (1987) 479.
30 J. K. Cha, W. J. Christ, Y. Kishi, Tetrahedron Lett., 24 (1983) 3943; W. J. Christ, J. K. Cha, Y. Kishi, Tetrahedron Lett., 24 (1983) 3947.
31 K. R. Kopecky, J. Am. Chem. Soc., 81 (1959) 2748; D. J. Cram, D. R. Wilson, J. Am. Chem. Soc., 85 (1963) 1245; J. D. Morrison, H. S. Mosher, Asymmetric Organic Reactions, Prentice-Hall, Englewood Cliffs, NJ, 1971, pp. 84.
32 W. C. Still, J. H. McDonald, Tetrahedron Lett., 21 (1980) 1032.
33 A. Klemer, G. Rodemeyer, Chem. Ber., 107 (1974) 2612.
34 R. Tsang, B. Fraser-Reid, Chem. Comm., (1984) 60.
35 Y. Chapleur, J. Chem. Soc., Chem. Comm., (1983) 141.
36 R. Tsang, B. Fraser-Reid, J. Am. Chem. Soc., 108 (1986) 2116.
37 For examples, see refs. 5 and 6.
38 R. U. Lemieux, S. Kato, D. Voisin, D., in: The Anomeric Effect: Origin and Consequences, W. A. Szarek and D. Horton (Eds.), Am. Chem. Soc., Symposium Series, no. 81, Washington, 1979.
39 S. Handa, R. Tsang, A. T. McPhail, B. Fraser-Reid, J. Org. Chem., 52 (1987) 3489.
40 J. Jurczak, S. Pikul, T. Bauer, Tetrahedron, 42 (1986) 447.
41 D. B. Tulshian, R. Tsang, B. Fraser-Reid, J. Org. Chem., 49 (1984) 2347; B. Fraser-Reid, D. B. Tulshian, R. Tsang, D. Lowe, V. G. S. Box, Tetrahedron Lett., 25 (1984) 4579.
42 D. R. Mootoo, V. Date, B. Fraser-Reid, J. Chem. Soc., Chem. Comm., (1987) 1461.

43 A. B. Reitz, S. O. Nortey, B. E. Maryanoff, D. Liotta, R. Monannhan, III, J. Org. Chem., 52 (1987) 4197.

44 (a) A. Basha, M. Lipton, S. M. Weinreb, Tetrahedron Lett., (1977) 4171; (b) S. Nahm, S. Weinreb, Tetrahedron Lett., 22 (1981) 3815.

45 M. Muraki, T. Mukaiyama, Chem. Lett., (1974) 1447.

New Reagents and Methods for the Synthesis of β-Lactams, Peptides and Oligonucleotides

I. Ugi*(Speaker), J. Achatz, M. Baumgartner-Rupnik, B. Danzer, C. Fleck, G. Glahsl, R. Herrmann, P. Jacob, C. Kambach, R. Karl, M. Klein, B. Landgraf, P. Lemmen, H. Martin, G. Neyer, R. Obrecht and B. Westinger.
*(Affiliation see page 133)

1. INTRODUCTION

This paper is a progress report on our continued effort to develop new reagents and methods for the synthesis of peptides, ß-lactams and oligonucleotides (ref. 1-4).

Our endeavours in peptide and ß-lactame chemistry concentrate on the use of four component condensations (4CC, Ugi reaction (refs. 1,2,5)) in the synthesis of peptides, ß-lactam antibiotics and related compounds.

The ß-halogenated protective groups that can be removed by Co(I)-phthalocyanine anion, $[PcCo]^{\ominus}$ (refs. 6,7) and the highly reactive five-membered cyclic phosphorylating reagents (refs. 3,4,8) are the chemical basis of our attempts to contribute to the preparative methodology of oligonucleotides.

2.THE SYNTHESIS OF PEPTIDE AND ß-LACTAM DERIVATIVES BY 4CC

The 4CC of carboxylic acids **1**, primary amines **2** or their imines of aldehydes **3** and isocyanides **4** to form **5** is a versatile synthetic principle for a variety of α-amino acid derivatives,

Scheme 1

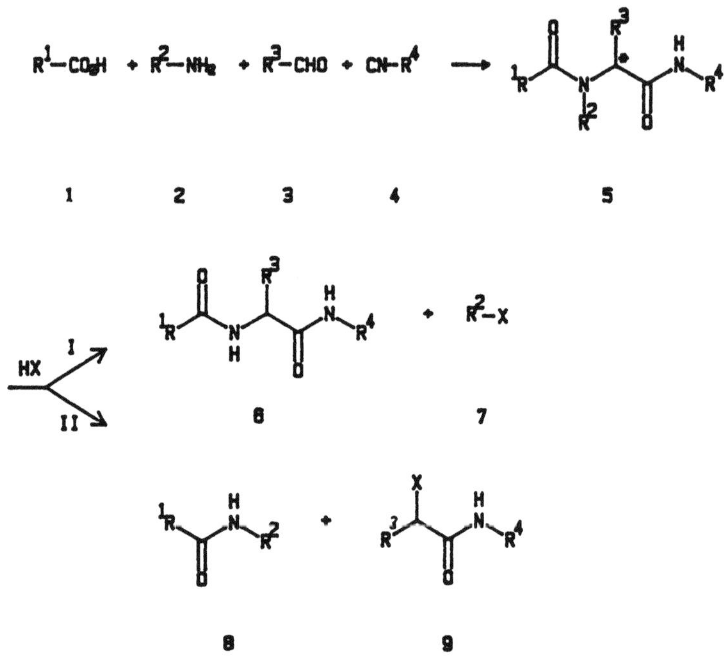

including many types of peptide and ß-lactam derivatives. In general, the synthetic results of a convenient one-pot 4CC can only be achieved through multistep syntheses by other reactions.

2.1 PEPTIDE SEGMENT COUPLING BY 4CC

About 50 aldehydes 3 have been tested as the carbonyl components of peptide segment coupling (SC) according to Scheme 1,II (ref. 1), but no satisfactory results have been obtained.

Our latest and possibly last attempt to find a satisfactory aldehyde 3 for 4CC-SC (ref.1) was an investigation, whether or not 3-formyl indene or its readily available enol acetate (ref. 9) is suitable.

Scheme 2

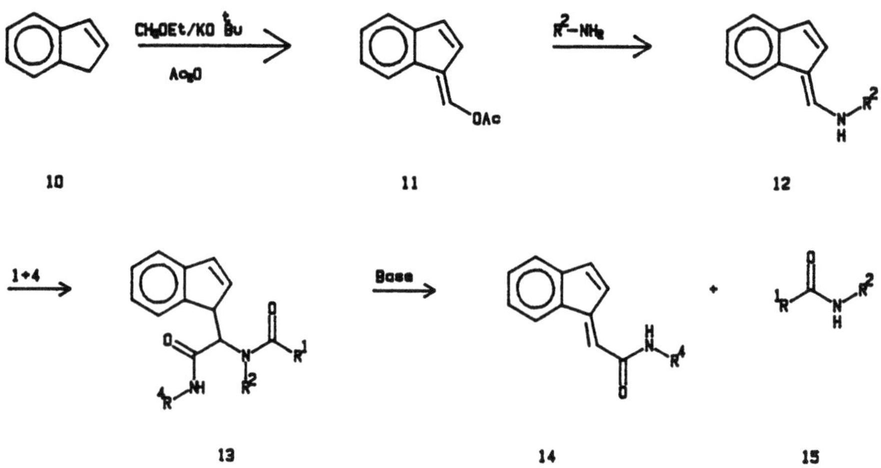

In a model experiment **12**, R^2=nPr was reacted with **1**,R^1 =Ph and **4**,R^4=tBu. According to our present model results 3-formyl indene will not outperform the previously tested aldehydes **3** in 4CC-SC.

2.2 THE SYNTHESIS OF PEPTIDE SEGMENTS BY STEREOSELECTIVE 4CC

Our previous work on 4CC-SSS (SSS=stereoselective segment synthesis) (ref. 1) emphasized stereoselective 4CC according to Scheme 1 I at low temperatures ($t \leq -40$ °C), and the use of α-ferrocenyl isobutylamine and analogs as the asymmetrically inducing amine components whose 4CC products are acidolytically cleavable. Sometimes this approach led to excellent results, but these were not reliably reproducible, because in some cases the 4CC took place in supersaturated solution at the chosen low reaction temperature, whereas in other cases the reactants precipitated at the low temperature, and the 4CC ensued on warming up for work-up (ref. 1).

Recently a study of the stereoselective acylation of α-aryl alkylamines **17** by phenyl trifluoromethylphenylketene **16** was done (ref. 10,11). The temperature dependence of the relative amounts of the diastereomeric products **18** and **19** indicated that for pairs of corresponding reactions (ref. 12,13) the zero selectivity temperature (ref. 11) varies considerably with the structural features of the amine **17**.

Scheme 3

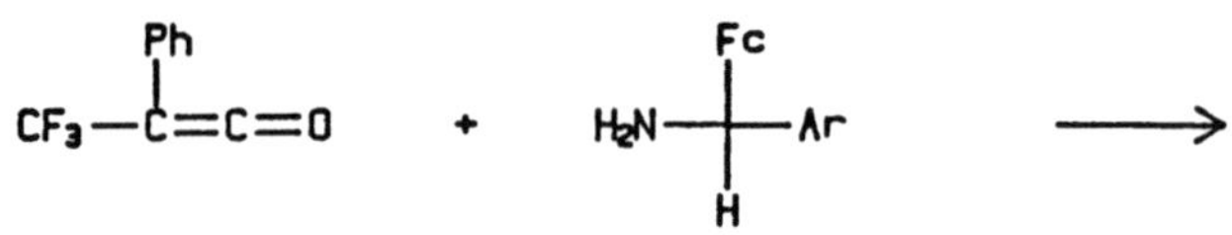

16 17 . (Ar= Ph, Fc= Ferrocenyl)

18 19

Since zero selectivity temperatures $t_0 << 0$ °C are found to occur, and since the degree of stereoselectivity increases with the difference between the reaction temperature chosen and the zero selectivity temperature, we realized that it is worthwhile to investigate the stereoselectivity of 4CC at t>0 °C under variation of the chiral amine component.

We found that the 4CC according to Scheme 4 are well-suited as model reactions in the search for amine components **2** and reaction conditions that guarantee satisfactory results in peptide syntheses by 4CC at "high temperatures"

Scheme 4

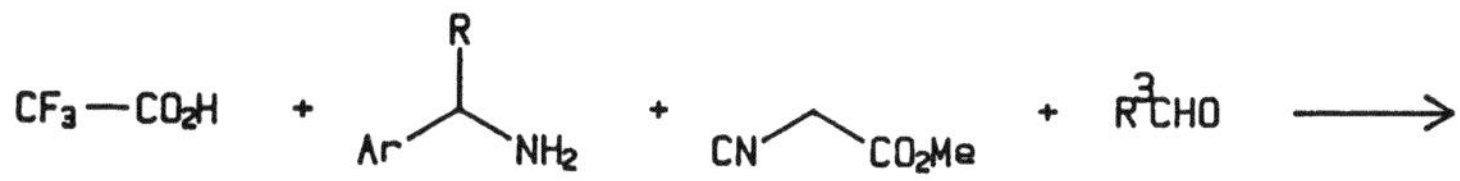

20 17 21 3

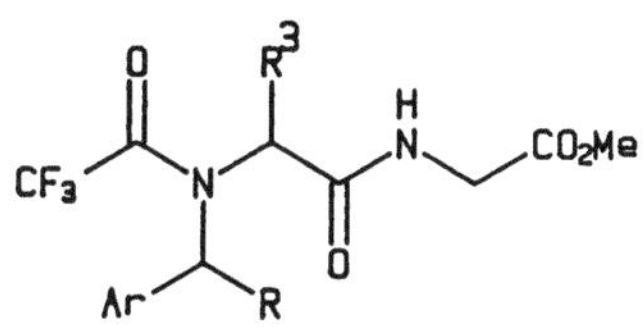

22

In contrast to the N-unsubstituted peptide derivatives (ref. 14) the ^{19}F-chemical shifts of the diastereomeric 4CC products of type **22** show relatively large differences in (see Table)

Table.

Table. The ^{19}F chemical shift differences of 22

22	Ar	R	R^3	δ(ppm) off the dia-stereoisomer of 22	Δδ(ppm)
a	Ph	Me	Me	-13,35;-13,41	0,06
b	Fc	Me	Me	-13,11;-13,12	0,01
c	Fc	iPr	Me	-15,63;-15,89	0,26
d	Ph	Me	iPr	-13,89;-14,36	0,47
e	Fc	Me	Me	-14,07;-14,36	0,29
f	Fc	iPr	Me	-15,41;-15,81	0,41
g	Ph	Me	Bz	-13,57;-13,67	0,10
h	Fc	Me	Bz	-13,34;-13,67	0,14
i	Fc	iPr	Bz	-15,99;-16,12	0,13

Here the degree of asymmetric induction of 4CC that form **22** can be conveniently determined with sufficient accuracy by high field ^{19}F-NMR spectroscopy. With this method the chiral amines **17** can be used as their racemates. Besides the α-ferrocenyl alkylamines **17**, Ar = Fc, R = alkyl the 1-amino monosaccharide derivatives will be included in a comparative study, since Kunz and Sager (ref. 15) have recently discovered that these amines have promising properties as asymmetrically inducing chiral components of 4CC with an acidolytically cleavable C-N-bond, like the α-ferrocenyl-alkylamines.

Our ongoing analysis of the dependence of the stereoselectivity and yield of 4CC on the amine component and the reaction conditions is guided by a computer assisted composite factorial design study. We hope to achieve in the near future a renaissance of the 4CC-SSS by this work.

2.3. THE SYNTHESIS OF β-LACTAMS BY 4CC

Due to their excellent bactericidal and bacteriostatic properties and their general lack of serious side-effects (ref. 16) the β-lactam antibiotics still belong to the most favoured antibacterial drugs that are avaible. The research activities in the field of β-lactam antibiotics, including the chemical synthesis of β-lactam antibiotics and related compounds, have increased up to now (ref. 16).

The centerpiece of any synthesis of this type of substances is the construction of the β-lactam moiety, because generally the strained, readily cleaved β-lactams are relatively difficult to build up from their acyclic precursors, e.g. the β-amino acids and their derivatives.

The 4CC is an exception. The β-lactams are readily produced by 4CC from the derivatives of β-amino acids according to schemes 5 and 6 (ref. 2).

Scheme 5 and 6

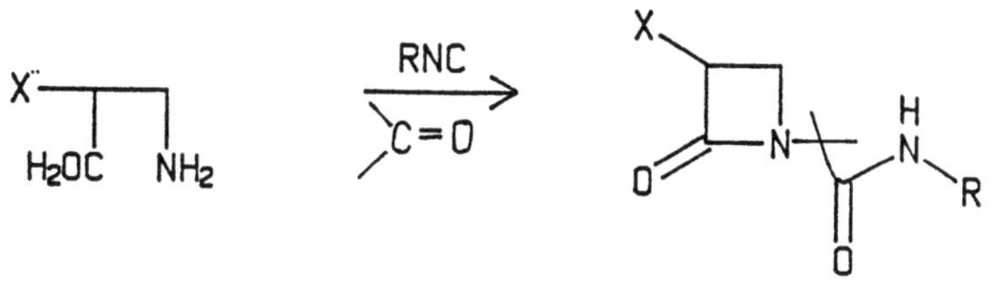

23 24

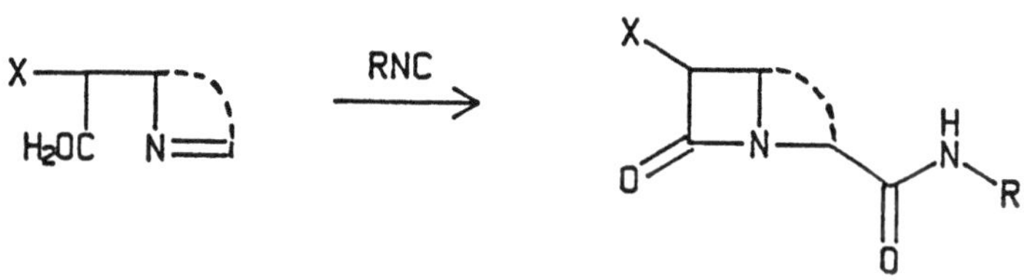

25 26

When ß-lactams are prepared by 4CC, the synthesis of the precursors is often more difficult than the otherwise critical formation of the ß-lactam system (ref. 2).

In our previous syntheses of penam (ref. 17) and cepham derivatives (ref. 18) one of the less satisfactory features was the use of the methyl group as carboxy protection. In a recent modification of our penam synthesis (ref. 17) (Scheme 7) the allyl group has been shown to be superior to the methyl group, because of its very mild deblocking conditions (ref. 19).

Scheme 7

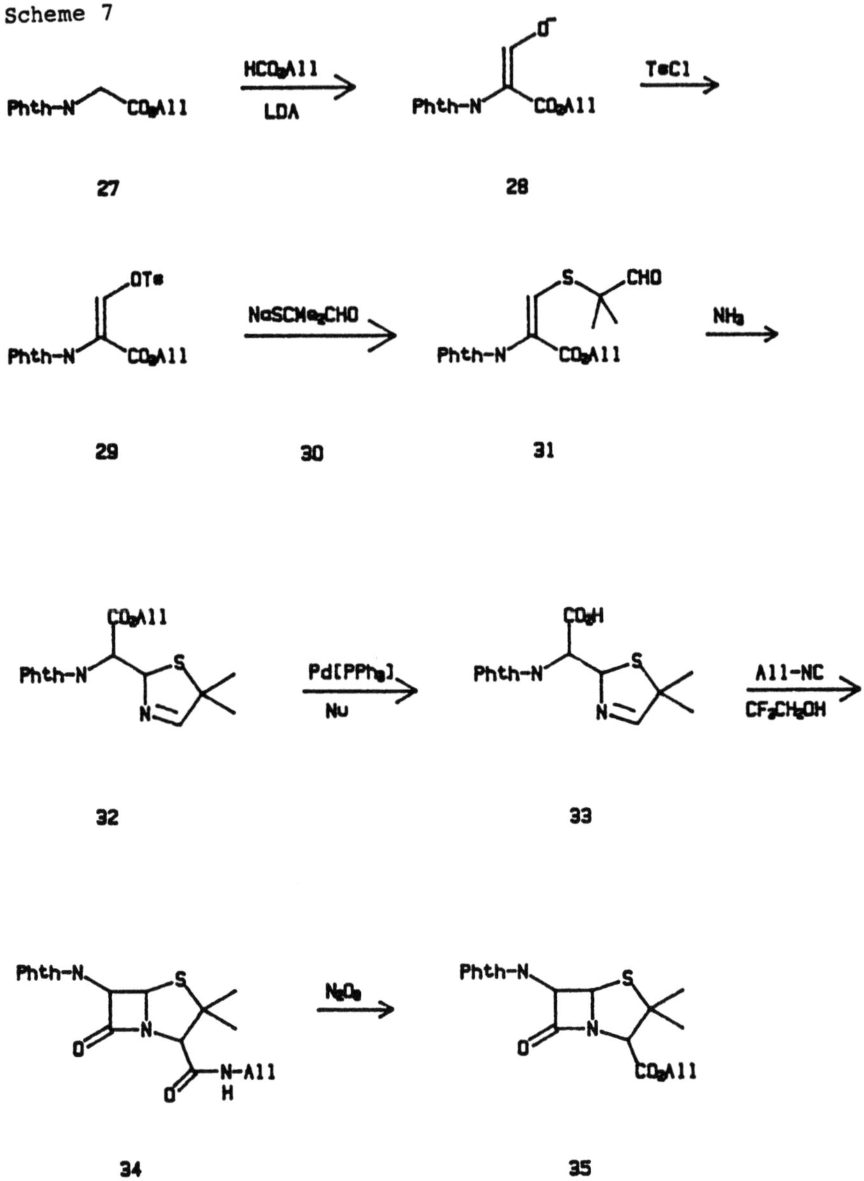

In this synthesis the formation of the thiazoline system **31** → **32** is carried out at 0 °C in ether. Thus a good yield and a high degree of stereoselectivity ($\geq$ 4:1) is achieved.

The cyclization by 4CC proceeds particularly well in tri-fluoroethanol as a solvent. The latter seems to be a "magic solvent" in the synthesis of ß-lactams by 4CC (ref. 20).

The conversion of the amide group in **34** into the carboxy group via **35** seems to be a promising way to solve the "amide problem" of ß-lactam syntheses by 4CC (refs. 2,21). The conversion of ß-lactam amides into the carboxylic acids via the esters was very successful in the hands of Hofheinz and Isenring (ref. 22) who worked with the N-benzhydryl amides of nocardicin derivatives. The stereoselective and regioselective 1,3-dipolar cycloaddition of chiral nitrones to benzyl crotonate is a key step in our approach to thienamycine (ref. 23) and related carbapenam derivatives. Besides chiral 2-phenylethyl hydroxylamine (ref. 24) the 1-hydroxylamino derivative of 2,3-5,6-diacetone mannose (ref. 25) is a promising source of chiral nitrones for such enantioselective cycloadditions.

Since we were unable to obtain the succinaldehyde derivatives, e.g. **38** (ref. 26) that are needed for the above carbapenam derivatives, we confined ourselves to model experiments with malonaldehyde dimethyl acetal (ref. 23). In the meantime, we have found a convenient synthesis for succinaldehyde diethyl acetal **38** (Scheme 8). Thus we now are able to continue our carbapenem syntheses via the cycloadducts **40**.

Scheme 8

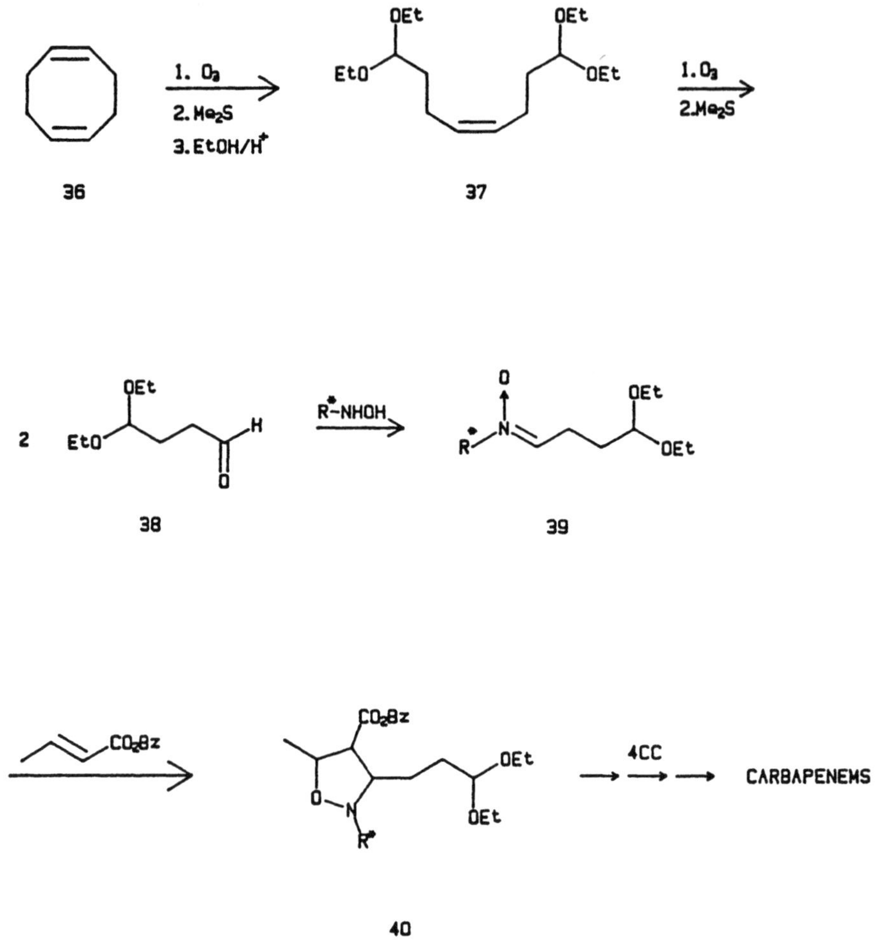

The fact that **43** is readily available the through the photolysis of **44** (ref. 28) suggests the idea to synthesize carbapenam derivatives **41** by 4CC of **42**.

After a systematic investigation of many potential routes, e.g. via 6-methyl-1.3-pyrimidine-2.4-dione derivatives, we favor a route in analogy to Scheme 10 (ref. 29) for the synthesis of **42** from the diethylacetal of **43**.

Scheme 9

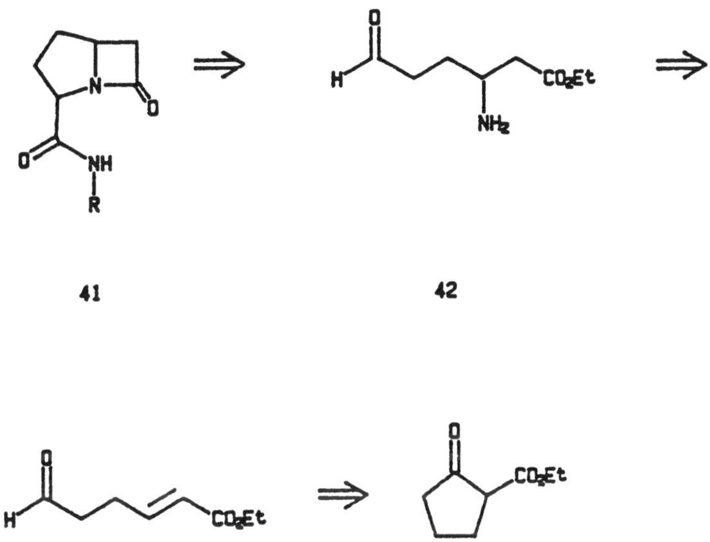

41 **42** **43** **44**

As a model for the conversion of **43** into a suitable derivative of **42** we studied the conversion of ethyl crotonate **45** into **49**.

Scheme 10

45 **46**

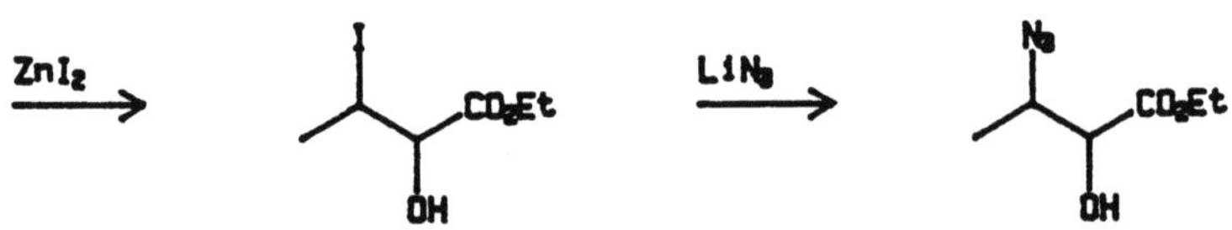

47 **48**

49

3. NEW REAGENTS AND METHODS FOR THE SYNTHESIS OF OLIGONUCLEOTIDES

3.1 PROTECTIVE GROUP TECHNIQUES

The Trichloro-tert.-butoxy-carbonyl (TCBOC) group survives the cleavage conditions of most other protective groups, and is cleaved under very mild conditions by $PcCo^{\ominus}$, a reagent that does not affect the other customary blocking groups (ref. 6).

Scheme 11

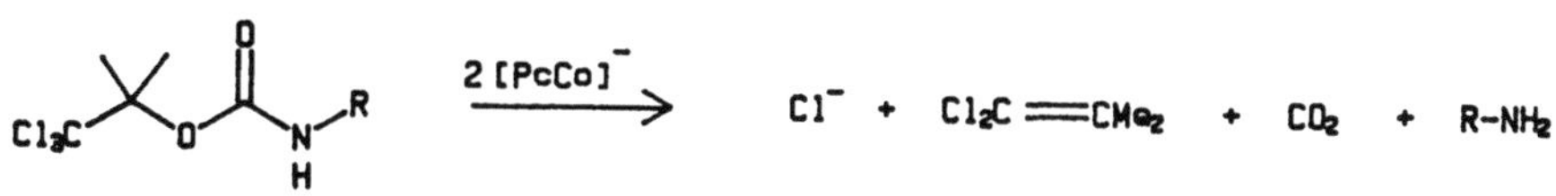

50 (TCBOC-NHR) 51

The TCBOC group (ref. 7) (Scheme 11) was first introduced for the "orthogonal" protection (ref. 6) of amino groups in peptide syntheses. Subsequently it was established that the TCBOC group is also usable in the protection of the nitrogen functions of the heterobases of nucleosides (ref. 30). Despite an intense search for a ß-halogenated protective group for oxygen and sulfur functions, no such group was found (ref. 30), e.g. for the protection of the alcoholic and phenolic hydroxy groups, carboxyl groups and thiol groups of amino acid derivatives, as well as the 5'- or the 2'-hydroxy groups of nucleoside derivatives.

The 4-trichloromethyl-trityl group was investigated as a potential protective group of the 5'- hydroxy function of nucleosides (see Scheme 12).

Scheme 12

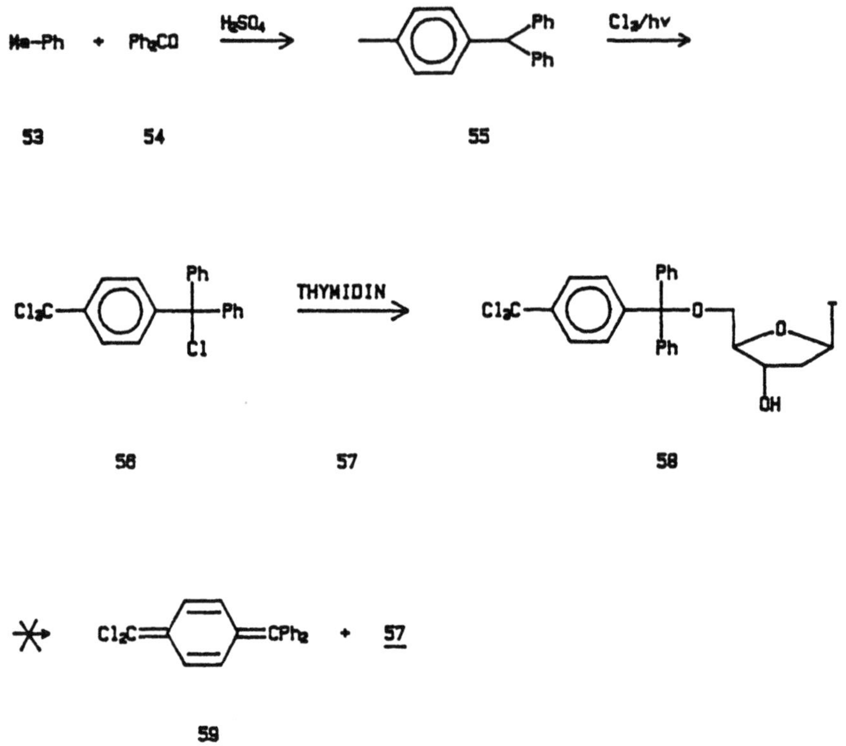

The reagent **56** is readily available from **53** and **54** via 55, and it can be attached to the 5'-OH group of nucleosides, just like the trityl group. However the cleavage into **59** and the deprotected nucleoside by $CoPc^{\theta}$ is not possible. The 4-trichloromethyl tritylated nucleoside reacts with $PcCo^{\theta}$ to form a mixture of products, that contain only minor amounts of the deprotected nucleoside.

The DATE (1,1-dianisyl-2,2,2-trichlorethyl) group has all the structural features that are needed in a universally usable protection by a ß-halogenated alkyl group (see Scheme 13).

Scheme 13

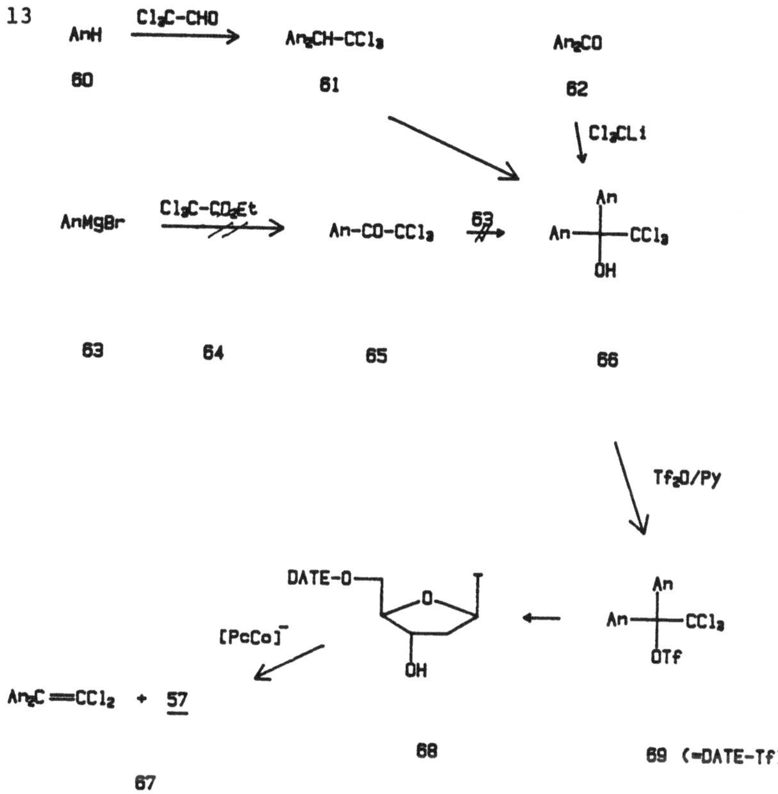

The two anisyl groups (An) and the trichloromethyl group provide the steric bulk that is needed for the selective protection of the 5'-hydroxyl group of nucleosides, à la trityl. The trichloromethyl group makes the DATE group cleavable by $PcCo^{\theta}$(ref. 32). Without the 4-methoxy groups of the anisyl moieties, **69** would not alkylate hydroxy groups; neither the phenyl analog of **69** is reactive enough in the sense of S_N1, nor DATE-Cl. The triflate group (Tf = CF_3-SO_2-) of **69** is roughly 10^8 times more reactive as a leaving group than chloride (ref. 33).

The precursor **66** of the DATE triflate reagent **69** can only be synthetized from **61** by bromination with NBS and subsequent hydrolysis, or from **62** by addition of the trichloromethyl carbanion. The attempted preparation of **66** from **63** and **64** via **65** does not succeed. Presently we are exploring the use of the DATE group as a protective group for other functions than the 5 '-hydroxy group of nucleosides (see above).

In the currently popular versions of the automated synthesis of deoxynucleosides by the phosphoramidite methods

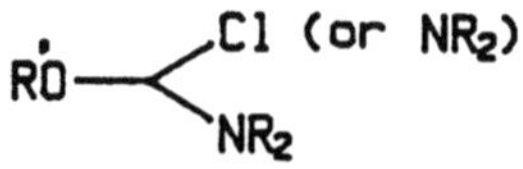

70

R' = Me-, NC-CH_2CH_2-, Cl_3C-CH_2-(=TCE), Cl_3C-CMe_2-(=TCB)

R = Me, ^{i}Pr

that are based on the reagents **70**, the methyl group and the ß-cyanoethyl group are favoured as protective groups of the phospite/phosphate function (refs. 34,35). The p-nitrophenyl-ethyl group shows also great potential for the above purpose (ref. 36), in particular, because the latter group is also usable in the protection of other functions, thus affording uniform deprotection conditions for all protected functions.

With uniform deprotection conditions in mind, we introduced the trichloro-tert.-butyl (TCB) group (ref. 35) for the protection of phosphite/phosphate (refs. 37-39) as a part of a phosphor amidite method. The TCB group imparts also enhanced stability to the intermediates of the types **71** and **74** (Scheme 14)

Scheme 14

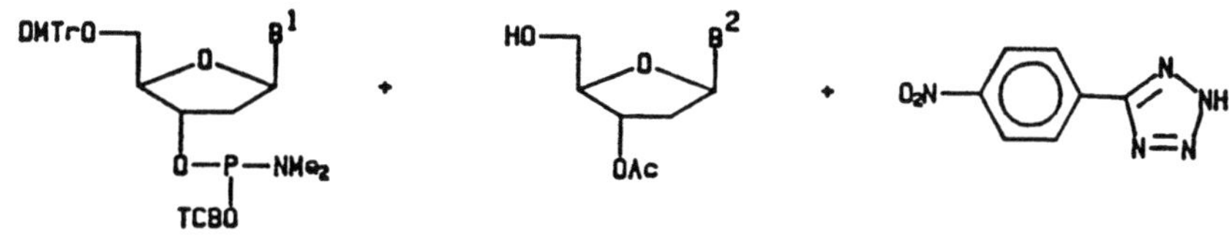

71 72 73

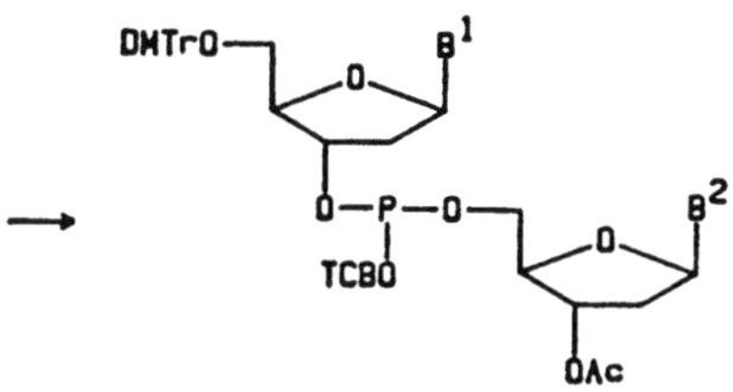

74

The TCB phosphoramidites are however less reactive than their methyl or ß-cyanoethyl counterparts. Therefore they need a more powerful activating reagent than the customary tetrazole. We first tried to use 5-trifluoromethyl tetrazole (ref. 37) but the latter was not only a more potent activating agent, it also removed partially the DMTr protection from the 5'oxygen. Thus an activating reagent of intermediate acidity was called for. An activating reagent that is sufficently potent but does not attack the DMTr group is **73**. As in the methyl case (ref. 40), it is an almost ideally suited activator for TCB phosphoamidites.

Diagram

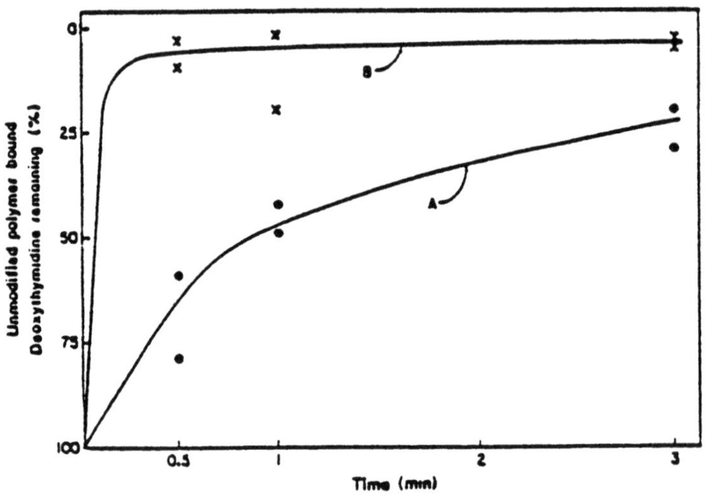

A) 5-H-tetrazole B) 5-p-NO_2 phenyl tetrazole

Diagram: The activation phosphoamidite by tetrazole and p-nitrophenyltetrazoles.

The use of TCB as the phosphite/phosphate protection and of TCBOC for the protection of the exocyclic amino functions of the bases enables an uniform deprotection strategy. Nevertheless care is necessary: a study of the influence of the electron withdrawing properties of guanosine 2-N-protecting groups on the stability of guanosine revealed relative destabilisation by TCBOC vs. p-tert.-butyl benzoyl.

3.2 FIVE MEMBERED CYCLIC PHOSPHORYLATING REAGENTS

The use of phosphorylating reagents of P(V) instead of the currently popular P(III) reagents would have the advantage that more stable intermediates could be expected, and that no oxidation step would be needed.

The P(V) reagents are generally not suitable for automated syntheses, because they are less reactive than the P(III) reagents. However, the five-membered cyclic phosphate derivatives are 10^5-10^7 times more reactive than their acyclic counterparts (ref. 8).

The 278 conceivable types of five-membered cyclic phosphorylating reagents **76** (Scheme 15) have been generated with the aid of the computer program IGOR (refs. 4,41).

Scheme 15

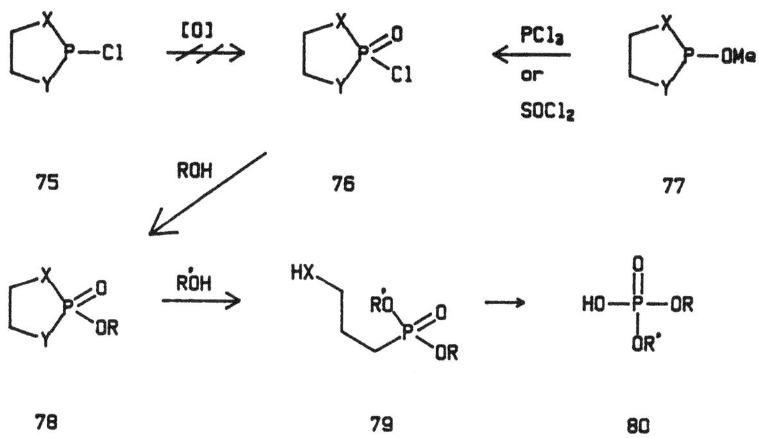

Since the thiol group has favorable protective group properties for phosphates (ref. 4), and since electron poor amido phosphates are reactive towards hydroxyl groups and are converted into systems that are fairly inert towards phosphorylation, we devoted our attention primarily on those of the 278 candidates that have X = S and Y = N.

With few exceptions, the reagents **76** cannot be formed by direct phosphorylating cyclization. The synthesis of **76** via **77** is presently the method of choice. The route via **75** is possible in the case of CEP-Cl (ref. 42), but not generally usable.

The cyclic phosphorylating reagents **85** and **92** were synthesized according to Schemes 16 and 17

Scheme 16

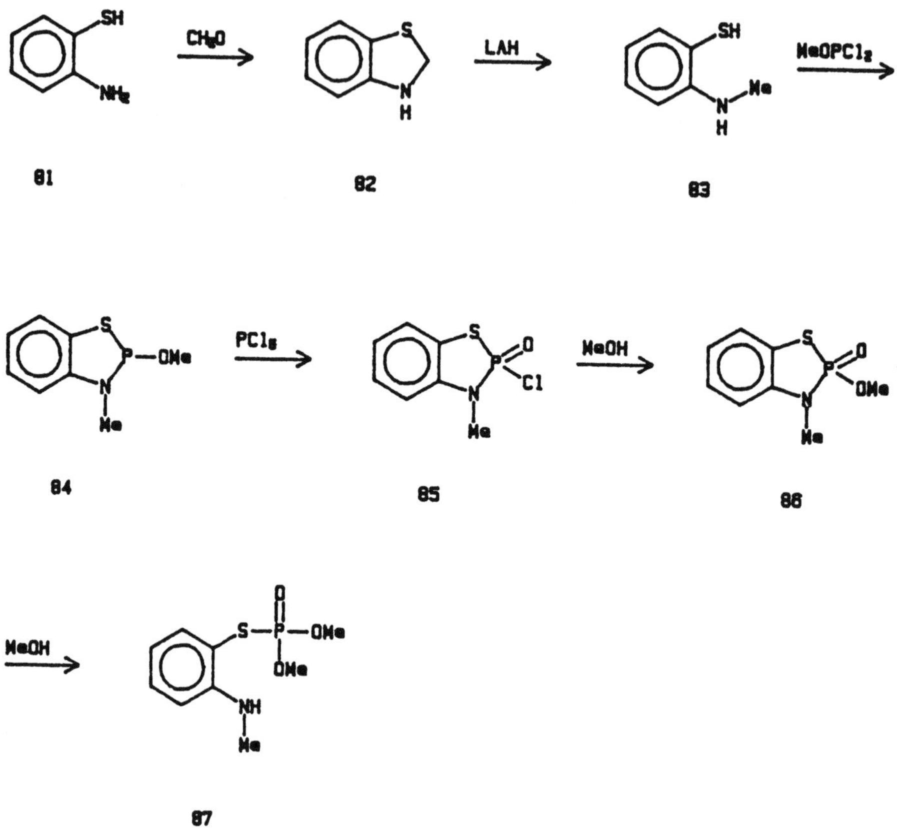

Scheme 17

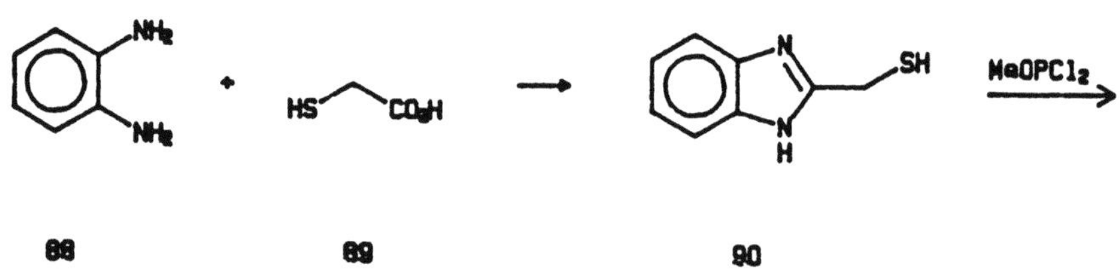

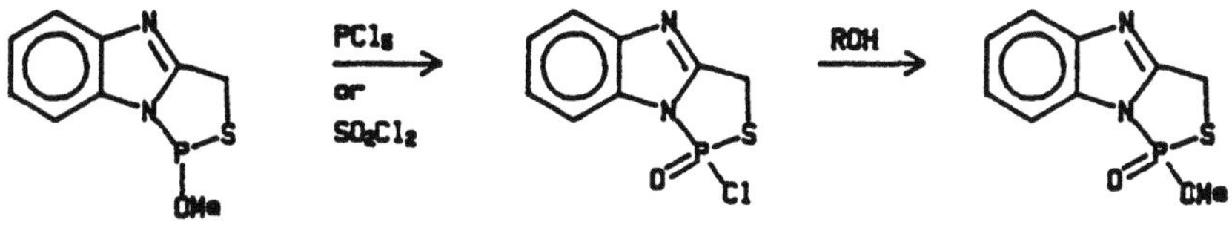

91 92 93

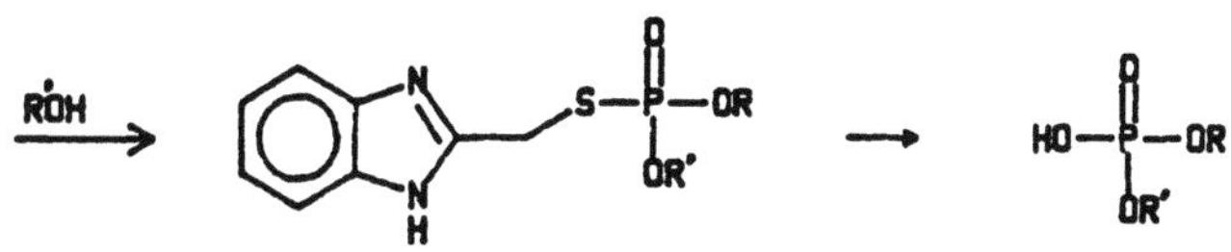

94 80

The reagent **85** is very reactive toward alcohols in the first phosphorylation step, e.g. **85** ⟶ **86**, but the second phosphorylation step **86**→**87** is too slow; so **85** is not usable in the synthesis of oligonucleotides. The reagent **92** has been made from **91** by reaction with PCl_5 or SO_2Cl_2, it has, however, not yet been obtained in a sufficiently pure state for being tested in phosphorylation according to **92** ⟶ **93** ⟶ **94**, although our preliminary results are encouraging.

The oxidation of P(III) compounds into their P(V) counterparts is a problem in phosphorus chemistry that poses itsself again and again, e.g. in the synthesis of **76** and it has not yet found a general answer. Since long we have searched for widely usable oxidizing reagents for the conversion P(III)→P(V) (refs. 41-43). We have now found, that the oxaziridines (ref. 44) achieve the conversion P(III) →P(V) smoothly, also in anhydrous solution . An investigation of the relative oxidizing power of the oxaziridines **95** as a function of the substituent R^1 and R^2, it was shown that electron pulling groups R^1 and R^2 enhance the relative rate of oxidation.

Scheme 18

95 **96** **97** **98**

The rate-increasing influence of electron withdrawal is stronger at R^1 than at R^2. For example, the rate constants of the oxidation of triethyl phosphite and triphenyl phosphine by **95**, R^1=p-nitrophenyl, R^2=iPr are ca. 10 times higher than the corresponding data for **95**, R^1=Ph,R^2=Tosyl.

ACKNOWLEDGEMENT

We acknowledge gratefully the financial support of our work by Deutsche Forschungsgemeinschaft, SFB 145, and Fonds der chemischen Industrie.

REFERENCES

1 I. Ugi, The Four Component Synthesis, in: M. Meienhofer, E. Gross, (Eds.); The Peptides, Academic Press, New York, 1979, Vol. 2, pp. 365-381; I. Ugi, D. Marquarding and R. Urban, Synthesis of Peptides by Four-Component Condensation, in: B. Weinstein (Ed.), Chemistry and Biochemistry of Amino Acids, Peptides and Proteins, Marcel Dekker, New York, 1982, pp. 245-289.

2 I. Ugi, Von Isocyaniden via Vierkomponenten-Kondensationen zu Antibiotika-Synthesen, Angew. Chem. **94** (1982), 826-835; Angew. Chem. Int. Ed. Engl. **21** (1982) 810-819.

3 I. Ugi, H. Aigner, G. Glahsl, R. Herges, G. Hering, R. Herrmann, G. Hübener, P. Lemmen, R. Obrecht, D. Rehn, R. Schwarz, K. Seeholzer, F. Siglmüller, J. Stackebrandt, R. Stöcklein-Schneiderwind, S.-T. Amadou and H. v. Zychlinski, New Reagents and Methods for the Synthesis of Peptides, ß-Lactams and Oligonucleotides, in: Atta-ur-Rahman (Ed.), Natural Produkt Chemistry, Springer-Verlag, Heidelberg, 1986, pp. 457-484.

4 I. Ugi, J. Bauer, E. Fontain, J. Götz, G. Hering, P. Jacob, B. Landgraf, R. Karl, P. Lemmen, R. Schneiderwind-Stöcklein, R. Schwarz, P. Sluka, N. Balgobin, J. Chattopadhyaya, T. Pathak and X.-X. Zhou, New phosphorylating Reagents and protective Group Techniques for Oligonucleotide Synthesis, as well as Computer Assistance in the Design of Reagents, Chemica Scripta **26** (1986) 205-215.

5 I. Ugi, Neuere Methoden der präparativen organischen Chemie IV. Mit Sekundär-Reaktionen gekoppelte α-Additionen von Immonium-Ionen und Anionen an Isonitrile, Angew. Chem. **74** (1962) 9-22, Angew. Chem. Int. Ed. Engl. **1** (1962) 8-21; G. Gokel, G. Lüdke and I. Ugi, Four-Component Condensations and Related Reactions, in: I. Ugi, (Ed.), Isonitrile Chemistry, Academic Press, New York, 1971, pp. 145-199.

6 H. Eckert, W. Breuer, J. Geller, I. Lagerlund und M. Listl, D. Marquarding, S. Stüber, I. Ugi, S. Zahr and H. v. Zychlinski, New Methods in Peptide Synthesis based on Supernucleophiles, Pure Appl. Chem. **51** (1979) 1219-1233.

7 H. Eckert, M. Listl and I. Ugi, Der 2,2,2-Trichlortert.-butyloxycarbonyl-(TCBOC)-Rest, eine säure- und basestabile, schonend abspaltbare Schutzgruppe, Angew. Chem. **90** (1978) 388-389; Angew. Chem. Int. Engl. **17** (1978) 361-362.

8 F. Ramirez, S.L. Glaser, P. Stern, I. Ugi and P. Lemmen; Five-membered cyclic Acyl Phosphates, Tetrahedron **29** (1973) 3741-3752; F. Ramirez, S. Glaser, P. Stern, P.D. Gillespie and I. Ugi, Fünfgliedrige cyclische Acylphosphate, neue hochreaktive Phosphorylierungsmittel, Angew. Chem. **85** (1973) 39-40, Angew. Chem. Int. Ed. Engl. **12** (1973) 66-67; D. Marquarding, F. Ramirez, I.Ugi and P.D. Gillespie, Austausch-Reaktionen von Phosphor-(V)-Verbindungen und ihre pentakoordinierten Zwischenstufen, Angew. Chem. **85** (1973) 99-127; Angew. Chem. Int. Ed. Engl. **12** (1973) 91-118.; F. Ramirez and I. Ugi, New Syntheses of Unsymmetrical Phosphodiesters Based on the Oxyphosphorane Concept, Phosphorus & Sulfur **1** (1976) 231-244.

9 R. Huisgen, E. Laschtuvka, I. Ugi and A. Kammermeier, Syntheseversuche in der Oxepin- und Azepin-Reihe, Liebigs Ann. Chem. **630** (1960) 128-139.

10 Joo-Hack Youn, Chirale Schwefelverbindungen und Ferrocenderivate als Bezugssysteme für asymmetrische Synthesen, Doctoral Thesis, Technical University München, 1986.

11 I. Ugi, Logic and Order in Stereochemistry, Chimia **40** (1986) 340-350.

12 E. Ruch and I. Ugi, The Stereochemical Analogy Model - A Mathematical Theory of Dynamic Stereochemistry, N.L. Allinger and E.L. Eliel (Eds.), Topics in Stereochemistry, Wiley, New York, 1969, Vol IV, pp. 99-125.

13 E. Anders, E. Ruch and I. Ugi, Experimentelle Prüfung des stereochemischen Strukturmodells am Beispiel der stereoselektiven Acylierung racemischer sekundärer Alkohole mit Phenyltrifluormethylketen, Angew. Chem. **85** (1973) 16-20; Angew. Chem. Int. Ed. Engl. **12** (1973) 25-29.

14 R.E. Sievers, E. Bayer and P. Hunziker, Fluorine Nuclear Magnetic Resonance of Peptides and Amino Acids, Nature **223** (1969) 179-181; E. Bayer, P. Hunziker, M. Hutter, R.E. Sievers and R. Uhmann, ^{19}F Nuclear Magnetic Resonance for the Control of Peptide Synthesis, J. Amer. Chem. Soc. **94** (1972) 265-268; see also: W.E. Hull, K. Seeholzer, M. Baumeister and I. Ugi, A Modified Synthesis of Mosher's Acid and its Use in a sensitive Stereoisomer Analysis of Amino Acid Derivatives, Tetrahedron **42** (1986) 547-552.

15 H. Kunz and W. Pfrengle, Asymmetric Synthesis on Carbohydrate Templates: Stereoselective Ugi Synthesis of o-Amino Acid Derivatives, J. Amer. Chem. Soc. **110** (1988) 651-652; see also:

H. Kunz and W. Sager, Diastereoselektive Strecker-Synthese an Kohlehydrat-Matrices, Angew. Chem. **99** (1987) 595-597; Angew. Chem. Int. Ed. Engl. **26** (1987) 557-559.

16 R.B. Morin and M. Gorma (Eds.), Chemistry and Biology of ß-Lactam Antibiotics, Academic Press, New York, 1982, Vols. 1-3.

17 A. Schutz and I. Ugi, Synthesis of a Penicillin Derivative by a Four-Component Condensation, J. Chem. Res. (S) **1979**, 157; J. Chem. Res. (M) **1979**, 2064-2071.

18 A. Schutz, I. Ugi, Die Synthese eines Cephamderivats mittels Vierkomponenten-Kondensation, Z. Naturforschung **34b** (1979), 1301-1306.

19 B.M. Trost, New Rules of Selectivity: Alkylic Alkylations catalyzed by Palladium, Acc. Chem. Res. **13** (1980) 385-393; P.D. Joffrey and S.W. Mc Comble, Homogeneous Palladium (SO)-Catalyzed Exchange Deprotection of Allylic Esters, Carbonates and Carbamates, J. Org. Chem. **47** (1982) 587; H. Kunz, H. Waldmann and C. Unverzagt, Allyl Ester as temporary Protecting Group for the ß-Carboxy Function of Aspartic Acid, Int. J. Peptide Protein Res. **26** (1985) 493-497; J. Tsuji, M. Nisar and I. Minami, A general Synthetic Method for α-Methylene Compounds by the Palladium-Catalyzed Decarboxylation-Deace-

toxylation of Allyl α-Acetoxymethylcarboxylates Substituted by an Electron -withdrawing Group at α-Position, Chem. Letters **1987**, 23-24; R. Deziel, Mild Palladium(0)-Catalyzed Deprotection of Allyl Esters. A useful Application in the Synthesis of Carbapenems and other ß-Lactam-Derivatives, Tetrahedron Lett. **28** (1987) 4371-4372.

20 R. Obrecht, Neue Methoden zur Synthese von ß-Lactamen mittels Vierkomponentenkondensationen, Doctoral Thesis, Technical University München, 1986.

21 I. Ugi and J. Geller, The selective Cleavage of Carbonamide Groups, and the Synthesis of ß-Lactam Antibiotics by Four Component Condensation, Chem. Scripta **22** (1983) 85-89.

22 H.P. Isenring and W. Hofheinz, A simple Two-Step-Synthesis of Diphenylmethyl Esters of 2-Oxo-1-azetidine Acetic Acid, Synthesis **1981**, 385-387; H. Martin, R. Herrmann and I. Ugi, Synthese einer chiralen Thienamycin-Vorstufe, Z. Naturforsch. (in press). see also:

23 H. Martin, Verbesserung der Totalsynthese von Thienamycin durch diastereoselektive 1,3-dipolare Cycloaddition chiraler Nitrone, Doctoral Thesis, Technical University of Munich, 1987.

24 T. Polonski and A. Chimiak, Oxydation of Amino Acid Esters into N-Hydroxylamino Acid Derivatives, Tetrahedron Lett. **1974**, 2453-2456.

25 A. Vasella, Stereoselektivität und Reaktivität bei der 1.3-dipolaren Cycloaddition chiraler N-(Alkoxyalkyl)nitrone, Helv. Chim. Acta **60** (1977) 1273-1295.

26 C. Botteghi, L. Lardicci and R. Menicagli, Optically active Heteroaromatic Compounds VI; 3-substituted Furans and Thiophenes from α,ß-unsaturated Aldehydes, J. Org. Chem. **38** (1973) 2361-1365; C.P. Forbes, W.J. Schoeman, H.F. Strauss,E.M.M. Venter and A. Wiechers, Applications of substituted Aryl Acetaldehydes in the Total Synthesis of seco-Mesembrane Alkaloids Part I. The Total Synthesis of 1-O-Methyljoubertiamine, J. Chem.Soc. Perkin I **1980**, 906-910.

27 C. Botteghi and F. Soccolini, Malonaldehyde, Succindialdehyde and Glutaraldehyde Monoacetals: Syntheses and Applications, Synthesis **1985**, 592-664.

28 J. Kassanyi, J. Perales, A. Laachach, I. Kawenoki and J.P. Morizur, Preparation of ω-Formylalkenoates from cyclic ß-Ketoesters, Synthesis **1979**, 279-281.

29 G. Neyer, Modellstudien für einen neuen Syntheseweg für

Carbapenam-Derivate, M.S. Thesis, Technical University München, 1987.

30 R.G.K. Schneiderwind and I.Ugi, Die 2,2,2-Trichlor-tert-butyloxycarbonyl-Gruppe als N-Schutzgruppe bei Oligonukleotidsynthesen, Z. Naturforsch. **36b** (1981) 1173-1175; T. Kamimura, T. Masegi and T. Hata, Protection of Imide Group of Uracil Moiety by Means of 2,2,2-Trichloro-tert.-butoxycarbonylchloride: A selective Synthesis of 2'-O-Methyluridine, Chem. Lett **1982**, 965-968; I. Ugi and R.G.K. Schneiderwind, Die 2,2,2-Trichlor-tert-butyloxycarbonyl-Gruppe, eine neue N-Schutzgruppe für Oligonukleotidsynthesen, Tetrahedron **30** (1983) 2207-2210; X.X. Zhou, I. Ugi and J. Chattopadhyaya, A Convenient Preparation of N-protected Nucleosides with the 2,2,2-Trichloro-terrt-butyloxycarbonyl (TCBOC) Group. Structural Assignment of N,N-bis-TCBOC-Guanosine and its Deoxy Analogue, Acta Chem. Scand, **B39** (1985) 761-765.

31 P. Seidel, I. Ugi, Die Photoaddition von Bromtrichlormethan an Allyl-Systeme, Z. Naturforsch. **37b** (1982) 376-379.

32 H. Eckert, I. Lagerlund and I. Ugi, Fragmentation with Supernucleophiles - VI. ß-Haloalkyl Groups as Functional Protection in Peptide Synthesis. A Kinetic Study of the Reaction of the Cobalt(I)phthalocyanine Anion with Organic Halides, Tetrahedron **32** (1977) 2243-2247.

33 P.J. Stang, M. Hanack and L.R. Subramanian, Perfluoroalkanesulfonic Esters: Methods of Preparation and Application in Organic Chemistry, Synthesis **1982**, 85-126.

34 M.D. Matteucci and M.H. Caruthers, Synthesis of Deoxynucleotides on a Polymer Support, J. Amer. Chem. Soc. **103** (1982) 3185-3191.

35 R.L. Letsinger und K.K. Ogilvie, Synthesis of Oligothymidilates via Phosphotriester Intermediates, J. Amer. Chem. Soc. **91** (1969) 3350-3355; J.C. Catlin and F. Cramer, Deoxyoligonucleotide Synthesis via the Triester Method, J. Org. Chem. **38** (1973) 245-250.

36 F. Himmelsbach, B.S. Schulz, T. Trichtinger, R. Charubala and W. Pfleiderer, The p-Nitrophenylethyl (NPE) Group, Tetrahedron **40** (1984) 59-72; W. Pfleiderer, M. Schwarz and H. Schirmeister, New Developments in Nucleotide Chemistry, Chem. Scripta **26** (1986) 147-154.

37 G. Hering, Neue Phosphorylierungsmittel für die Oligonukleotidsynthese, Doctoral Thesis, Technical University München, 1987.

38 G. Hering, R. Stöcklein-Schneiderwind, I. Ugi, T. Pathak, N. Balgobin and J. Chattopadhyaya, Preparation and

Properties of Chloro-N,N-dialkylamino-2,2,2-trichlorethoxy- and Chloro-dialkylamine 2,2,2-Trichloro-dimethyl-ethoxyphosphines and their Deoxynucleoside Phosphiteamidates, Nucleosides and Nucleotides **4** (1985) 169-171.

39 R.L. Letsinger, E.P. Groody and T. Tanaka, J. Amer. Chem. Soc. **104** (1982) 6805-6806.

40 B.C. Froehler and Matteucci, Substituted 5-Phenyltetrazoles: Improved Activators of Deoxynucleoside Phosphoamidites in Deoxyoligonucleotide Synthesis, Tetrahedron Lett. **24** (1983) 3171-3174.

41 J. Bauer and I. Ugi, Chemical Reactions and Structures without Precedent generated by Computer Programs, J. Chem. Res. **1982**, (S) 298; **1982** (M) 3101-3196, 3201-3260; J. Bauer, R. Herges, E. Fontain and I. Ugi, IGOR and Computer assisted Innovation in Chemistry, Chimia **39** (1985) 43-53.

42 I. Ugi and I. Götz, The Synthesis of 2-Chloro-2-oxo-4.5-dimethyl-1.3.2Δ^5-dioxaphospholene (CEP Chloride), Heterocycles **21** (1984) 265-269.

43 H. Kellner, Neue ß-Halogenalkyl-chlorphosphate als selektive Phosphorylierungsreagentien für die Nukleotidsynthese, Doctoral Thesis, Technical University München, 1980.

44 M.J. Haddadin and J.P. Freeman, Oxaziridines, in: A. Hassner (Ed.), Heterocyclic Compounds, Vol. 42,3, Interscience, New York, 1985, pp. 283-350; G. Glahsl and R. Herrmann, (+)3-Oxo-camphorsulfonyl-oxaziridine, a highly stereoselective Reagent for the Oxidation of Sulfides to chiral Sulfoxides, J. Chem. Soc. Perkin I (in press), and ref. therein.

Affiliation

Prof. I. Ugi, Institut für Organische Chemie,
Technische Universität München, 8046 Garching, FRG

General Strategies for the Asymmetric Synthesis of Oxygenated Natural Products

Stephen F. Martin†
Department of Chemistry, The University of Texas at Austin, Austin, TX 78712 USA

Introduction. Several years ago we initiated a program that was directed toward the design and development of general strategies that could be expeditiously applied to the formulation of concise, total syntheses of enantiomerically pure, oxygenated natural products. Some targets of considerable interest and obvious importance in this regard include: KDO (3 deoxy-D-*manno*-octulopyranosate) (**1**), which is a higher monosaccharide found in the cell wall of Gram-negative bacteria;[1,2] the antibiotic tirandamycin A (**2**) as well as its degradation product tirandamycic acid (**3**);[3,4] Prelog-Djerassi lactone (**4**), which is obtained from the degradation of several macrolide antibiotics including methymycin (**5**);[5,6] other medicinally important macrolide antibiotics such as the erythromycins A (**6**) and B (**7**);[5,7] and the ansa antibiotics macbecin (**8**) and herbimycin A (**9**), which are potential anticancer agents.[8,9]

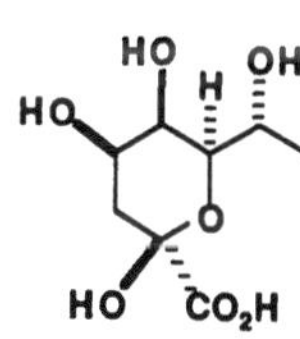

1

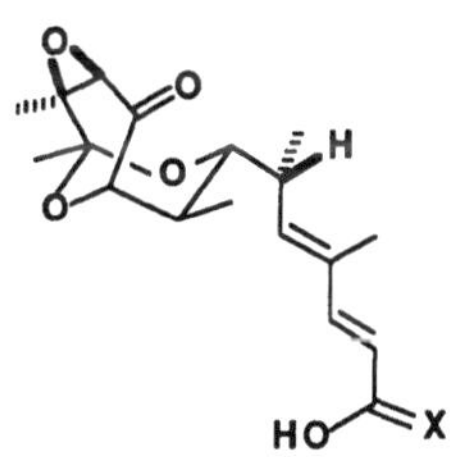

2: X =

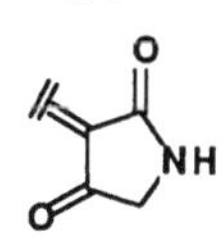

3: X = O

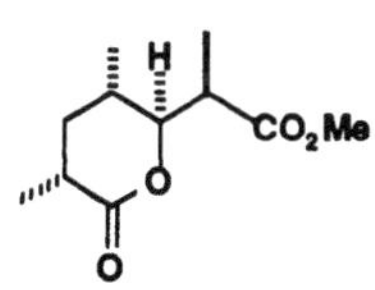

4

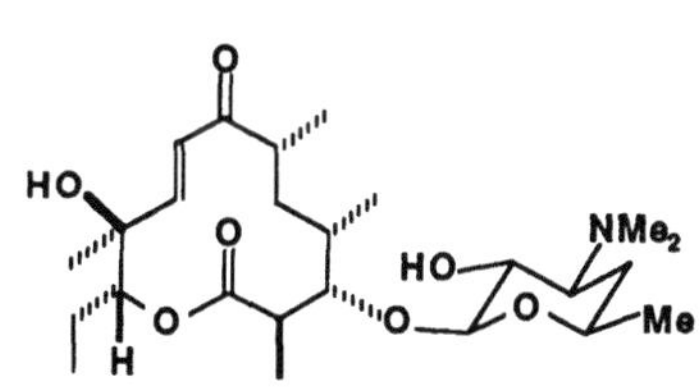

5

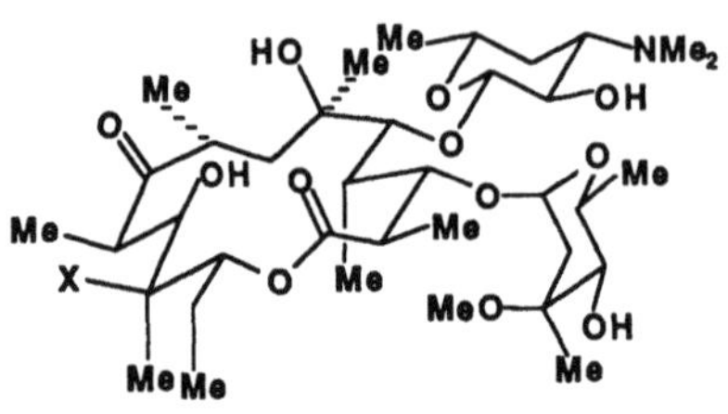

6: X = OH
7: X = H

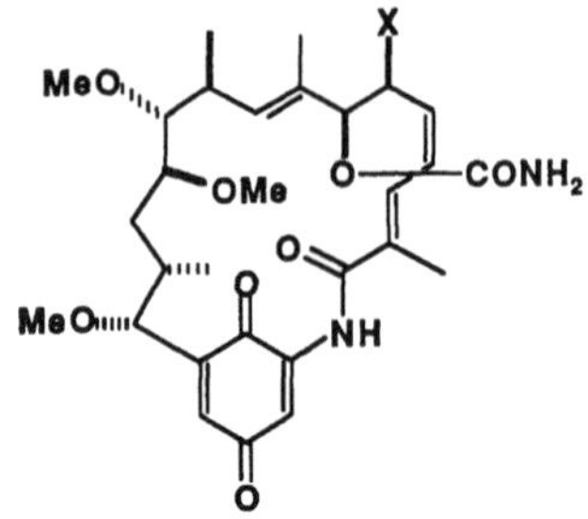

8: X = Me
9: X = OMe

On the methodological front of these broadly based endeavors, we have exploited pericyclic processes such as the dipolar cycloadditions of nitrile oxides together with the aldol reaction and related constructions as tactical devices for the formation of new carbon-carbon bonds with high levels of stereochemical control. Another important focus of these explorations has been upon the development of techniques for the manipulation and refunctionalization of hydropyrans, since this structural subunit is not only common to a variety of natural substances, but it may also be effectively exploited as a conformationally-biased template for the stereoselective construction of various skeletal arrays present in numerous natural products. In this context, we have devised a novel and highly effective strategy for the asymmetric syntheses of oxygenated natural products. The fundamental approach features the intermediacy of the hydro-3-pyranones **12**, which may be accessed from the chiral furfuryl carbinols **10** via the hydroxy enediones **11** by well-established oxidative techniques (Scheme 1).[10] A critical element of this overall plan[11] is that the hydro-3-pyranones **12** are admirably endowed with differentiated functionality that is suitable for further elaboration by reaction with selected nucleophiles

Scheme 1

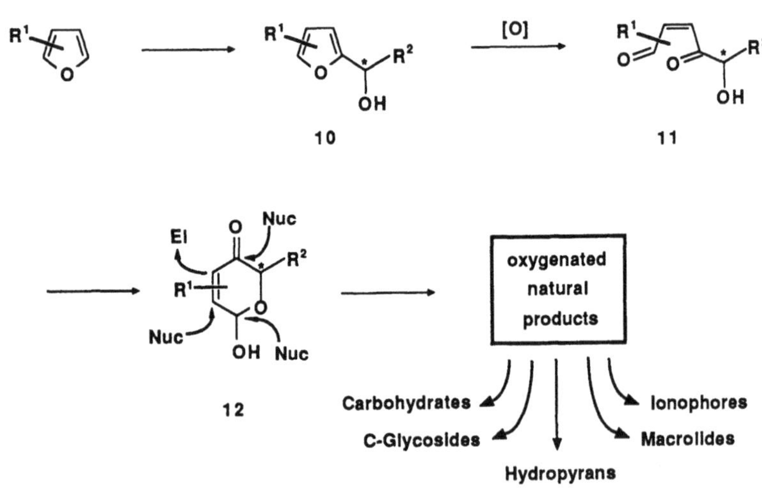

and electrophiles. These manipulations thus allow the facile introduction of other functional groups and/or alkyl residues onto the basic framework. Significantly, the stereochemical courses of these operations are directed in a highly predictable fashion by the preexisting center at C*, which may be fixed in an absolute sense via the asymmetric synthesis of the furfuryl carbinols **10**, and a variety of standard tactics for the convenient preparation of these chiral alcohols in enantiomerically pure form from substituted furans may be envisaged.

We now wish to record some of the details of our recent investigations in this area that have culminated in the design of highly concise entries to (+)-KDO (**1**), (-)-tirandamycin A (**2**) and (+)-tirandamycic acid (**3**), (+)-Prelog-Djerassi lactone (**4**), and the *seco*-acids of the aglycones of the erythromycins A (**6**) and B (**7**); our basic plan for the total syntheses of macbecin (**8**) and the closely related substance herbimycin A (**9**) will also be outlined.

Higher Monosaccharides: A Novel Approach to (+)-KDO (1). (+)-KDO (**1**) links the hydrophilic carbohydrate polymer region to the hydrophobic lipid A moiety in the cell wall of Gram-negative bacteria, and it has recently been postulated that analogs of **1** might serve as useful antibiotics by inhibiting the biosynthesis of lipopolysaccharides.[12] Although there has been some synthetic activity in this arena,[2,12] we are presently aware of no accounts of the *total* synthesis of the optically pure substance. Inasmuch as **1** bears close structural resemblance to the generalized hydropyranone **12**, it occurred to us that the strategy outlined in Scheme 1 might be readily applicable to its synthesis.

The initial move in this modest undertaking involved the addition of 2-lithiofuran (**13**) to isopropylidene-D-glyceraldehyde (**14**) in the presence of zinc bromide[11c,13] followed by protection of the intermediate alkoxide *in situ* as a silyl ether to give **15**, thereby establishing the crucial stereochemical relationship at C(6) and C(7) of **1** (Scheme 2). Metalation of **15** at C(2) of the furan ring and sequential alkylation with benzylchloromethylether and deprotection of the C(6) hydroxyl group provided the furfuryl carbinol **16**. The oxidation of the furan moiety was readily achieved by reaction of **16** with *tert*-butylhydroperoxide in the presence of a vanadium catalyst,[10i] and subsequent *O*-methylation of the resulting lactols delivered an easily separable mixture (>5:1) of the hydropyranone **17** together with its C(2) anomer. Since (+)-KDO (**1**) itself possesses a carboxyl function at C(1), we also prepared **18** in anticipation that it might provide a more direct route to **1**. However, to date all attempts to oxidize the electronically deactivated furan ring of **18** without the concomitant and deleterious cleavage of the acetonide moiety were unsuccessful, and this route has been temporarily placed on the sideline as another pathway has proven more promising.

With **17** now secured, it remained only to refunctionalize the hydropyran ring, to adjust the oxidation level at C(1), and to deprotect the hydroxyl groups at C(7) and C(8). To this end, the reduction of **17** with various hydride reagents was explored. When **17** was treated with either diisobutylaluminum hydride or sodium borohydride, a mixture of the epimeric allylic alcohols **19** and **20** was obtained in ratios that varied from 5-10:1 in which the undesired diastereoisomer **19** dominated. Alternatively, when **17** was allowed to react with K-Selectride, reduction proceeded with a high degree of selectivity from the more accessible equatorial face to give **20** as the major product (>10:1). The original synthetic plan called for the stereoselective introduction of the C(4) oxygen substituent via

Scheme 2

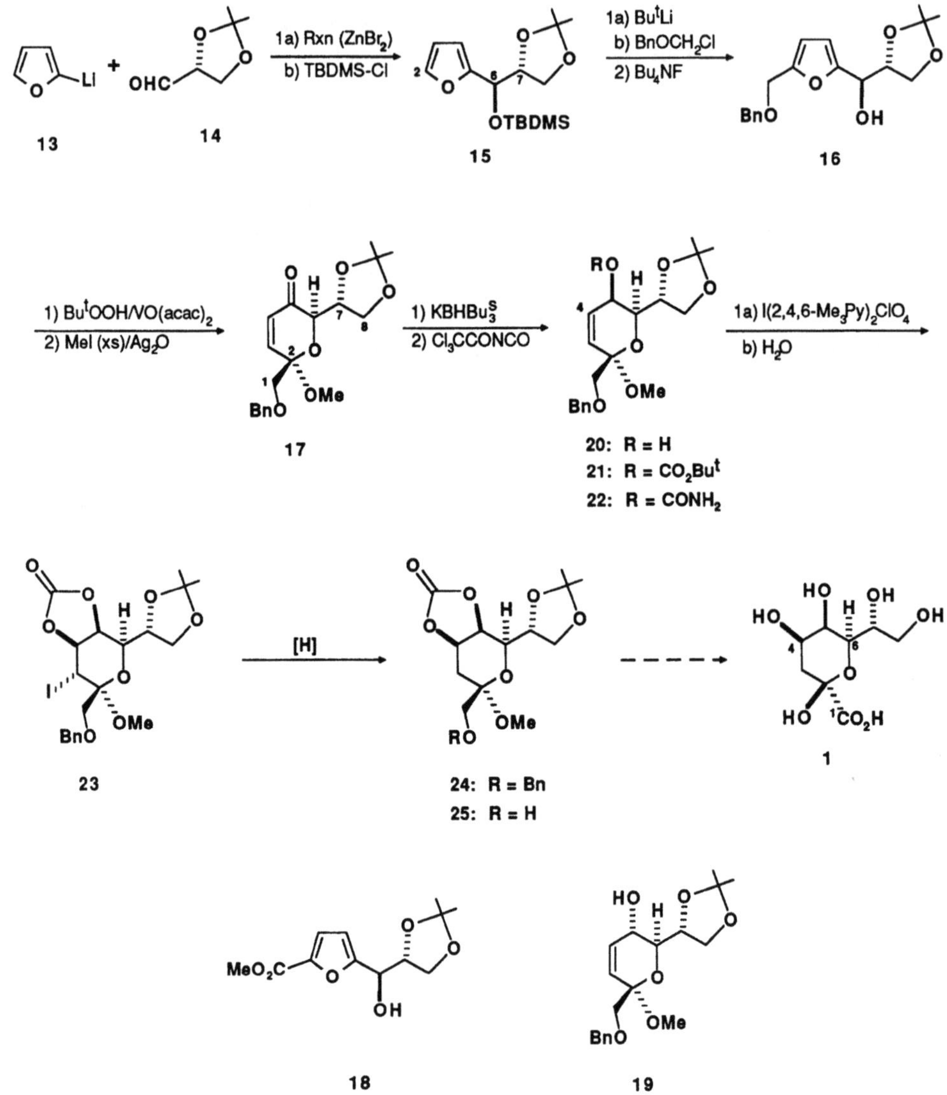

an intramolecular process through the agency of the electrophilic cyclization of a suitable derivative of the allylic alcohol **20**. However, preliminary attempts to effect the cyclization of the carbonate **21** and related substances were unsuccessful. Nevertheless, we discovered that **20** could be easily converted into the carbamate **22** according to a technique previously employed by Kishi.[14] Although the electrophilic cyclization of allylic alcohol derivatives related to **22** was known to be rather sluggish, the application of methodology developed by Fraser-Reid[15] was brought to bear on the problem at hand to provide **23** in good overall yield.

The next task required the removal of the 3-iodo substituent and the *O*-benzyl group from the alcohol at C(1). Somewhat surprisingly, when we attempted to effect the radical deiodination of **23** by the action of tri-*n*-butyltin hydride in the presence of AIBN in refluxing toluene, a mixture of the expected product **24** *and the debenzylated substance* **25** was obtained. The *O*-benzyl protecting group may then be cleaved from **24** by catalytic hydrogenolysis. We are presently exploring methods that would allow the efficient and simultaneous removal of both the iodine and the benzyl group from **23** to give **25** in a single step. In order to complete this short synthesis of **1**, the oxidation of the primary alcohol at C(1) and deprotection of the various hydroxyl substituents is required, and we anticipate that these processes may be induced without untoward difficulty.

A Concise Approach to (-)-Tirandamycin A (2) and (+)-Tirandamycic Acid (3). Tirandamycin A (**2**)[3] is a representative member of a novel class of naturally-occurring antibiotics that are characterized by the presence of an enolized 3-dienoyl tetramic acid moiety coupled with the unusual dioxabicyclo[3.3.1]nonane ring system. In addition to its antimicrobial activity, **2** inhibits bacterial DNA-directed RNA polymerase, and it interferes with oxidative phosphorylation. Early synthetic work in this area focused upon the preparation of (+)-tirandamycic acid (**3**),[4a-d] which is a degradation product of tirandamycin A (**2**). The first major achievement in this area was Ireland's synthesis of **3** from D-glucose,[4a] but more recently **2** has also been prepared by total synthesis.[4e-h] Our own investigations in this area culminated in a facile, asymmetric synthesis of **26**, which played a pivotal role in Ireland's synthesis of **2**[4a] and Schlessinger's synthesis of **3**.[4f]

The essential features of the strategy which evolved in our original plan that was directed toward the synthesis of **3** are summarized by the logical disconnections depicted in Scheme 3, wherein **26** was envisaged as the key intermediate and primary subgoal of the project. A Wittig construction

Scheme 3

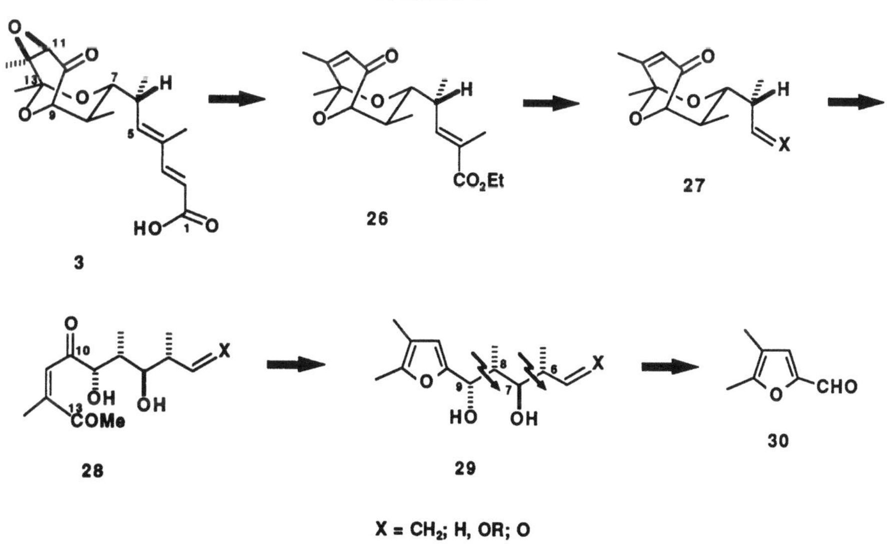

was envisioned to allow facile access to **26** via the aldehyde **27** (X = O), which may in turn be unraveled by hydrolysis of the bicyclic ketal moiety to provide the densely functionalized intermediate **28**. Since the synthetic equivalencies of the 1,4-dione and ene-1,4-dione arrays with a furan ring have now been well established,[10,11] it occurred to us and others[4b,c,e] that a furan might admirably serve as the essential C(10)-C(13) core of **3**. In view of this analysis, it naturally followed that the substituted furan **29** should become the initial target of our investigations, and a variety of entries to this substance from 4,5-dimethylfurfuraldehyde (**30**) may be formulated that involve stereoselective aldol or related constructions for the formation of the C(6)-C(7) and the C(8)-C(9) bonds.

In order to probe the underlying feasibility of the crucial cyclization **28** → **27**, we elected to engage first in a brief and illustrative model study (Scheme 4). Thus, the racemic ß-keto ester **32a** was converted into the *syn*-diol **33a** by the highly stereoselective, chelation-controlled reduction of the ketone group with zinc borohydride[16] followed by the hydride reduction of the remaining ester function. Sequential oxidation of **33a** with bromine in methanol and treatment of the intermediate 2,5-dimethoxy-2,5-dihydrofurans with aqueous acid afforded the hydropyranone **34a**, but all attempts to induce its cyclization under either acidic or basic reaction conditions failed to give detectable quantities of **35a**. In view of these discouraging results, we speculated that the bicycloketalization of the related hydropyranone **34b** should be more favorable, since this process would involve the cyclization of a more stable, and hence more readily formed, tertiary, oxygen-stabilized carbocation. Moreover, the presence of the ß-methyl group on the conjugated olefin would be expected to hinder the 1,4-addition of nucleophiles to the enone array. In support of this hypothesis, when **33b** was treated with bromine and methanol followed by hydrolysis of the intermediate ketals in aqueous acid, the desired bicycle **35b** was obtained as the sole product in excellent yield. Interestingly, there was no evidence for the formation of the intermediate hydropyranone **34b**.

Scheme 4

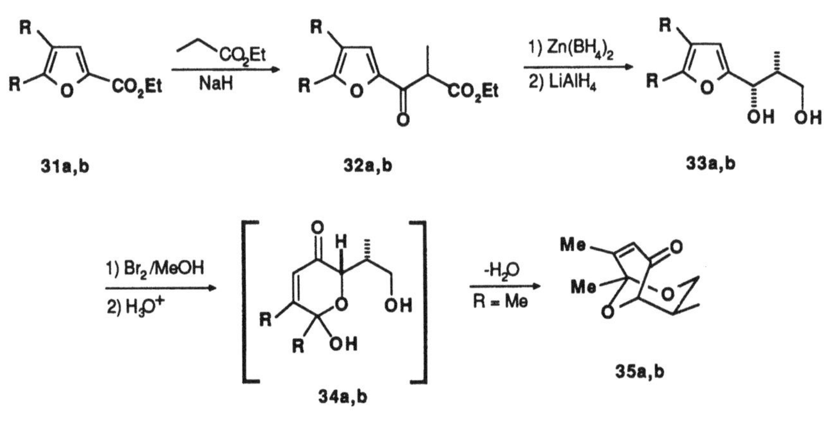

Series a: R = H

Series b: R = Me

Thus, encouraged by these preliminary findings, we set to the more challenging task of preparing a substituted, enantiomerically pure furan of the general type **29** (X = CH_2). We envisioned that the two stereocenters at C(8) and C(9) could be established with the correct absolute configuration via a diastereoselective *syn*-aldol[17] process. The remaining stereocenters at C(6) and C(7) bearing the pendant methyl and hydroxyl groups in an *anti* orientation together with the terminal vinyl group at C(5) might then be installed by the nucleophilic addition of a crotyl organometallic reagent[18] to a suitably protected aldehyde. The major uncertainty in this new undertaking would likely arise during the construction of the second carbon-carbon bond, since this event entails a substrate-based reaction and requires controlling the relative stereochemistry between C(7) and C(8) by a nucleophilic addition to the carbonyl group in an *anti*-Cram sense. The terminal vinyl function would serve nicely as a latent aldehyde moiety that could then be subsequently exploited by selected olefination procedures for the further elaboration of the dienoyl side chain in the targeted antibiotics.

The synthesis commenced with the reaction of the known 4,5-dimethylfuraldehyde (**36**)[19] with the boron enolate **37** according to the elegant methodology originally developed by Evans,[17,20] and the subsequent removal of the chiral auxiliary afforded the enantiomerically pure *syn*-ß-hydroxy ester **38** in excellent yield (Scheme 5). Protection of the free hydroxyl group followed by the partial reduction of the ester afforded the aldehyde **39**, which underwent a chromium (II) mediated addition of crotyl bromide[21] to give a separable mixture (1:1.5) of the homoallylic alcohols **40** and **41**. Unfortunately, all attempts to improve this ratio in favor of the desired stereoisomer **40** were unsuccessful, and those efforts included the use of other hydroxyl protecting groups on derivatives of **39** that might be more

Scheme 5

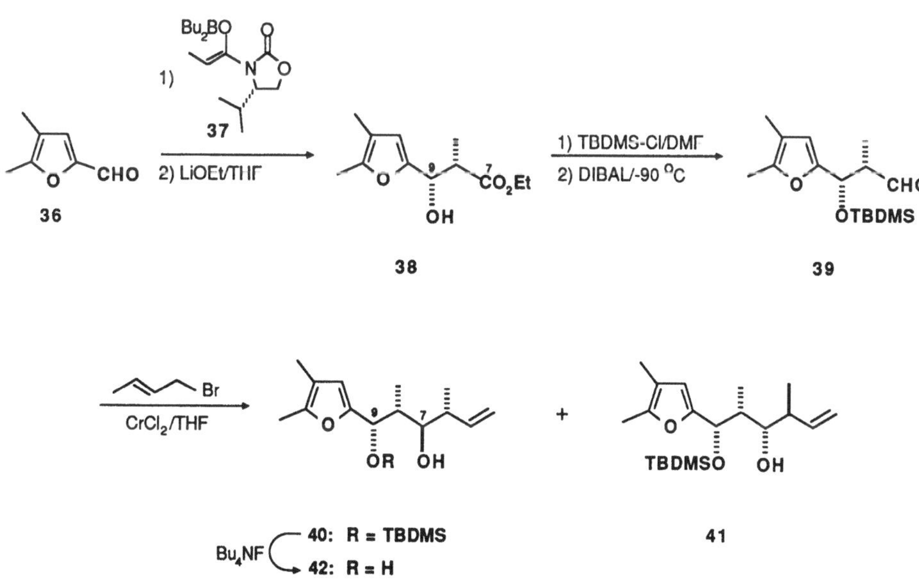

prone to favor the chelation controlled addition to the aldehyde function. In retrospect, it appears that the use of chiral crotyl boronates as the nucleophilic partners, which has been nicely developed by Roush,[18e] might constitute a reasonable solution to this stereochemical problem. Subsequent fluoride-induced removal of the *tert*-butyldimethylsilyl group from **40** then gave **42**, and the stage was then set for oxidative transformation of the furan ring.

Although preliminary experiments to effect the efficient conversion of **42** into **44** by oxidation with bromine in methanol followed by treatment with aqueous acid according to the protocol previously established in our model studies (Scheme 4) were unsuccessful, we soon discovered that the reaction of **42** with *m*-chloroperbenzoic acid (MCPBA) followed by the direct subjection of the intermediate hydropyranones **43** to aqueous hydroiodic acid delivered the desired bicyclic ketal **44** (Scheme 6). It should be noted that it was necessary to exercise careful control over the experimental conditions, since prolonged exposure of **44** to aqueous acid led to the production of a mixture **44** together with other substances. Selective ozonolysis of the terminal vinyl group present in **44** then proceeded without difficulty to afford the unstable aldehyde **45**, which was then subjected without further purification to a Wittig olefination according to the precise conditions originally defined by Ireland[4a] to deliver the known unsaturated ester **26**. Since enantiomerically pure **26** had been converted previously in five steps by Ireland into (+)-tirandamycic acid (**3**)[4a] and more recently by Schlessinger[4f] into (-)-tirandamycin A (**2**), the extraordinarily concise (nine steps from the known aldehyde **36**) preparation of **26** outlined herein constitutes, in a strictly formal sense, the total syntheses of these substances.

Scheme 6

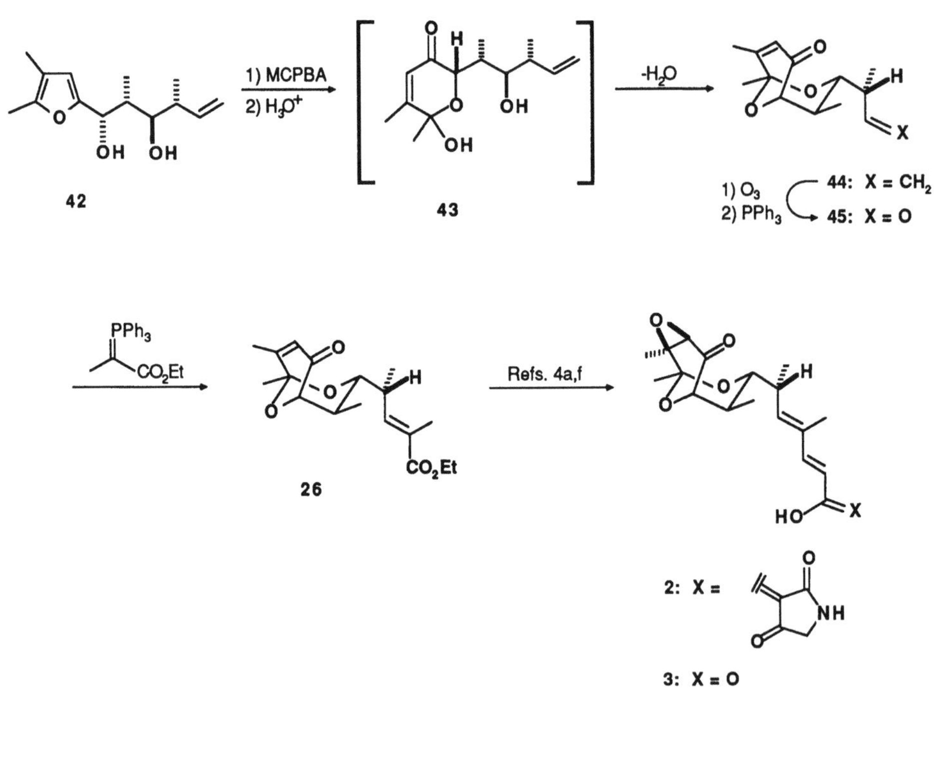

Total Synthesis of (+)-Prelog-Djerassi Lactone Methyl Ester (4). The methyl ester of Prelog-Djerassi lactone **4**, which incorporates the C(1)-C(7) fragment of the macrocyclic antibiotic methymycin (**5**), has served as an important focal point for the design and application of new synthetic methods that result in the stereoselective elaboration of carbon frameworks bearing a number of contiguous and/or alternating stereocenters. The stereochemical relationships present in **4** are also found in other macrolide and ionophore antibiotics, and consequently the tactics that have been developed for its assemblage may oftentimes be extended to the syntheses of these and other architecturally complex natural products. Indeed, **4** has emerged as one of the standards against which new methodology for effecting stereoselective constructions and functionalizations may be measured. Our own success in this area exploits the general strategy for the synthesis of oxygenated natural products that is embodied in Scheme 1 and is summarized in Scheme 7.[6]

The stereocenters at C(2) and C(3) of **4** were conveniently established with the correct relative and absolute configurations once again through the utilization of the aldol methodology developed by Evans.[17,20] Thus, reaction of furaldehyde (**46**) with the boron enolate **47** followed by methanolysis afforded the β-hydroxy ester **48** in enantiomerically pure form. The sequential reaction of **48** with bromine in methanol and then aqueous acid followed by protection of the intermediate anomeric lactols

Scheme 7

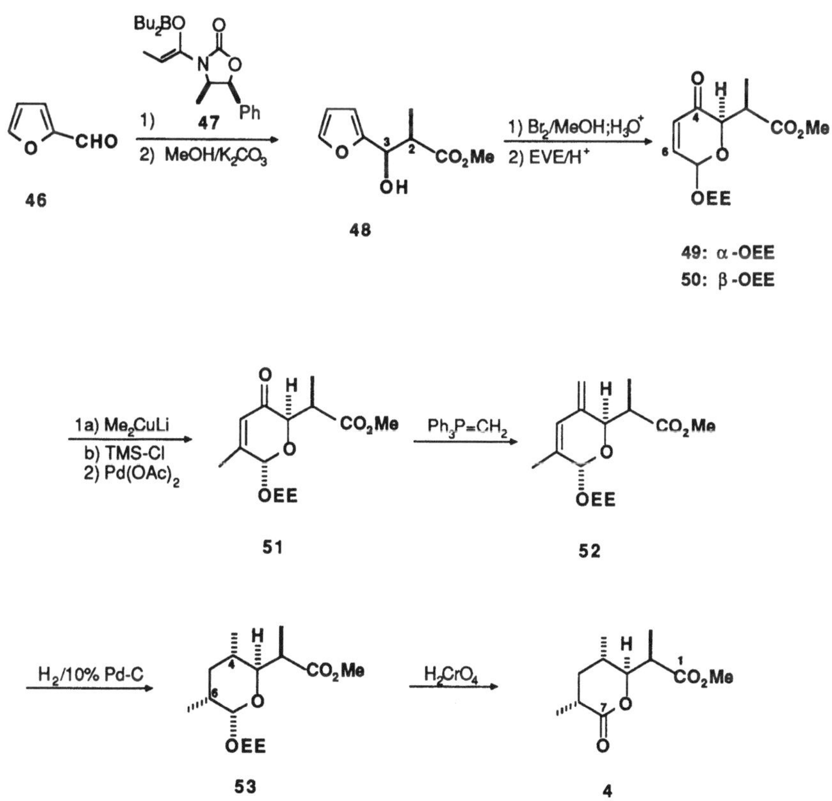

with the bulky ethoxyethyl group provided a separable mixture (approximately 4:1) of the hydropyranones **49** and **50**, respectively. It now remained to introduce the methyl groups at C(4) and C(6) with the proper relative stereochemistry. The conjugate addition of lithium dimethylcuprate to **49** proceeded as expected[22] virtually exclusively from the ß-face to give the incorrect stereochemistry at C(6). However, reversal of this stereochemical flaw was readily realized via a straightforward sequence of reactions that commenced with trapping the intermediate enolate formed after the cuprate addition as its silyl enol ether which was oxidized with excess palladium acetate according to the Ito-Saegusa protocol[23] to give **51**. Subsequent Wittig olefination of **51** furnished the diene **52**, which presumably underwent stereoselective reduction from the ß-face owing to the presence of the sterically demanding α-ethoxyethyl group,[24] to deliver a mixture (approximately 9:1) of **53** together with a substance that appeared to be the corresponding C(4) epimer. The final transformation of **53** into **4** was then achieved in a single operation upon treatment of **53** with aqueous chromic acid.

This concise entry (nine chemical operations from **46**) to **4** in enantiomerically pure form convincingly establishes the viability of the approach outlined in Scheme 1 for the asymmetric synthesis of certain key structural subunits that are present in members of the family of macrolide antibiotics. However, an even more compelling case in support of the utility of this strategy is evident from its application to the total synthesis of the antibiotics of the erythromycin class.

Total Synthesis of the Seco Acid of Erythronolide B. The basic strategy for the challenging assault on the clinically useful erythromycin antibiotics is summarized in the antithetic format in Scheme 8 for the aglycones of the erythromycins A and B, **54** and **55** respectively. Thus, the initial carbon-heteroatom bond disconnection (Step ***a***) of the 14-membered ring of the lactone functions of **54** and **55** renders the protected forms of the corresponding seco acid derivatives **56** and **57**. Subsequent scission of the C(10)-C(11) bond (Step ***b***) via a retro-aldol construction then leads to the ketone **60** that must be coupled with either of the aldehydes **58** or **59** depending upon whether **54** or **55**, respectively, is the ultimate target. A crucial question at this juncture entails the stereochemical outcome of this bond forming step. In this regard, it is noteworthy that prior to the initiation of this study, Masamune had convincingly demonstrated in his elegant synthesis of 6-deoxyerythronolide B[7b] that a closely related aldol reaction ensued with a high level (17:1) of diastereoselection (*vide infra*); more recently, Kochetkov observed a similar preference in a related process.[7n] There exist two fundamental options for the preparation of **60** that feature the combination according to standard methodology of the keto aldehyde **61**, or its functional equivalent, with either an allylic organometallic reagent or an enolate as the C(1)-C(2) synthon.[17,18] A *syn*-selective process that occurs with a high level of diastereofacial selectivity in the Cram sense is mandated in this step. Refunctionalization of **61** in a retrosynthetic sense might then be envisioned to lead *inter alia* to the bicyclic ketal **62**, which would be accessible from a substituted furan **63**.

The bicyclic ketal **62** incorporates the intact C(3)-C(10) portion of the backbone of the erythromycin antibiotics, and its construction was viewed as the first objective. In the event, this undertaking was initiated with the straightforward formylation of commercially available 2-ethylfuran (**64**) according to the Vilsmeier-Haack protocol to provide **65** (Scheme 9). The diastereoselective

Scheme 8

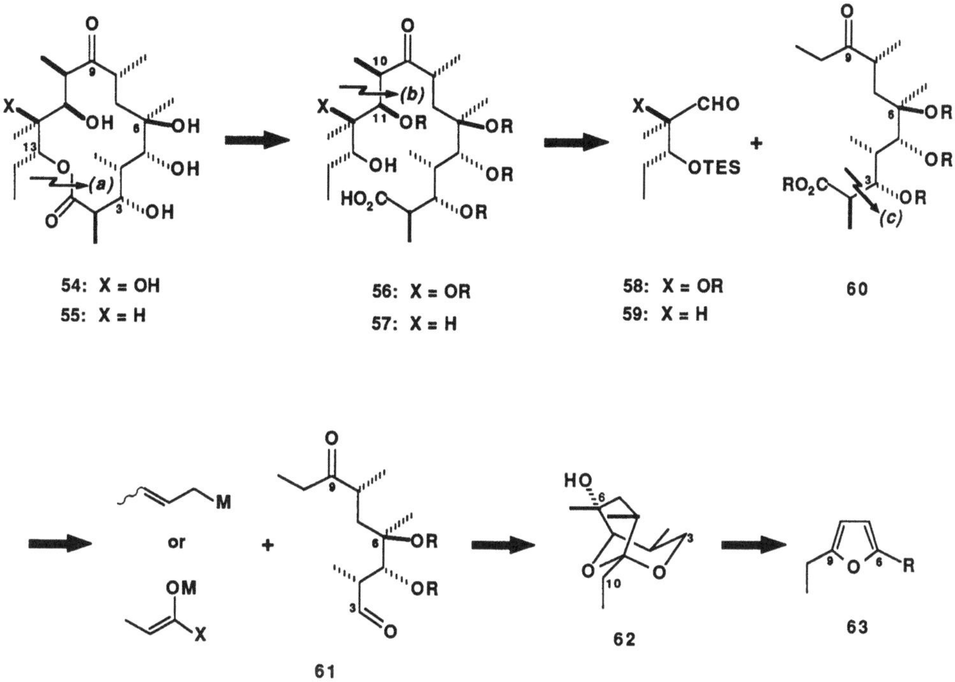

aldol reaction[17,20] of **65** with the boron enolate **47** then provided **66**, which was readily converted into the *syn*-diol **67** upon reduction with lithium borohydride. In close analogy with chemistry that had been previously developed in efforts directed toward the synthesis of (+)-tirandamycic acid (**3**), we discovered that treatment of **67** with bromine in aqueous acetonitrile under rigorously defined experimental conditions led directly to the bicyclic enone **68**, a substance that is generously endowed with desirable structural features. For example, whenever protective measures must be employed to ensure that only specific functional groups enter into chemical reactions, two additional chemical steps are typically required. Thus, in order to develop a truly concise approach to polyfunctional, complex natural products, it is imperative to minimize the total number of such synthetically unproductive but often necessary operations. In the particular case at hand, the potentially interfering functional groups are the hydroxy substituents at C(3) and C(5) together with the ketone at C(9), and these are nicely protected internally by the formation of the bicyclic ketal. Furthermore, the bicyclic nature of **68** presents incoming nucleophilic reagents with two highly stereodifferentiated faces of the enone moiety. Consequently, the addition of lithium dimethylcuprate to **68** at C(8) ensued with virtually complete (>98%) diastereofacial selectivity, and when the intermediate, readily enolizable ketone thus produced was allowed to react with methyllithium in the presence of cerium trichloride,[25] the addition to the carbonyl group at C(6) also proceeded with high (>20:1) stereoselection to provide **62**.

Scheme 9

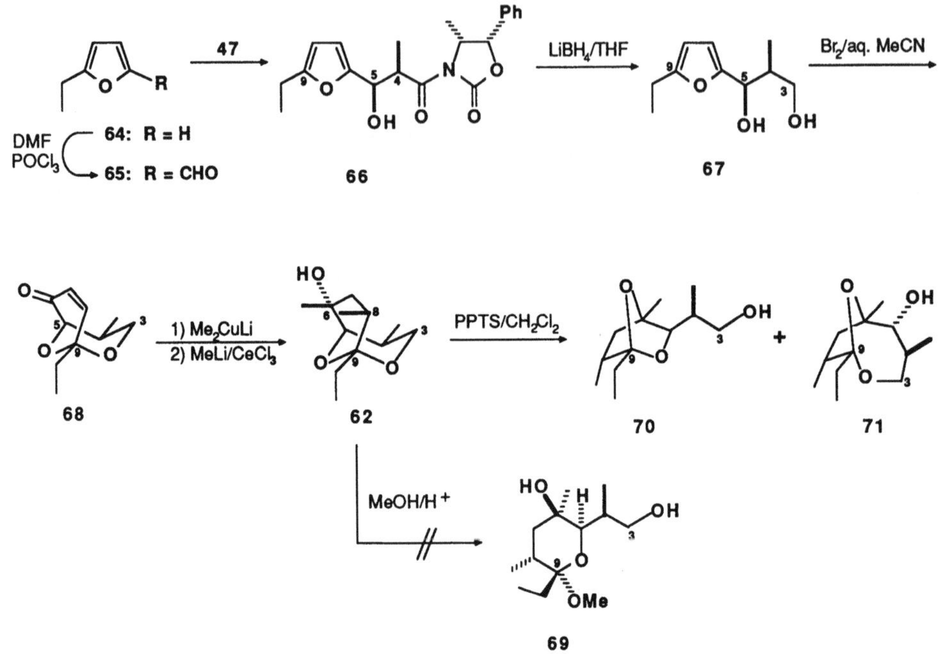

We had originally hoped to exploit the internal strain present in **62** to provide a significant driving force to effect its facile conversion to the monocyclic hydropyran **69**, but we have been unable to achieve this important transformation under an extensive variety of experimental conditions. Rather, the acid-catalyzed reorganization of **62** proceeded to give **70** as the kinetic product when non-nucleophilic solvents were employed; however, prolonged exposure of **70** to these acidic conditions led to the formation of **71** together with other isomeric substances.

With the three ketals **62**, **70**, and **71** thus readily available, the problem of selecting the optimal route to the macrolide skeleton had to be addressed. Of the latter two possibilities, a preliminary analysis revealed that more synthetic manipulations would be required for the elaboration of the ketal **71** toward this final goal than would be necessary for **70**, which appeared well suited to the task (Scheme 10). For example, the Swern oxidation of **70** afforded the aldehyde **72**, which underwent a highly diastereoselective, directed aldol reaction[17,20] with the boron enolate **47** to provide **73**, the structure of which was unequivocally verified by the single crystal X-ray analysis of its *tert*-butyldimethylsilyl ether. It is noteworthy that **73** incorporates the complete C(1)-C(10) segment of the erythromycin backbone with the proper absolute and relative stereochemistry at each of the stereocenters; efforts to transform **73** via a suitably short sequence of chemical operations into the target macrolides is under careful scrutiny. A different method for adding the C(1)-C(2) fragment to **72** by the Lewis acid-catalyzed addition of a crotyl stannane to **72** was also briefly evaluated.[18a-c,26] Unfortunately, under the Lewis-acidic conditions that were required to effect this nucleophilic addition,

Scheme 10

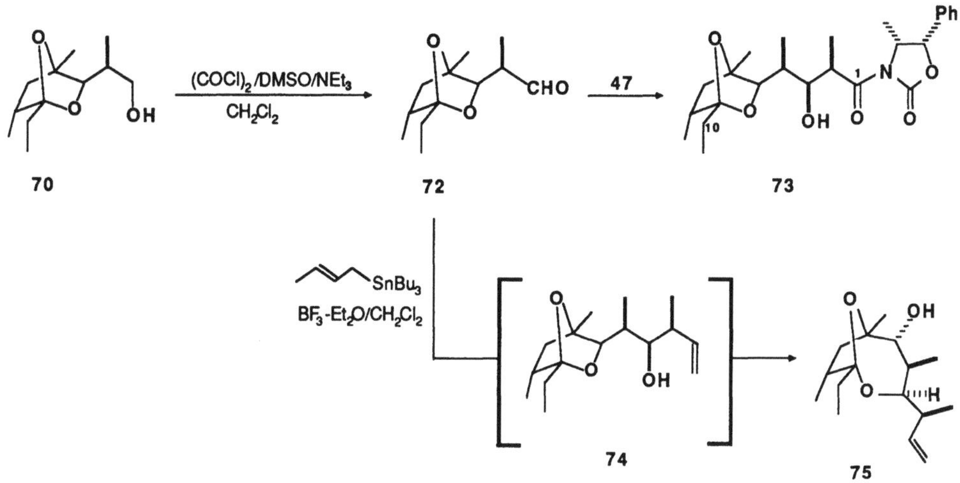

the major adduct **74** that was initially produced underwent spontaneous isomerization to give a substance that was tentatively identified as **75** in an unavoidable and damaging side reaction.

During the course of the aformentioned studies, an alternative series of reactions was being explored, and these investigations have recently resulted in the development of an extraordinarily concise route to a protected derivative of the seco acid of erythronolide B (Scheme 11) that promises to offer ready access to the seco acid of erythronolide A as well. Treatment of **62** with ethanedithiol in the presence of boron trifluoride etherate resulted in a ketal exchange to give the protected keto triol **76**. Selective protection of the vicinal diol array at C(5) and C(6) was achieved under equilibrating conditions by the action of camphorsulfonic acid (CSA) in acetone, and the thioketal protecting group at C(9) was then removed through the agency of *N*-bromosuccinimide in aqueous acetone to give **77**. Efforts to effect the transformation of **62** into **77** by more direct pathways have been unavailing thus far. Swern oxidation of the primary alcohol at C(3) delivered the keto aldehyde **78**, which underwent a highly chemoselective, Lewis acid-catalyzed addition of crotyl stannane[26] to the aldehyde function to give a mixture (approximately 3-4:1) of the homoallylic alcohol **79** together with another diastereomer that is presumed to be epimeric at both C(2) and C(3). The structure of **79** was secured by single crystal X-ray analysis. Despite several efforts to date, we have been unable to improve upon the diastereomeric ratio in this substrate-based process, and it appears that it may be necessary to engage methodology involving a reagent-based aldol reaction to resolve this stereochemical problem.

In anticipation of the final carbon-carbon bond construction that is required to prepare the intact seco acid of erythronolide B using a directed aldol reaction to form the C(10)-C(11) bond, it was first necessary to prepare the requisite chiral aldehyde **80**. Although the synthesis of **80** had been previously reported,[7b] we elected to devise an alternative route to access this material that commenced with the addition of the chiral boron enolate **37**[20] to propionaldehyde to furnish **81** (Scheme 12). The sequential protection of the secondary hydroxyl group and removal of the chiral auxiliary gave **82**, which was then converted to **80** by over-reduction followed by reoxidation under Swern conditions.

Scheme 11

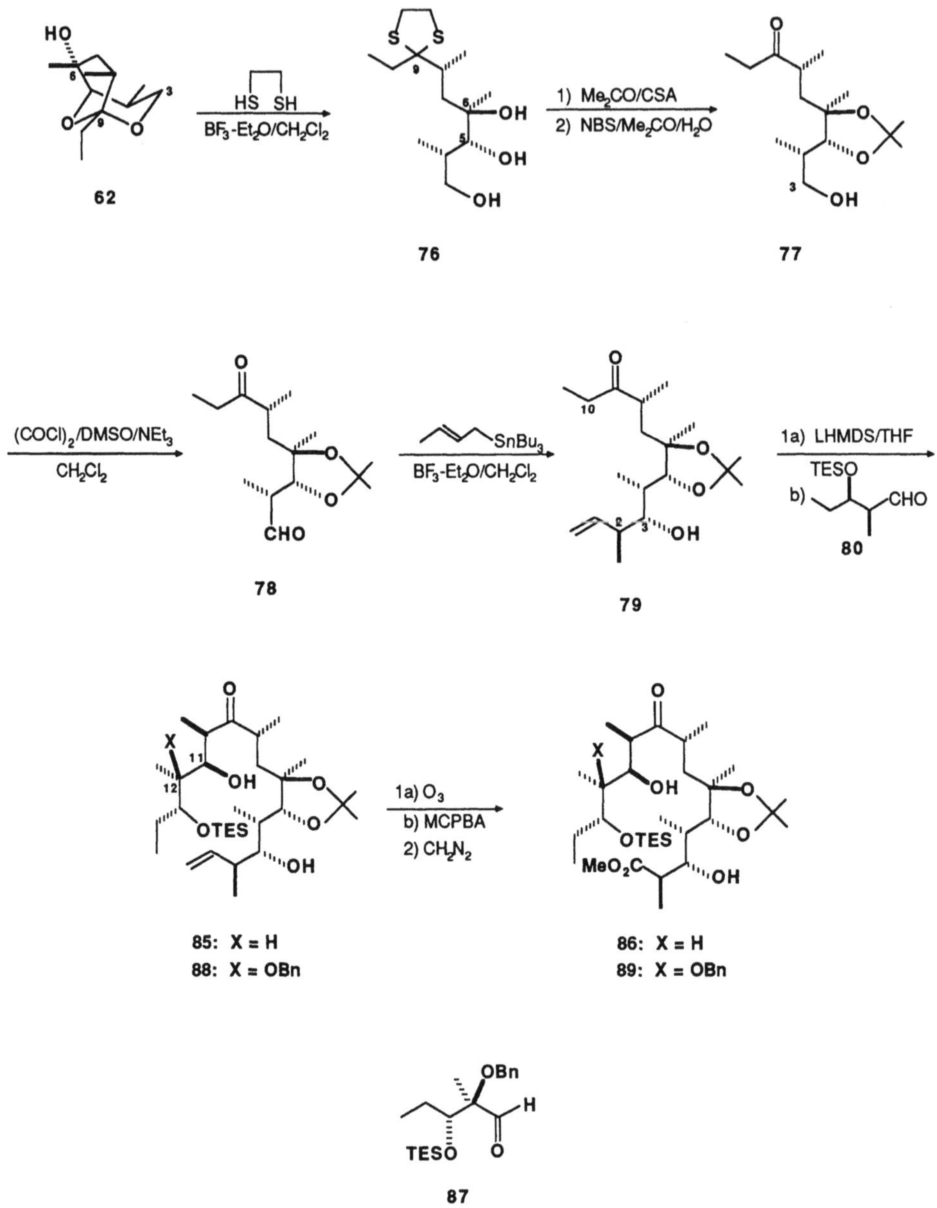
HO
6
9
3
O
O
62
HS SH
BF3-Et2O/CH2Cl2
S S
9
6
5
OH
OH
OH
76
1) Me2CO/CSA
2) NBS/Me2CO/H2O
O
O
O
3
OH
77
(COCl)2/DMSO/NEt3
CH2Cl2
CHO
78
SnBu3
BF3-Et2O/CH2Cl2
10
2
3
OH
79
1a) LHMDS/THF
TESO
b)
CHO
80
X
11
12
OH
OTES
85: X = H
88: X = OBn
1a) O3
b) MCPBA
2) CH2N2
MeO2C
86: X = H
89: X = OBn
OBn
H
TESO
87

Scheme 12

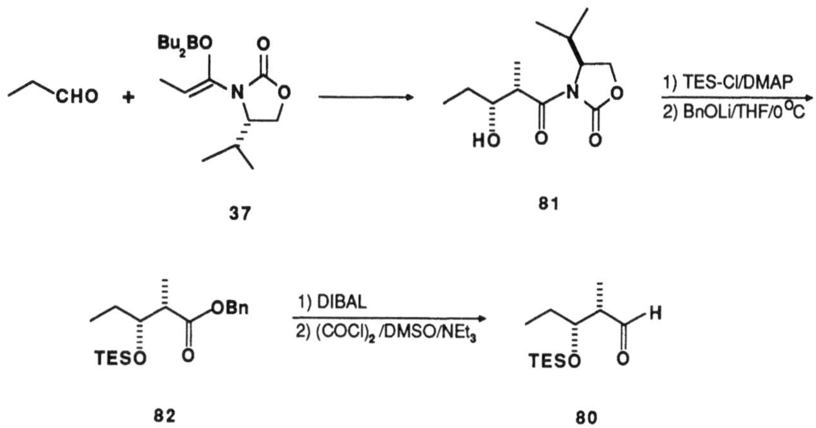

With the key molecular subunits **79** and **80** now in hand, attention was focused upon the critical aldol reaction (Scheme 11). Following closely the protocol previously developed by Masamune for the conversion of **83** into **84**, the less substituted *Z*-enolate generated from **79** was allowed to react with the aldehyde **80** to provide a mixture (approximately 2-3:1) of the desired adduct **85** together with

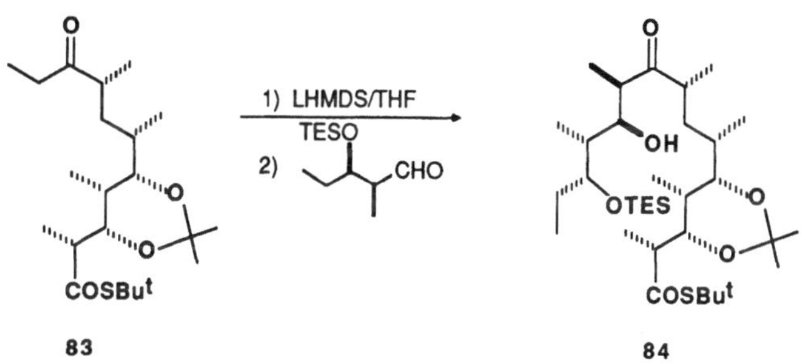

another diastereoisomer that is presently assumed to be the isomeric *syn*-product epimeric at both C(11) and C(12).[7b,n] Since we have thus far been unable to improve upon this diastereomeric ratio by making several straightforward adjustments in the reaction conditions, it presently seems likely that it may be necessary to explore the alternative of utilizing boron enolates derived from chiral boranes[27] to override the inherent diastereofacial selectivity in this substrate-controlled reaction. Inasmuch as such a tactic might afford a convenient and practical solution to this stereochemical dilemma, we are currently in the process of probing this intriguing possibility. In any case, the synthesis of the protected seco acid derivative of erythronolide B **86** was completed by the oxidative scission of the terminal vinyl group of **85** followed by esterification with diazomethane.

Upon examination of the reactions outlined in Schemes 9, 11, and 12, it is evident that the present route to the seco acid of erythronolide B is remarkably concise, involving a longest linear sequence of a mere 14 chemical operations from the commercially available 2-ethylfuran (**64**). Even counting the steps required for the preparation of the aldehyde **80**, the total number of operations is only 19. We anticipate that the analogous aldol reaction of the enolate derived from **79** with the electrophilic partner **87**, which should be readily available by application of standard methods,[28] will also provide an expedient pathway to the erythronolide A seco acid derivative **89** via the adduct **88**. Future investigations in this area will focus upon the elaboration of **86** and **89** into the naturally-occurring erythromycins B (**7**) and A (**6**), respectively, by a strategic device that entails the macrocyclization of suitable analogs of the fully glycosylated seco acids.

Concluding Remarks. From the foregoing discussions, it is obvious that the general strategy outlined in Scheme 1 is eminently suitable for the efficacious asymmetric syntheses of certain highly oxygenated and stereochemically complex natural products. Additional tests of this strategy and the associated tactics will be encountered as we challenge other naturally-occurring substances. For example, one major project that is currently under close scrutiny is directed toward the total syntheses of the ansa antibiotics macbecin (**8**) and herbimycin A (**9**) according to the general plan that is

Scheme 13

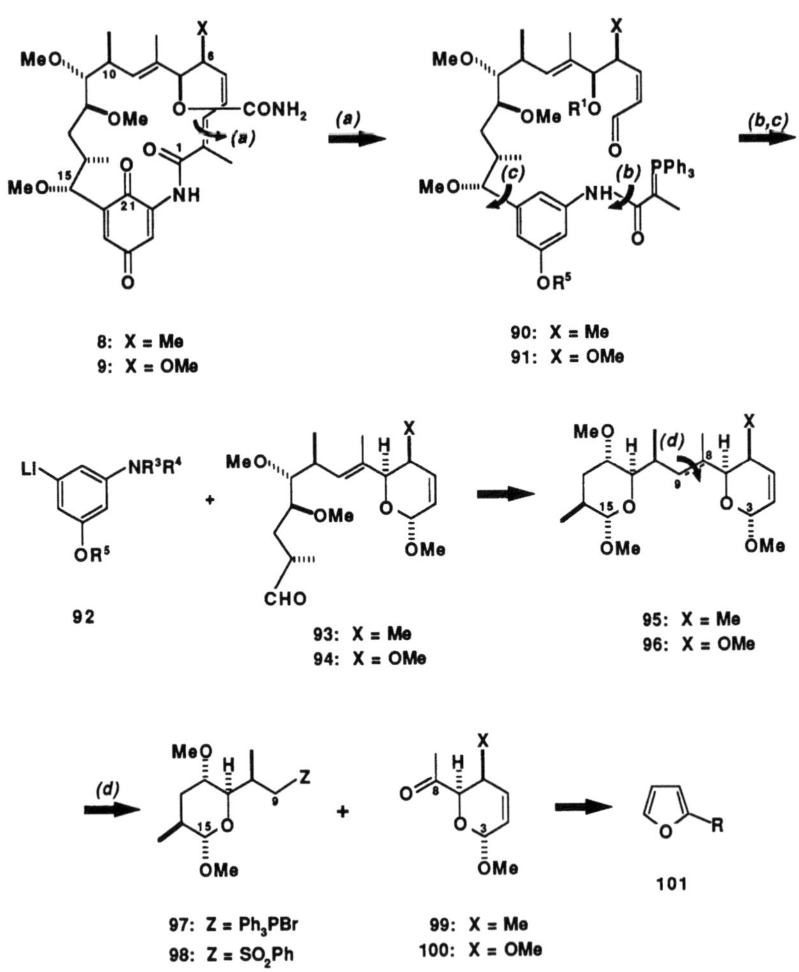

adumbrated in the retrosynthetic format in Scheme 13, wherein the key hydropyran intermediates **97** - **100** would be prepared from the substituted furans **101**. The details of these investigations will be recorded in due course.

Acknowledgment. I wish to thank my highly dedicated group of coworkers who made the critical and extensive intellectual, creative, and technical contributions that were required to complete the synthetic work described herein. These individuals include: L. E. Burgess, C. L. Campbell, Dr. R. C. Chapman, Dr. C. Gluchowski, R. P. Gist, D. E. Guinn, Dr. T. A. Mulhern, Dr. G. J. Pacofsky, and Dr. P. W. Zinke. I am also grateful to the National Institutes of Health (GM 31077) and the Robert A. Welch Foundation for their generous support of this research.

REFERENCES AND NOTES

†. Recipient of a National Institutes of Health (National Cancer Institute) Research Career Development Award (1980-1985).

1. Unger, F. M. *Adv. Carbohydr. Chem. Biochem.* **1981**, *38*, 323.
2. (a) Schmidt, R. R.; Betz, R. *Angew. Chem. Int. Ed. Engl.* **1984**, *23*, 430. (b) Danishefsky, S. J.; Pearson, W. H.; Segmuller, B. E. *J. Am. Chem. Soc.* **1985**, *107*, 1280.
3. (a) Meyer, C. E. *J. Antibiot.* **1971**, *24*, 558. (b) MacKellar, F. A.; Grostic, M. F.; Olson, E. C.; Wnuk, R. J.; Branfman, A. R.; Rinehart, Jr., K. L. *J. Am. Chem. Soc.* **1971**, *93*, 4943. (c) Reusser, F. *Infect. Immun.* **1970**, *2*, 77, 82.
4. For summary accounts of leading synthetic efforts, see: (a) Ireland, R. E.; Wuts, P. G. M.; Ernst, B. *J. Am. Chem. Soc.* **1981**, *103*, 3205. (b) Ziegler, F. E.; Wester, R. T. *Tetrahedron Lett.* **1984**, *25*, 617. (c) DeShong, P.; Ramesh, S.; Perez, J. J. *J. Org. Chem.* **1983**, *48*, 2117. (d) Martin, S. F.; Gluchowski, C.; Campbell, C. L.; Chapman, R. C. *J. Org. Chem.* **1984**, *49*, 2512; *Tetrahedron*, in press. (e) DeShong, P.; Ramesh, S.; Elango, V.; Perez, J. J. *J. Am. Chem. Soc.* **1985**, *107*, 5219. (f) Schlessinger, R. H.; Berbernitz, G. R.; Lin, P.; Poss, A. J. *J. Am. Chem. Soc.* **1985**, *107*, 1777. (g) Boeckman, Jr., R. K.; Starrett, Jr., J. E.; Nickell, D.G.; Sum, P.-E. J. *Am. Chem. Soc.* **1986**, *108*, 5549. (h) Neukom, C.; Richardson, D. P.; Myerson, J. H.; Bartlett, P. A. *J. Am. Chem. Soc.* **1986**, *108*, 5559.
5. For reviews of macrolide antibiotics, see: (a) Masamune, S.; Bates, G. S.; Corcoran, J. W. *Angew. Chem. Int. Ed. Engl.* **1977**, *16*, 585. (b) Nicolaou, K. C. *Tetrahedron*, **1977**, *33*, 683. (c) Back, T. G. *Ibid.* **1977**, *33*, 3041. (d) Paterson, I.; Mansuri, M. M. *Ibid.* **1985**, *41*, 3569.
6. For recent syntheses of Prelog-Djerassi lactone, see: Martin, S. F.; Guinn, D. E. *J. Org. Chem.* **1987**, *52*, 5588 and references cited therein.
7. For some synthetic work in the erythromycin field, see: (a) Corey, E. J. et al. *J. Am. Chem. Soc.* **1978**, *100*, 4618, 4620; *Ibid.* **1979**, *101*, 7131. (b) Masamune, S.; Hirama, M.; Mori, S.; Ali, S. A.; Garvey, D. S. *Ibid.* **1981**, *103*, 1568. (c) Woodward, R. B. et al. *Ibid.* **1981**, *103*, 3210. (d) Stork, G.; Paterson, I.; Lee, F. K. C. *Ibid.* **1982**, *104*, 4686. (e) Vedejs, E.; Dolphin, J. M.; Mastalerz, H. *Ibid.* **1983**, *105*, 127. (f) Oikawa, Y.; Nishi, T.; Yonemitsu, O. *J. Chem. Soc. Perkin Trans. I* **1985**, 1, 7, 19, 27. (g) Heathcock, C. H.; Young, S. D.; Hagen, J. P.; Pilli, R.; Badertscher, U. *J. Org. Chem.* **1985**, *50*, 2095. (h) Danishefsky, S.

J.; Larson, E.; Askin, D.; Kato, N. *J. Am. Chem. Soc.* **1985**, *107*, 1246. (i) Kobayashi, Y.; Uchiyama, H.; Kanbara, H.; Sato, F. *Ibid.* **1985**, *107*, 5541. (j) Bernet, B.; Bishop, P. M.; Caron, M.; Kawamata, T.; Roy, B. L.; Ruest, L.; Sauve, G.; Soucy, P.; Deslongchamps, P. *Can. J. Chem.* **1985**, *63*, 2810, 2814, 2818. (k) Burke, S. D.; Schoenen, F. J.; Murtiashaw, C. W. *Tetrahedron Lett.* **1986**, *27*, 449. (l) Kinoshita, M.; Arai, M.; Tomooka, K.; Nakata, M. *Ibid.* **1986**, *27*, 1811. (m) Kinoshita, M.; Arai, M.; Ohsawa, N.; Nakata, M. *Ibid.* **1986**, *27*, 1815. (n) Sviridov, A. F.; Ermolenko, M. S.; Yashunsky, D. V.; Borodkin, V. S.; Kochetkov, N. K. *Ibid.* **1987**, *28*, 3835, 3839. (o) Burke, S. D.; Schoenen, F. J.; Nair, M. S. *Ibid.* **1987**, *28*, 4143. (p) Burke, S. D.; Chandler, III, A. C.; Nair, M. S.; Campopiano, O. *Ibid.* **1987**, *28*, 4147. (q) Tone, H.; Nishi, T.; Oikawa, Y.; Hikota, M.; Yonemitsu, O. *Ibid.* **1987**, *28*, 4569. (r) Stork, G.; Rychnovsky, S. D. *J. Am. Chem. Soc.* **1987**, *109*, 1564.

8. (a) Tanida, S.; Hasegawa, T.; Higashide, E. *J. Antibiotics* **1980**, *33*, 199. (b) Muroi, M.; Izawa, M.; Kosai, Y.; Asai, M. *Ibid.* **1980**, *33*, 201. (c) Muroi, M.; Haibara, K.; Asai, M.; Kamiya, K.; Kishi, T. *Tetrahedron* **1981**, *37*, 1123. (d) Omura, S.; Nakagawa, A.; Sadakane, N. *Tetrahedron Lett.* **1979**, 4323. (e) Turusaki, A.; Matsumoto, T.; Nakagawa, A.; Omura, S. *J.Antibiotics*, **1980**, *33*, 781.
9. For synthetic approaches to macbecin, see: (a) Evans, D. A.; Ornstein, P. L.; Ennis, M. D. 187th American Chemical Society National Meeting, April 8-13, St. Louis, 1984, MEDI-154. (b) Baker, R.; Cummings, W. J.; Hayes, J. F.; Kumar, A. *J. Chem. Soc. Chem. Commun.* **1986**, 1237.
10. For some leading references, see: (a) Elming, N. in "Advances in Organic Chemistry," Raphael, R. A.; Taylor, E. C.; Wynberg, H., Eds., Interscience: New York, 1960; Vol. II, p 67. (b) Achmatowicz, Jr., O.; Bukowski, P.; Szechner, B.; Zwierzchowska, Z.; Zamojski, A. *Tetrahedron* **1971**, *27*, 1973. (c) Achmatowicz, Jr., O.; Bielski, R. *Carbohydr. Res.* **1977**, *55*, 165. (d) Piancatelli, G.; Scettri, A.; D'Auria, M. *Tetrahedron* **1980**, *36*, 661. (e) Weeks, P. D.; Brennan, T. M.; Brannegan, D. P.; Kuhla, D. E.; Elliot, M. L.; Watson, H. A.; Wlodecki, B.; Breitenbach, R. *J. Org. Chem.* **1980**, *45*, 1109. (f) Shono, T.; Matsumura, Y.; Tsubata, K.; Takata, J. *Chem. Lett.* **1981**, 1121. (g) Georgiadis, M. P.; Couladouros, E. A.; Polissiou, M. K.; Filippakis, S. E.; Mentzafos, D.; Terzis, A. *J. Org. Chem.* **1982**, *47*, 3054. (h) Graziano, M. L.; Iesce, M. R.; Carli, B.; Scarpati, R. *Synthesis* **1983**, 125. (i) Ho, T.-L; Sapp, S. G. *Synth. Commun.* **1983**, *13*, 207.
11. For applications of similar strategies to natural product synthesis, see: (a) References 4b-d, 6, 10b,c. (b) Achmatowicz, Jr., O.; Grynkiewicz, G.; Szechner, B. *Tetrahedron* **1976**, *32*, 1051. (c) Suzuki, K.; Yuki, Y.; Mukaiyama, T. *Chem. Lett.* **1981**, 1529. (d) Szechner, B. *Tetrahedron* **1981**, *37*, 949. (e) Dziewiszek, K.; Chmielewski, M.; Zamojski, A. *Carbohydr. Res.* **1982**, *104*, C1. (f) Jurczak, J.; Pikul, S. *Tetrahedron Lett.* **1984**, *25*, 3107. (g) Bromidge, S.M.; Sammes, P.G.; Street, L. J. *J. Chem. Soc. Perkin Trans. I* **1985**, 1725. (h) Sammes, P. G.; Street, L. J.; Whitby, R. J. *Ibid.* **1986**, 281. (i) Gunn, B.P. *Heterocycles* **1985**, *23*, 3061. (j) DeShong, P.; Lin, M.-T.; Perez, J. J. *Tetrahedron Lett.* **1986**, *27*, 2091. (k) Mori, K.; Kisida, H. *Tetrahedron* **1986**, *42*, 5281. (l) Pikul, S.; Raczko, J.; Ankner, K.; Jurczak, J. *J. Am. Chem. Soc.* **1987**, *109*, 3981.
12. (a) Molin, H.; Pring, B. G. *Tetrahedron Lett.* **1985**, *26*, 677. (b) Higgins, C. *Nature* **1987**, *327*, 655. (c) Hammond, S. M.; Claesson, A.; Jansson, A. M.; Larsson, L. -G.; Pring, B. G.; Town, C. M.; Ekstrom, B. *Ibid.* **1987**, *327*, 730. (d) Claesson, A.; Jansson, A. M.; Pring, B. G.; Hammond, S. M.; Ekstrom, B. *J. Med. Chem.* **1987**, *30*, 2309. (e) Goldman, R.; Kohlbrenner, W.; Lartey, P.; Pernet, A. *Nature* **1987**, *329*, 162.
13. Jurczak, J.; Pikul, S.; Ankner, K. *Tetrahedron Lett.* **1986**, *27*, 1711.
14. Minami, N.; Ko, S. S.; Kishi, Y. *J. Am. Chem. Soc.* **1982**, *104*, 1109.

15. Pauls, H. W.; Fraser-Reid, B. *J. Carbohydr. Chem.* **1985**, *4*, 1.
16. Nakata, T.; Oishi, T. *Tetrahedron Lett.* **1980**, *26*, 1641.
17. For reviews, see: (a) Evans, D. A.; Nelson, J. V.; Taber, T. R. in "Topics in Stereochemistry;" Allinger, N. L.; Eliel, E. L.; Wilen, S. H., Eds.; Interscience: New York; 1982, Vol. 13, p 1. (b) Heathcock, C. H. in "Asymmetric Synthesis, Stereodifferentiating Addition Reactions, Part B," Morrison, J. D., Ed.; Academic Press: New York; 1984, Vol. 3, p 111.
18. For reviews, see: (a) Hoffmann, R. W. *Angew. Chem. Int. Ed. Engl.* **1982**, *21* 555. (b) Yamamoto, Y.; Maruyama, K. *Heterocycles* **1982**, *18*, 357. (c) Yamamoto, Y. *Aldrichimica Acta* **1987**, *20*, 45. See also (d) Roush, W. R.; Adam, M. A.; Walts, A. E.; Harris, D. J. *J. Am. Chem. Soc.* **1986**, *108*, 3422. (e) Roush, W. R.; Palkowitz, A. D.; Palmer, M. A. *J. Org. Chem.* **1987**, *52*, 316. (f) Brown, H. C.; Bhat, K. S. *J. Am. Chem. Soc.* **1986**, *108*, 293. (g) Garcia, J.; Kim, B. M.; Masamune, S. *J. Org. Chem.* **1987**, *52*, 4831.
19. Rice, K. C.; Dyer, J. R. *J. Heterocyclic Chem.* **1975**, *12*, 1325.
20. Evans, D. A.; Bartroli, J.; Shih, T. L. *J. Am. Chem. Soc.* **1981**, *103*, 2127.
21. (a) Buse, C. T.; Heathcock, C. H. *Tetrahedron Lett.* **1978**, 1685. (b) Kishi, Y.; Nagaoka, H. *Tetrahedron* **1981**, *37*, 3873. (c) Lewis, M. D.; Kishi, Y. *Tetrahedron Lett.* **1982**, *23*, 2343. (d) Hiyama, T.; Okude, Y.; Kimura, K.; Nozaki, H. *Bull. Chem. Soc. Jpn.* **1982**, *55*, 561 and previous work.
22. (a) Paulsen, H.; Koebernick, W.; Koebernick, H. *Tetrahedron Lett.* **1976**, 2297. (b) Plaumann, D. E.; Fitzsimmons, B. J.; Ritchie, B. M.; Fraser-Reid, B. *J. Org. Chem.* **1982**, *47*, 941.
23. Ito, Y.; Hirao, T.; Saegusa, T. *J. Org. Chem.* **1978**, *43*, 1011.
24. For examples of the importance of the steric bulk of the anomeric substituent in directing the stereochemical course of catalytic hydrogenations in related systems, compare: (a) Jarosz, S.; Fraser-Reid, B. *Tetrahedron Lett.* **1981**, *22*, 2533. (b) Isobe, M.; Ichikawa, Y.; Goto, T. *Ibid.* **1981**, *22*, 4287. (c) Hanessian, S.; Demailly, G.; Chapleur, Y.; Leger, S. *J. Chem. Soc. Chem. Commun.* **1981**, 1125. (d) Schlessinger, R. H.; Poss, M. A. *J. Am. Chem. Soc.* **1982**, *104*, 357 and reference 13 therein.
25. (a) Imamoto, T.; Kusumoto, T.; Tawarayama, Y.; Sugiura, Y.; Mita, T.; Hatanaka, Y.; Yokoyama, M. *J. Org. Chem.* **1984**, *49*, 3904. (b) Imamoto, T.; Takiyama, N.; Nakamura, K. *Tetrahedron Lett.* **1985**, *26*, 4763.
26. (a) Yamamoto, Y.; Yatagai, H.; Ishihara, Y.; Maeda, N.; Maruyama, K. *Tetrahedron* **1984**, *40*, 2239. (b) Keck, G. E.; Boden, E. P. *Tetrahedron Lett.* **1984**, *25*, 1879. (c) Nakajima, N.; Hamada, T.; Tanaka, T.; Oikawa, Y.; Yonemitsu, O. *J. Am. Chem. Soc.* **1986**, *108*, 4645.
27. (a) Masamune, S.; Sato, T.; Kim, B. M.; Wollmann , T. A. *J. Am. Chem. Soc.* **1986**, *108*, 8279. (b) Paterson, I.; McClurc, C. K. *Tetrahedron Lett.* **1987**, *28*, 1229.
28. See for example in Heathcock, C. H.; Hagen, J. D.; Jarvi, E. T.; Pirrung, M. C.; Young, S. D. *J. Am. Chem. Soc.* **1981**, *103*, 4972.

Total Synthesis of Nitrogen-Containing Natural Products via Nitroso Diels-Alder Reaction

CHIHIRO KIBAYASHI
Tokyo College of Pharmacy, 1432-1 Horinouchi, Hachioji, Tokyo 192-03, Japan

1 INTRODUCTION

Hetero Diels-Alder reaction in which the C-nitroso moiety (R-N=O) functions as a heterodienophile provides cyclic derivatives of hydroxylamine namely 3,6-dihydro-1,2-oxazines 1 (ref. 1). A synthetically important feature of the products is introduction of nitrogen and oxygen functionality at the same time with a high degree of regiochemical and stereochemical control. Subsequent reduction of these cycloadducts can easily be achieved by using a wide variety of reducing agents, such as zinc in acetic acid, sodium or aluminum amalgam in ethanol, or hydrogen over catalysts, and normally result in cleavage of the N-O bond leading to the formation of 4-amino alcohols.

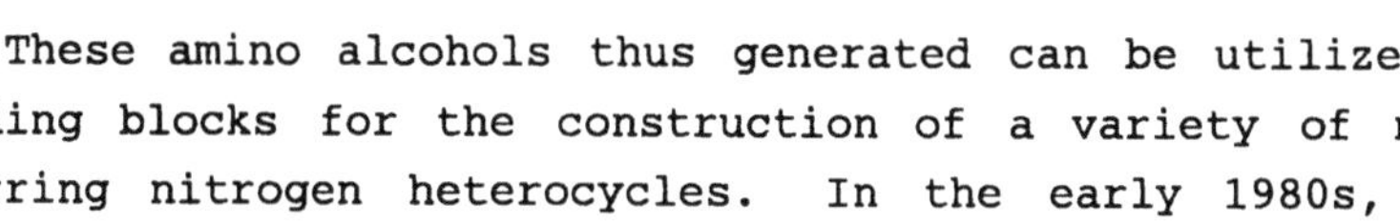

These amino alcohols thus generated can be utilized as key building blocks for the construction of a variety of naturally occurring nitrogen heterocycles. In the early 1980s, when we began a program to develop general utility of the cycloaddition-N-O bond cleavage approach to natural product synthesis, this approach had only received isolated attention although nitroso cycloaddition itself was a well known reaction.

Reviewed here are contributions from our laboratories to the development of nitroso methodology for preparing nitrogen-containing natural products including alkaloids and a pheromone.

2 TROPANE ALKALOIDS

The tropane alkaloids occur as esters of relatively simple organic carboxylic acids with amino alcohols (alkamines) which are all hydroxylated derivatives of tropane (2), i.e., tropine (3), pseudotropine (4), and nortropine (5) as monohydroxylated alkamines. Because of their pharmaceutical significance and the presence of an unusual ring system, this class of alkaloids have

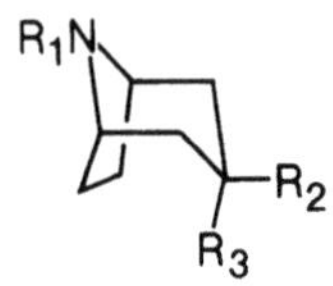

2 R_1 = Me; R_2 = R_3 = H
3 R_1 = Me; R_2 = H; R_3 = OH
4 R_1 = Me; R_2 = OH; R_3 = H
5 R_1 = R_2 = H; R_3 = OH
6 R_1 = Me; R_2, R_3 = O

been the subject of intensive stereochemical, biogenetical, and synthetic activities. In particular, a great deal of synthetic work on natural and nonnatural tropane bases has been carried out with the aim of investigating their pharmacological activity. The earliest synthetic approach to a tropane base was described by Willstätter (ref. 2). This approach to tropinone (6) in a multistage synthesis was followed by a more lucid and practical Robinson synthesis (ref. 3). Since these classical syntheses of tropinone (6), a number of general synthetic methods for the preparation of some tropanes have been reported. However, except for two instances of new approaches to tropane alkaloids via [3 + 2] cycloaddition by Tufariello (ref. 4) and [3 + 4] cycloaddition by Noyori (ref. 5), respectively, efficient methods for the preparation of naturally occurring tropane alkaloids are limited.

We anticipated that a hetero Diels-Alder cycloaddition using nitroso compounds as dienophiles could be a successful new approach in accessing heterocyclic natural products. Our initial investigations focused on this area, which had been little explored at the time our work began (ref. 6). Thus attention was turned to developing a facile new route for the elaboration of the tropane ring system by utilizing a [4 + 2] nitroso

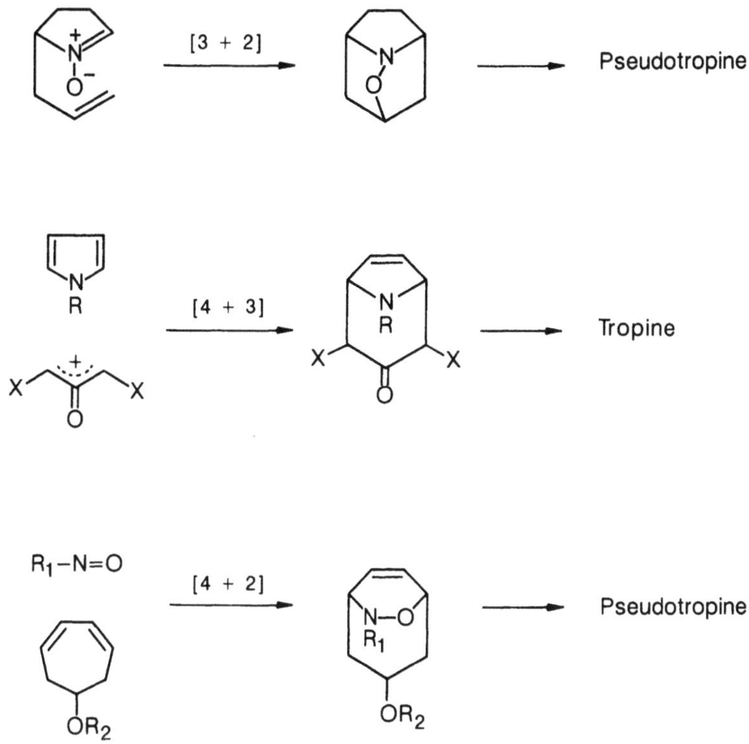

cycloaddition and its application to the synthesis of the naturally occurring tropane alkaloids such as pseudeotropine and tropacocaine.

2.1 N-Benzoylnortropane

As our first model we chose N-benzoylnortropane (12) to investigate construction of the tropane ring system based on [4 + 2] nitroso cycloaddition with a 1,3-cycloheptadiene. A search of the literature indicated that only one example of a Diels-Alder cycloaddition of a nitroso compound with a seven-membered ring diene has been reported (ref. 8). In view of this, the present study of tropane synthesis was initiated by the examination of the nitroso Diels-Alder reaction of 1,3-cycloheptadiene (7) (Fig. 1). Thus reaction of 7 with the acylnitroso compound 8 generated in situ from benzohydroxamic acid by oxidation with tetrapropylammonium metaperiodate (ref. 9) was carried out at room temperature, affording the [4 + 2] cycloadduct 9 in 85% yield. Reduc-

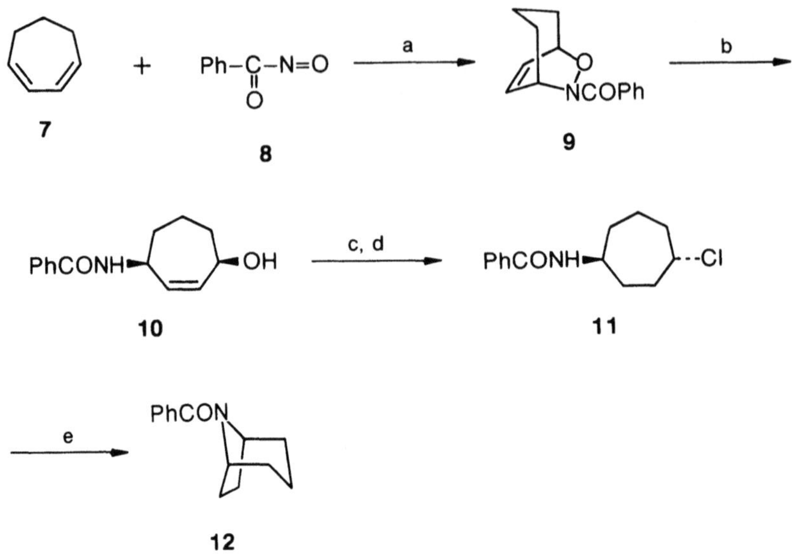

Fig. 1 (a) DMF-$CHCl_3$, r.t.; (b) Na-Hg, EtOH, Na_2HPO_4, 0 °C → r.t.; (c) H_2, Pd/C, MeOH; (d) $SOCl_2$, Et_3N, $CHCl_3$, r.t.; (e) t-BuOK, HMPA-benzene, r.t.

tive N–O bond cleavage of **9** with sodium amalgam in ethanol gave the amide alcohol 10 (77%), which was converted to the trans chloride 11 by hydrogenation followed by treatment with thionyl chloride and triethylamine (64%). The desired N-benzoylnortropane (12) was obtained in 87% yield when the chloride 11 was treated with potassium tert-butoxide in HMPA–benzene (1:1).

2.2 Tropane

An alternative approach for the constructing the tropane ring system according to [4 + 2] nitroso cycloaddition utilizing 1-chloro-1-nitrosocyclohexane (13) as a dienophile is as follows (Fig. 2). The cyclic diene 7 was reacted with 13 in an ethanol–carbon tetrachloride solution at -20 °C for 10 h then at -10 °C for 48 h to yield the cycloadduct as its hydrochloride 14 in 68% yield. Catalytic hydrogenation of 14 afforded the amino alcohol

15 (isolated as its hydrochloride) which was then subjected to selective acylation with benzyl or ethyl chloroformates to provide the corresponding carbamates **16** or **17** in 80% and 84% yields, respectively.

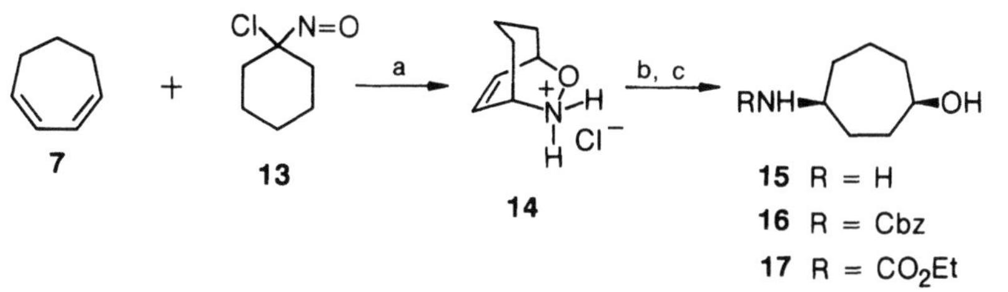

Fig. 2 (a) EtOH-CCl_4, -10 °C; (b) H_2, Pd/C, MeOH; (c) benzyl (for **16**) [or ethyl (for **17**)] chloroformate, aq. K_2CO_3, $CHCl_3$, 0 °C → r.t.

When the benzyl carbamate **16** was treated with thionyl chloride, the results varied widely depending upon the reaction conditions employed. Thus treatment with thionyl chloride in chloroform in the absence of a base gave a 1:1 mixture of the dehydration products **19a** and **20a** in 61% yield along with the chloride **18a** in 12% yield. When the reaction was carried out in the presence of 1.3 equiv of pyridine at room temperature the chloride **18a** was obtained in 32% yield along with the 1:1 olefin mixture of **19a** and **20a** (6%) and the dicycloheptylsulfite **21a** (18%). On heating in the presence of excess pyridine, **18a** was obtained in improved yield (55%), together with the olefin mix-

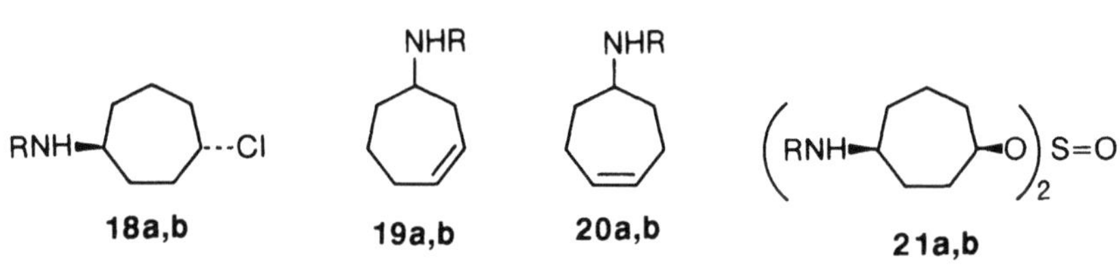

a R = Cbz; **b** R = CO_2Et

ture of **19a** and **20a** (4%) and **21a** (18%). Otherwise, the reaction of ethyl carbamate **17** with thionyl chloride and pyridine (1.9 equiv) in chloroform at reflux generated the chloride **18b** (65%) accompanied by the 1:1 mixture of olefins **19b** and **20b** (11%) and the dicycloheptyl sulfite **21b** (5%).

Recent studies have demonstrated (ref. 10) the potential synthetic utility of C–N bond formation via heteromercuration of carbamate derivatives of unsaturated amines. This prompted us to utilize the dehydration products **19a** and **20a** in the tropane synthesis via intramolecular heteromercuration. Thus the 1:1 mixture of **19a** and **20a** was treated with mercuric acetate followed by reduction with $NaBH_4$ to furnish N-carbobenzoxynortropane (**26**) in 45% yield (based on reacted starting material). The product **26** must result from reductive demercuration of both organomercurials **24** and **25** initially formed from the asymmetrical and symmetrical olefins **19a** and **20a**, respectively, via cationic intermediates **22** and **23** as depicted in Fig. 3. Thus separation of the olefin mixture is not necessary since both the olefins can equally be utilized as the reaction substrates in this cyclization.

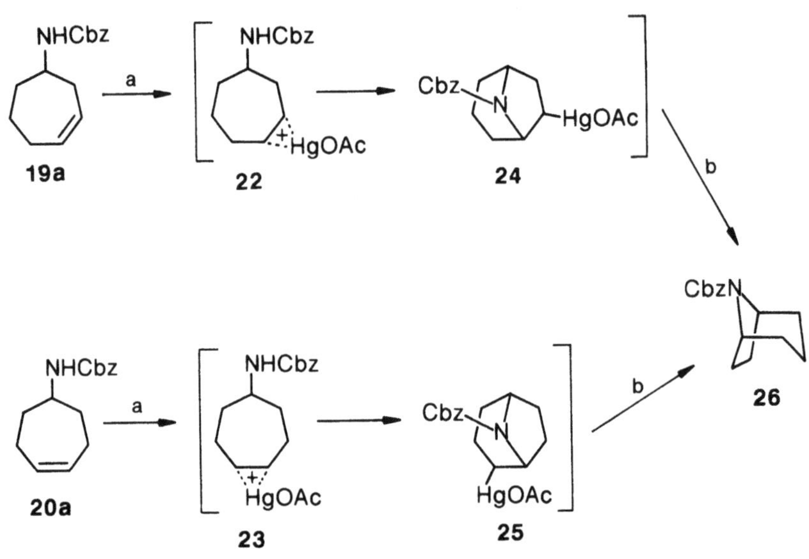

Fig. 3 (a) $Hg(OAc)_2$, aq. THF, r.t.; (b) $NaBH_4$, aq. NaOH, r.t.

Alternatively, the chloride **18a** was directly converted into N-carbobenzoxynortropane (**26**) in satisfactory yield (78%) when subjected to base-induced intramolecular "amidocyclization" as previously described for the preparation of N-benzoylnortropane (12) (Fig. 1). In a similar reaction involving the chloride **18b**, N-(ethoxycarbonyl)nortropane (**27**) was formed in 78% yield. The synthesis of tropane (2) was achieved by $LiAlH_4$ reduction of the benzyl carbamate **26** or ethyl carbamate **27** in 72% or 71% yield, respectively (Fig. 4).

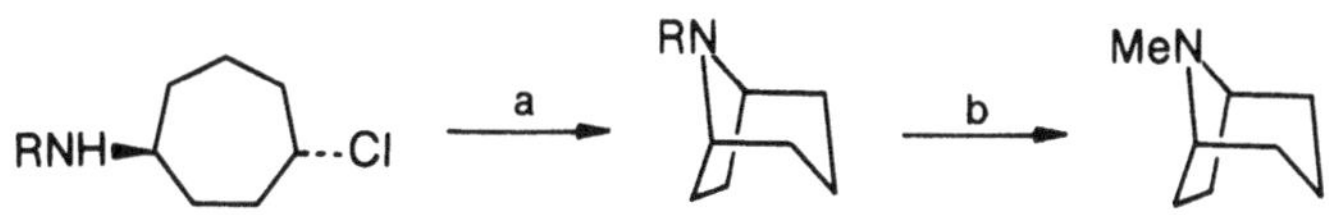

18a R = Cbz
18b R = CO_2Et

26 R = Cbz
27 R = CO_2Et

2

Fig. 4 (a) t-BuOK, HMPA-benzene, r.t.; (b) $LiAlH_4$, THF, reflux.

2.3 Pseudotropine and Tropacocaine

Having developed the method for the preparation of the tropanes based on a nitroso Diels-Alder approach, we sought to apply this to the synthesis of the naturally occurring tropane alkaloids pseudotropine (**4**) and tropacocaine (**41**) (ref. 11), which exhibit marked local anesthetic action.

From the above results, the most reliable approach to these alkaloids seemed to be the one involving cycloaddition with 1-chloro-1-nitrosocyclohexane (13) followed by base-induced amido-cyclization as outlined in Fig. 5. Thus the reaction of 3,5-cycloheptadienyl benzoate (**28**) with the nitroso compound 13 generated a 4:1 mixture of the oxazabicyclononene hydrochlorides favoring exo form **29** over endo form **30**. The stereochemistry of these cycloadducts was verified with their free bases on the basis of their ^1H NMR spectra. The signal attributed to the 3-endo proton adjacent to the benzoyloxy group in the spectrum of the major adduct (free base of **29**) occurs at significantly higher field (δ 4.95) than usual, indicating that it must lie within the shielding cone of the C-6—C-7 double bond, while the 3-exo proton signal (δ 5.57) suffers much less from a shielding effect of the double bond.

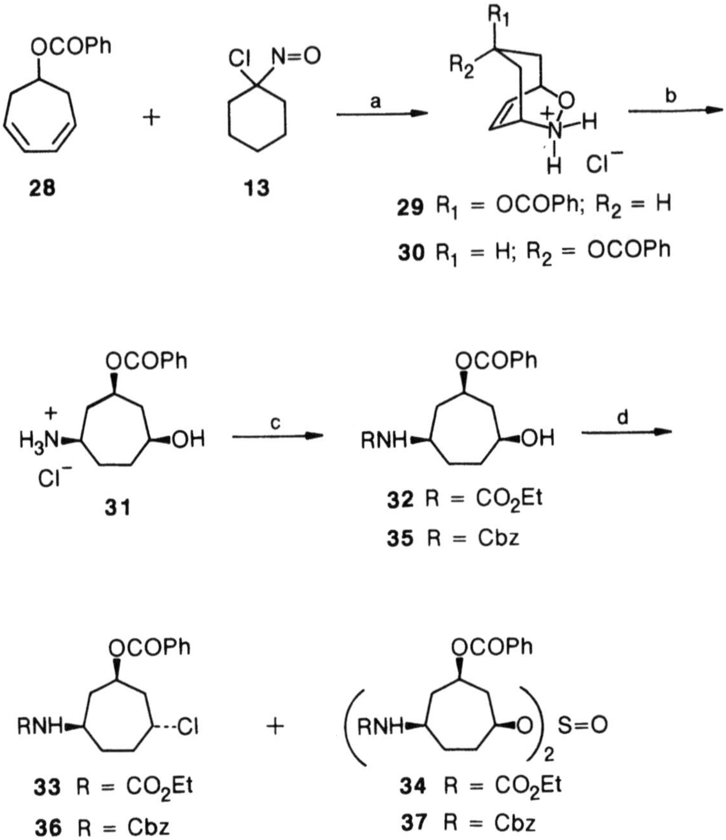

Fig. 5 (a) EtOH-CCl_4, -10 °C; (b) H_2, Pd/C, MeOH; (c) ethyl (for 32) [or benzyl (for 35)] chloroformate, aq. K_2CO_3, $CHCl_3$, 0 °C → r.t.; (d) $SOCl_2$, Py, $CHCl_3$, reflux.

The major cycloadduct 29, readily separable from 30 by recrystallization, underwent catalytic hydrogenation to give the amino alcohol hydrochloride 31 which was in turn selectively N-acylated with ethyl chloroformate affording the carbamate 32 in 84% overall yield. Chlorination of 32 with thionyl chloride and excess of pyridine in chloroform at reflux yielded the chloride 33 (55%) along with a byproduct assigned the dicycloheptyl sulfite **34** (28%). Similarly, treatment of 35, prepared from 31, with thionyl chloride and pyridine gave 36 (36%) and 37 (38%).

The consequent tropane ring elaboration was pursued via base-induced amidocyclization of the chloride 33 under the previously described conditions (t-BuOK, HMPA-benzene, room temperature), resulting in the formation of 38 (46%) (Fig. 6). Reduction with $LiAlH_4$ converted **38** to pseudotropine (**4**) (67%). Furthermore, the chloride **36** was subjected to amidocyclization as well to give **39**. Deprotection by catalytic hydrogenation converted **39** into N-nortropacocaine (**40**), which subsequently underwent the Eschweiler-Clarke reaction with formic acid and formaldehyde to provide desired tropacocaine (**41**) in 71% yield from **39** (Fig. 6).

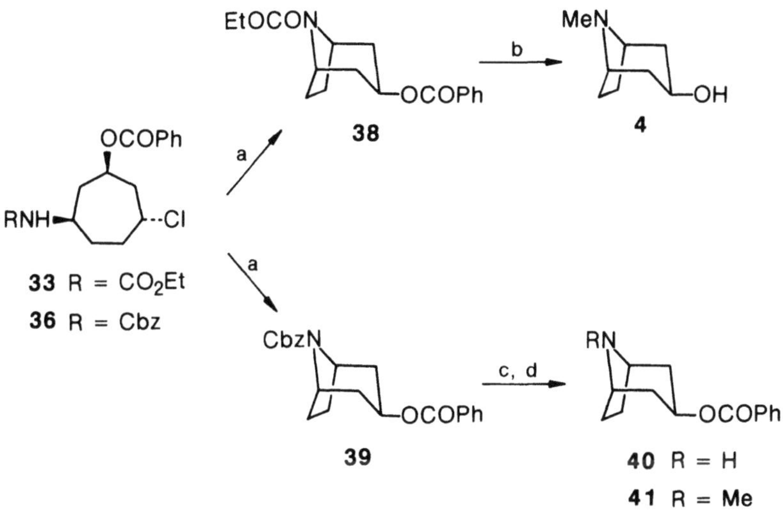

Fig. 6 (a) t-BuOK, HMPA-benzene, r.t.; (b) $LiAlH_4$, THF, reflux; (c) H_2, Pd/C, MeOH; (d) HCO_2H, HCHO, reflux.

3 INTRAMOLECULAR NITROSO DIELS-ALDER REACTION

The methodology, based on nitroso Diels-alder reaction, that proved useful in the synthesis of tropane alkaloids also seemed to open an attractive route to some other alkaloids. Compared to the intramolecular imino Diels-Alder reaction (ref. 1c), the intramolecular variant of the nitroso Diels-Alder reaction has received far less attention (refs. 6b, 12), despite the enormous potential it holds for alkaloid synthesis. With this in mind, we proceeded to examine the application of the intramolecular nitroso Diels-Alder cycloaddition in the synthesis of alkaloids possessing saturated nitrogen heterocyclic ring systems.

3.1 Gephyrotoxin 223AB

Gephyrotoxin (GTX) 223AB (**42**), one of the neurotoxin alkal-

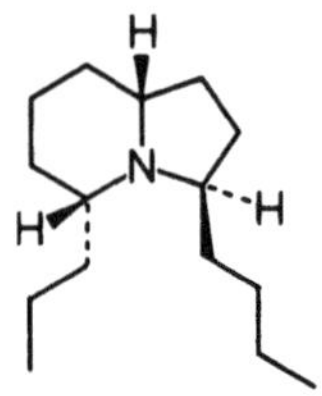

42

oids isolated in minute quantity from skin extracts of neotropical poison-dart frogs (family Dendrobatidae) (refs. 13, 14), has attracted a vast amount of interest because of its unusual biological characteristics. Numerous groups have thus been involved in the development of methodology for the total synthesis of this molecule. These efforts have recently resulted in the synthesis of **42** (ref. 15) and its stereoisomers (ref. 16). Our approach to **42** involving an acylnitroso Diels-Alder reaction, which took place in an intramolecular fashion (Fig. 7) (ref. 17), is distinguished from these previous efforts (ref. 18) by the

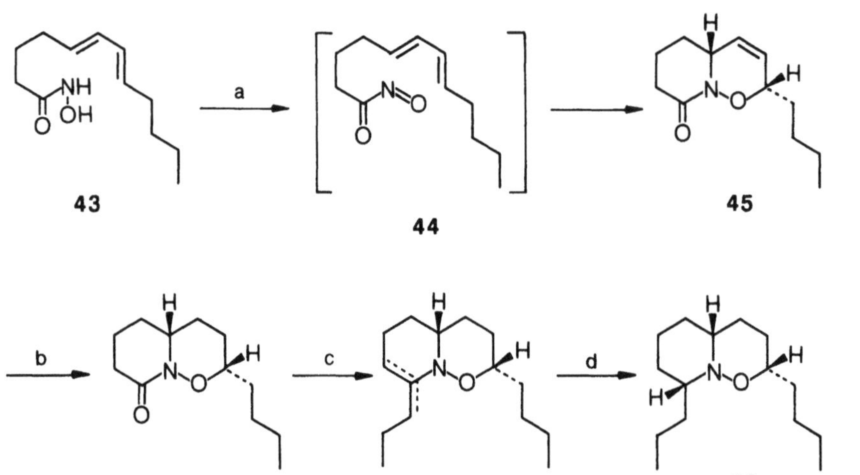

43 **44** **45** **46** **47** **48**

Fig. 7 (a) n-$Pr_4N(IO_4)$, $CHCl_3$, 0–5 °C; (b) H_2, Pd/C, MeOH; (c) n-PrMgBr, Et_2O, 0 °C → r.t.; (d) $NaCNBH_4$, HCl-MeOH (pH 3.8–5.4), 0 °C → r.t.

fact that the entire sequence provided single stereoisomers in the desired sense and thus met with no separation problems.

Oxidation of the hydroxamic acid **43** by the periodate in chloroform (0–5 °C) generated **44** in situ. Under the reaction conditions, **44** underwent intramolecular [4 + 2] cycloaddition to give as the single product the 1,2-oxazine derivative **45** in 82% yield. Hydrogenation of **45** provided **46** in 90% yield. Then **46** was treated with the Grignard reagent at ambient temperature to give the enamine **47** (89%), which was subsequently subjected to reduction with $NaCNBH_3$ in methanol at pH 3.8–5.4 resulting in the exclusive formation of **48** (70%).

The ^{13}C NMR spectrum of **48** at 24 °C showed pairs of resonances for each of the carbons in the molecule. The relative intensity of each doublet was nearly unit independent of temperature and the doublet collapsed into a single line at high temperature depending on its peak separation. These observations strongly indicate that **48** exists in conformational equilibrium between **48A** and **48B** due to nitrogen inversion with nearly equal population energy barrier ($\Delta G^{\ddagger}$), estimated to be about 8.0 kcal mol^{-1}.

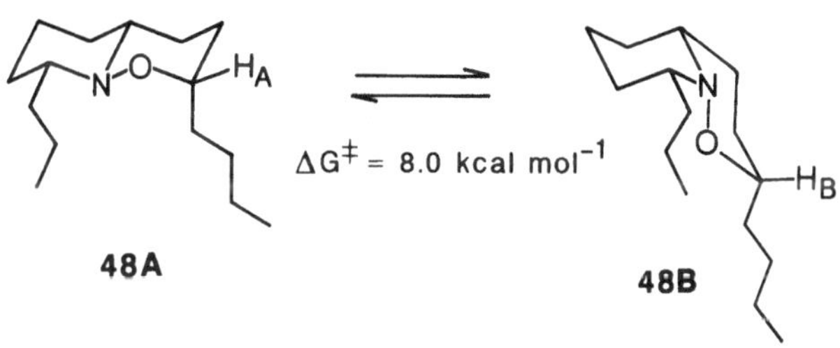

48A **48B**

The ^{1}H NMR spectra in pyridine-d_5 of **48** at 27.5 °C showed two sets of multiplets at δ 3.87 ($W_{1/4}$ = 23.9 Hz) and 3.72 ($W_{1/4}$ = 31.6 Hz) with an integration ratio of 10:9 for the C-2 proton, indicative of the equatorial hydrogen (H_A) in **48A** and the axial hydrogen (H_B) in **48B**, respectively. These C-2 proton signals converged to a single resonance centered at δ 3.88 at 100 °C.

The exclusive formation of **48** can be rationalized to be a result of stereoelectronically controlled reaction of the transient iminium salt generated from **47** under acidic conditions. Due to inversion at nitrogen, there are thus four possible transition states A–D (Fig. 8) which maintain maximum orbital overlap with respect to the approaching hydride ion and the developing nitrogen lone pair. Two of these transition states (A and D) are of more stable chair shape, with the latter being disfavored due to a strong peri interaction between the butyl group and the incoming hydride ion. On the other hand, A can accommodate the entering hydride ion without steric interference of the 2-alkyl side chain thereby leading to **48**.

With the required stereochemistry thus established, the only requirement in order to complete the synthesis was constructing the pyrrolidine moiety of the target molecule. Reductive cleavage of the N–O bond in **48** gave the amino alcohol **49** (85%). Exposure of this material to benzyl chloroformate in an alkaline solution furnished the hydroxy carbamate **50** (36%), along with **51** (24%) and **52** (25%), the latter two products of which could easily

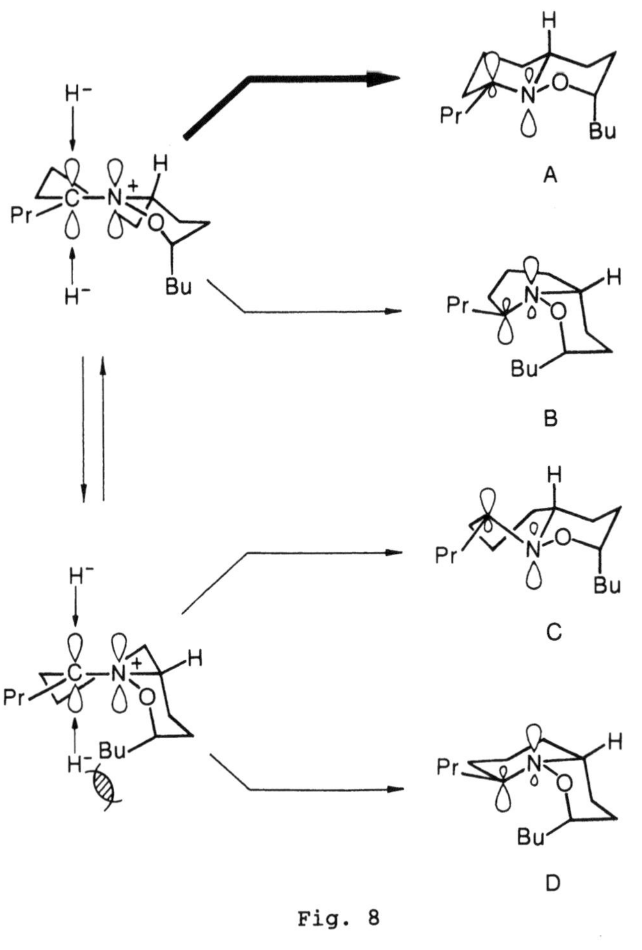

Fig. 8

be hydrogenated back to the amino alcohol **49**. Thus actual yield of **50** based on recovered **49** was 71%. Finally, the hydroxy carbamate **50** was converted to the mesylate **53** (83%) which upon hydrogenation provided GTX 223AB (**42**) in 81% yield as the sole product (Fig. 9).

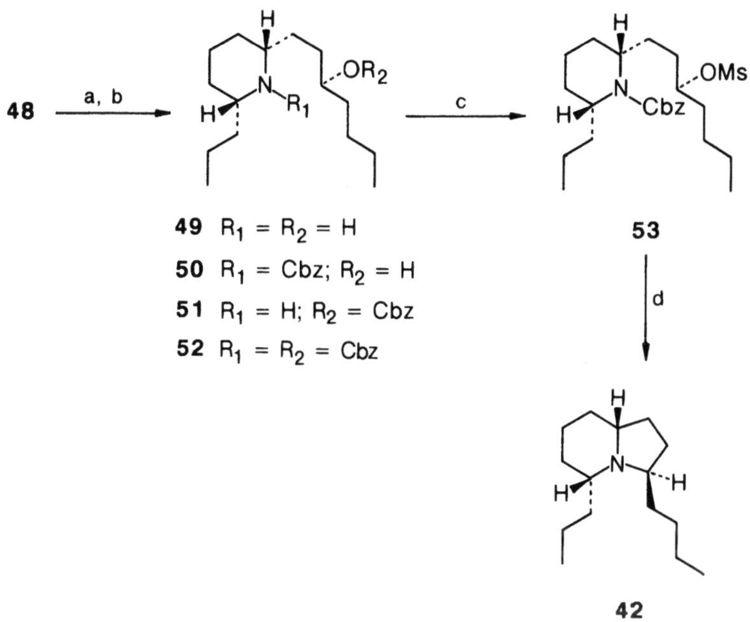

Fig. 9 (a) Zn, aq. AcOH, 60 °C; (b) benzyl chloroformate, aq. Na_2CO_3, $CHCl_3$, r.t.; (c) MsCl, Et_3N, CH_2Cl_2, -10 °C; (d) H_2, Pd/C, MeOH.

3.2 Monomorine I

From the above survey of results the synthetic potential of the oxazinolactam intermediate **46** in hand, we envisaged to synthesize monomorine I (**62**) which constitutes of an extension of the methodology based on intramolecular nitroso Diels-Alder cycloaddition. The relative stereochemistry of this substance, isolated as one of the trail pheromones from Pharaoh ants (Monomorium pharaonis L.) (ref. 19), has been determined its relative stereochemistry by nonstereoselective synthesis (ref. 20). More recently, a stereospecific synthesis of racemic **62** (ref. 21) and a chiral synthesis of the (-)-enantiomer of natural **62** (ref. 22) were reported.

Our synthesis was initiated with the Grignard reaction of the bicyclic 1,2-oxazine **46** using methylmagnesium bromide to generate the unstable enamine **54**, which was then immediately hydrogenated leading to a single isomer 55 in 70% overall yield (Fig. 10). Compound 55 was subjected to N–O bond cleavage by treatment with zinc in aqueous acetic acid to give the amino alcohol **56** in 68% yield. The cis relationship of C-2 and C-6 alkyl substituents in **56** was confirmed by the ^{13}C NMR spectrum showing 2-C and 6-C signals at δ 56.2 and 52.5, respectively, in good agreement with those reported for the fire-ant vernom iso-solenopsin A (57) rather than those reported for solenopsin A (**58**) (ref. 23). Treatment of **56** with benzyl chloroformate (1.5 equiv) afforded the benzyl carbamate 59 (53%), accompanied by the formation of the mixture of the O-acyl (6%) and N,O-diacyl (32%) derivatives of 56 which could readily be reverted to starting 56 by hydrogenolysis in quantitative yield. Thus the actual yield of 59 based on recovered 56 was 85%.

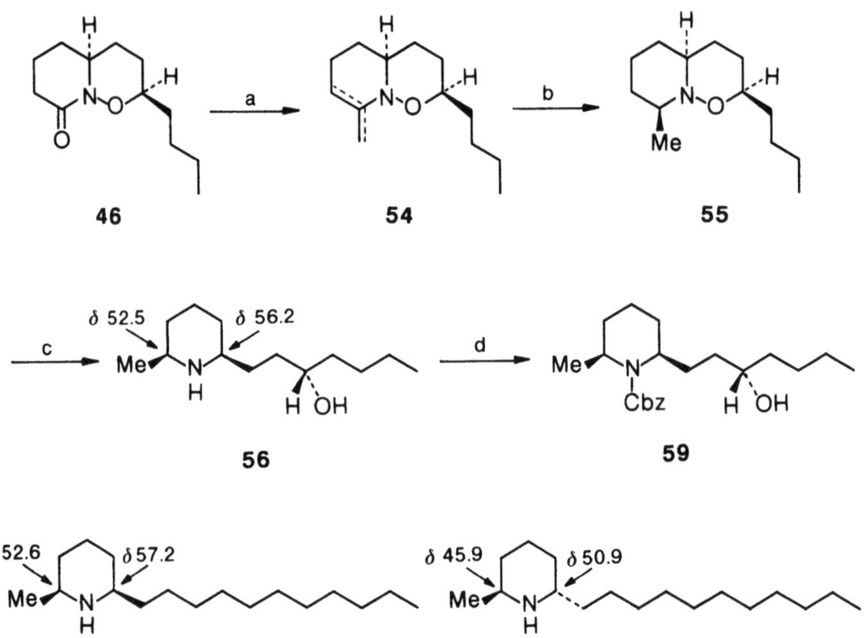

57 Isosolenopsin A

58 Solenopsin A

Fig. 10 (a) MeMgBr, Et_2O, 0 °C → r.t.; (b) H_2, Pd/C, MeOH; (c) Zn, aq. AcOH, 60 °C; (d) benzyl chloroformate, aq. Na_2CO_3, CH_2Cl_2, r.t.

When exposed to iodotrimethylsilane at room temperature, **59** underwent C–O bond cleavage along with iodination to produce the silyl ester **60** (Fig. 11). In situ cyclization was carried out by treating **60** in methanol at room temperature to provide (±)-monomorine I (**62**) and its C-3 epimer (**63**) in 42% and 40% yields from **59**, respectively. An appreciable improvement of the diastereoselectivity in the cyclization to monomorine I was obtained by the following sequence. With compound **59** in hand, the ketone **61** was prepared by Collins oxidation in 94% yield. On catalytic hydrogenation of **61**, (±)-monomorine I (**62**) was obtained in 71% yield, along with (±)-3-epimonomorine I (**63**) as a minor isomer (15%).

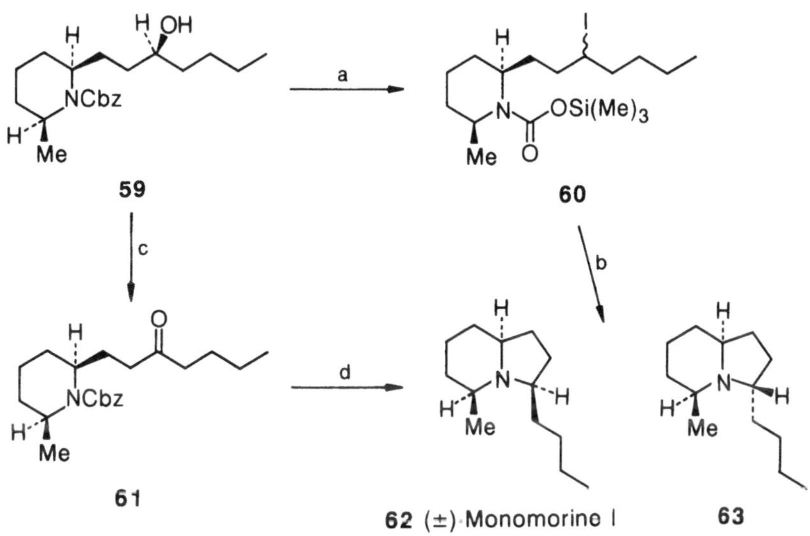

Fig. 11 (a) Me_3SiI, CH_2Cl_2, r.t.; (b) MeOH, r.t.; (c) $CrO_3 \cdot 2Py$, CH_2Cl_2, r.t.; (d) H_2, Pd/C, MeOH.

6.3 Dihydropinidine and Pinidine

In order to demonstrate further the utility of the above described sequence for the synthesis of gephyrotoxin 223AB (**48**) and monomorine I (**62**) involving intramolecular hetero Diels-Alder reaction followed by highly stereoselective introduction of the alkyl group to the amide we then explored an extension of these studies to the synthesis of dihydropinidine (**73**) (ref. 24).

The synthetic route utilizing this sequence for the elaboration of the cis-2,6-dialkylated piperidine is illustrated in Fig. 12. Intramolecular cycloaddition of the acylnitroso compound **65**, produced in situ from the corresponding hydroxamic acid **64** by periodate oxidation, yielded the oxazinolactam **66**. Compound **66** was converted to the enamine **68** by catalytic hydrogenation followed by treatment with methylmagnesium bromide. Catalytic hydrogenation of **68** resulted in both olefin hydrogenation and N-O bond cleavage providing the amino alcohol **69** as a single diastereomer. Confirmation that **69** bears the 2 and 6 alkyl substituents in a cis relationship was made by the ^{13}C NMR spectrum in the same manner as described above for the amino alcohol **56** (Fig. 9), a synthetic precursor to monomorine I. The signals for the 2-C and 6-C ring carbons appearing at δ 52.2 and 56.9 matched

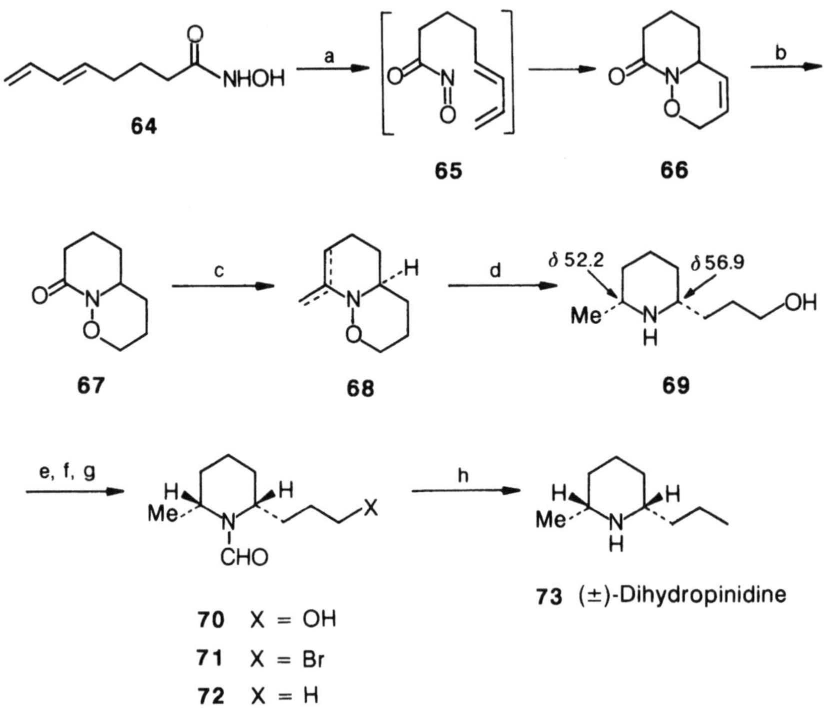

Fig. 12 (a) n-$Pr_4N(IO_4)$, $CHCl_3$, -10 °C; (b) H_2, Pd/C, MeOH; (c) MeMgBr, Et_2O, 0 °C → r.t.; (d) H_2, Pd/C, MeOH; (e) HCO_2H, toluene, reflux; (f) PBr_3, CH_2Cl_2, r.t.; (g) H_2, Pd/C, Et_3N, MeOH; (h) 10% HCl, reflux.

well with those reported for isosolenopsin A (**57**) than solenopsin A (**59**) (ref. 23).

The amino alcohol **69** was selectively N-formylated to give the formamide **70**, deoxygenation of which was achieved by bromination with PBr_3 followed by hydrogenation to yield 72. Removal of the protective group by acid treatment afforded (±)-dihydropinidine (**73**) (Fig. 12).

The nitroso Diels-Alder approach, so useful in synthesis of dihydropinidine (**73**), was applied to the synthesis of (±)-pinidine (**80**) as ontlined in Fig. 13 (ref. 24).

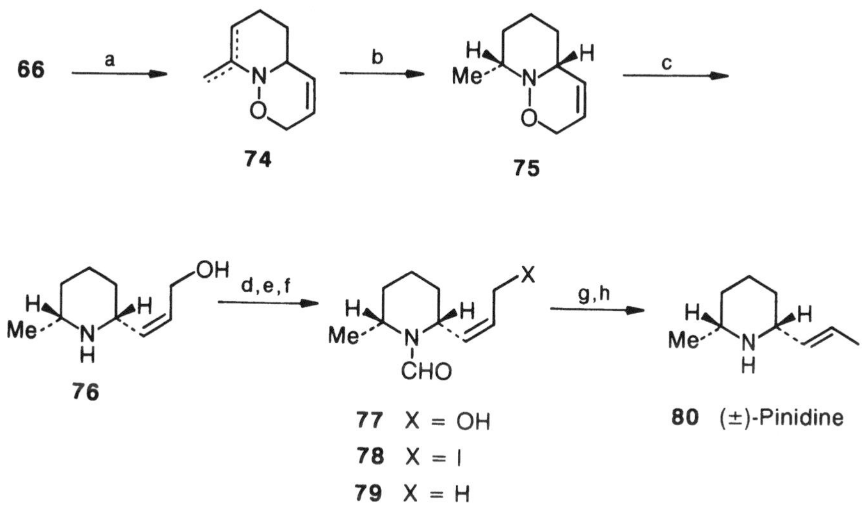

Fig. 13 (a) MeMgBr, Et_2O, 0 °C → r.t.; (b) $NaCNBH_4$, HCl-MeOH (pH 3.8–5.4), 0 °C → r.t. (c) Zn, aq. AcOH, 60 °C; (d) HCO_2H, toluene, reflux; (e) $(PhO)_3PMeI$, HMPA, r.t.; (f) $NaCNBH_4$, r.t.; (g) hν, I_2, benzene; (h) 10% HCl, reflux.

REFERENCES

1. For reviews of heterodinophile cycloadditions see: (a) S. M. Weinreb and J. I. Levin, Heterocycles, 12, 949 (1979). (b) S. M. Weinreb and R. S. Staib, Tetrahedron, 38 3127 (1982). (c) S. M. Weinreb, Acc. Chem. Res., 18, 16 (1985). (d) ref. 2c.
2. H. L. Holmes in "The Alkaloids," R. H. F. Manske, Ed., Academic Press, New York, 1950, Vol. I, Chapter 6.

3. R. Robinson, J. Chem. Soc., **111**, 762 (1917).
4. J. J. Tufariello and E. J. Trybulski, J. Chem. Soc., Chem. Commun., 720 (1973). J. J. Tufariello, G. B. Mullen, J. J. Tegeler, E. J. Trybulski, S. C. Wang, and Sk. Asrof Ali, J. Am. Chem. Soc., **101**, 2435 (1979).
5. R. Noyori, Y. Baba, and Y. Hayakawa, J. Am. Chem. Soc., **96**, 3336 (1974). Y. Hayakawa, Y. Baba, Y. Makino, and R. Noyori, Ibid., **100**, 1786 (1978).
6. For recent entries into natural products utilizing nitroso Diels-Alder reactions, see: (a) N. J. Leonard and A. J. Playtis, J. Chem. Soc., Chem. Commun., 133 (1972). (b) G. E. Keck and D. G. Nickell, J. Am. Chem. Soc., **102**, 3632 (1980). (c) G. E. Keck and R. R. Webb, II, J. Org. Chem., **47**, 1302 (1982). (d) J. E. Boldwin, P. D. Bailey, G. Gallacher, K. A. Singleton, and P. M. Wallace, J. Chem. Soc., Chem. Commun., 1049 (1983); J. E. Baldwin, P. D. Bailey, G. Gallacher, M. Otsuka, K. A. Singleton , and P. M. Wallace, Tetrahedron, **40**, 3695 (1984). (e) A. Defoin, H. Fritz, G. Geffroy, and J. Streith, Tetrahedron Lett., **27**, 4727 (1986).
7. H. Iida, Y. Watanabe, and C. Kibayashi, Tetrahedron Lett., 25, 5091 (1984); J. Org. Chem., **50**, 1818 (1985).
8. H. Hart, S. K. Ramaswami, and R. Willer, J. Org. Chem., **44**, 1 (1979).
9. G. E. Keck and S. A. Fleming, Tetrahedron Lett., 4736 (1978).
10. D. L. J. Clive, V. Farina, A. Singh, C. K. Wong, W. A. Kiel, and S. M. Menchen, J. Org. Chem., **45**, 2120 (1980). K. E. Harding and S. R. Burks, Ibid., **46**, 3920 (1981), **49**, 40 (1984). S. Danishefsky, E. Taniyama, and R. R. Webb, II, Tetrahedron Lett., **24**, 11 (1983).
11. S. R. Johns, J. A. Lamberton, A. A. Sioumis, Aus. J. Chem., **24**, 2399 (1971).
12. G. E. Keck, Tetrahedron Lett., 4767 (1978).
13. J. W. Daly, G. B. Brown, M. Mensah-Dwumah, and C. W. Myers, Toxicon, **16**, 163 (1978).
14. Th. F. Spande, J. W. Daly, D. J. Hart, Y.-M. Tsai, T. L. Macdonald, Experientia, **37**, 163 (1981).
15. For syntheses of gephyrotoxin 223AB, see: (a) T. L. Macdonald, J. Org. Chem., **45**, 193 (1980). (b) J. Royer and H.-P. Husson, Tetrahedron Lett., **26**, 1515 (1985). (c) C. A. Broka and K. K. Eng, J. Org. Chem., 51, 5045 (1986).
16. (a) R. V. Stevens and A. W. M. Lee, J. Chem. Soc., Chem.

Commun., 133 (1982). (b) D. J. Hart and Y.-M. Tsai, J. Org. Chem., **47**, 4403 (1982).

17. H. Iida, Y. Watanabe, and C.Kibayashi, J. Am. Chem. Soc., **107**, 5535 (1985).
18. Refs. 16 and 17. For unpublished syntheses of **42** and a complehensive review in this area, see: J. W. Daly and Th. F. Spande in "The Alkaloids II: Chemical and Biological Perspectives," S. W. Pelletier, Ed., Wiley-Interscience, New York, 1986, Vol. 4, Chapter 1.
19. F. J. Ritter, T. E. M. Rotgens, E. Talman, P. E. J. Verwiel, and F. Stein, Experientia, **29**, 530 (1973); F. J. Ritter and C. J. Persoons, Neth. J. Zool., **25**, 261 (1975).
20. (a) J. E. Oliver and P. E. Sonnet, J. Org. Chem., **39**, 2662 (1974). (b) P. E. Sonnet and J. E. Oliver, J. Heterocycl. Chem., **12**, 289 (1975). (c) P. E. Sonnet, D. A. Natzel, and R. Mendoza, Ibid., **16**, 1040 (1980).
21. (a) R. V. Stevens and A. W. M. Lee, J. Chem. Soc., Chem. Commun., 102 (1982). (b) R. Yamaguchi, E. Hata, T. Matsuki, and M. Kawanishi, J. Org. Chem., **52**, 2049 (1987).
22. J. Royer and H.-P. Husson, J. Org. Chem., **50**, 670 (1985).
23. Y. Moriyama, D.-D. Huynh, C. Monneret, and Q.-K. Huu, Tetrahedron Lett., 825 (1977).
24. Unpublished results from C. Kibayashi and Y. Watanabe.

Synthetic Studies in the Alkaloid Field

Csaba Szántay*, György Kalaus**, Lajos Szabó**,
Mária Incze*, Zsuzsanna Kardos-Balogh* and Ferenc Sóti*

*Central Research Institute for Chemistry of the Hungarian Academy of Sciences, H-1525 Budapest, POB 17, Hungary

**Institute for Organic Chemistry, Technical University, H-1521 Budapest, Hungary

It is a common knowledge that plants are useful sources of many valuable medicines. Among these medicines several indole alkaloids can be found, e.g. reserpine and deserpidine from *Rauwolfia serpentina*, ajmalicine and yohimbine from *Corynanthe yohimbe*, vinblastine and vincristine from *Catharanthus roseus*, just to mention a few of them.

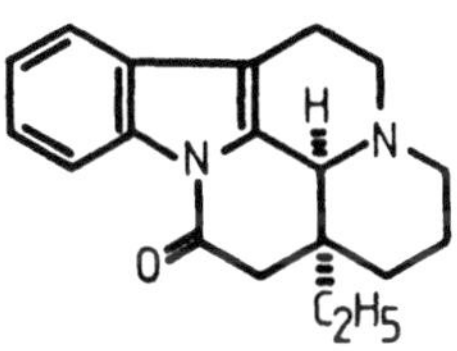

(-)-eburnamonine

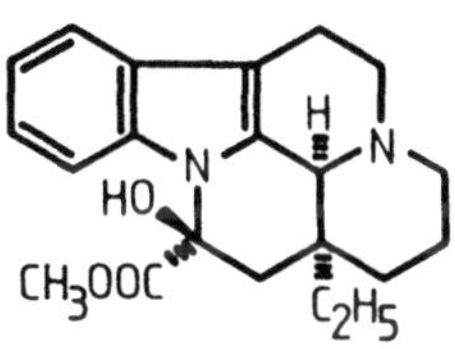

(+)-vincamine

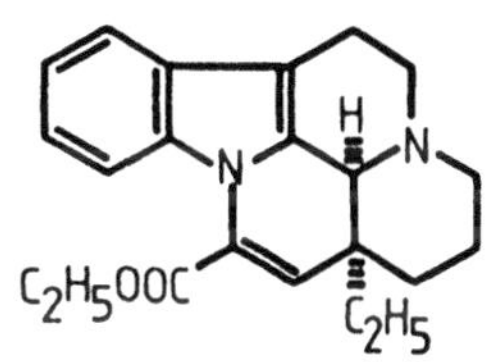

Cavinton

(+)-Vincamine, an alkaloid first isolated from *Vinca minor* has gained wide application in recent years as a specific cerebral vasodilator. Another alkaloid from the same species, (-)-eburnamonine is marketed with a similar indication. A semisynthetic derivative of vincamine, the (+)-apovincaminic acid ethyl ester is produced under the trade name CAVINTON by the Hungarian Pharmaceutical Company Gedeon Richter. The latter compound has the trade name CALAN in Japan with a substantial share of the market.

Of course when someone finds a successful molecule, the structural variation in order to establish structure-bioactivity relationship (SAR) is inevitable. In the course of that investigations we found some interesting relations in addition to SAR, between structure and chemical reactivity as well.

We have realized a strikingly different chemical behaviour of mesylates of epimers 1 on the one hand, and the homologues 2 on the other.

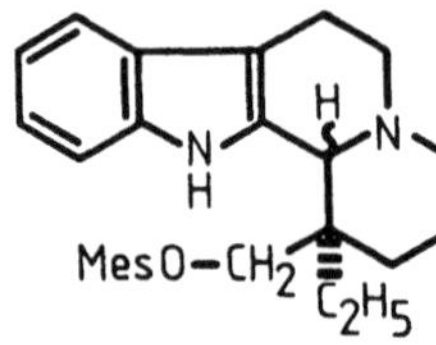

1a 12b Hα
1b 12b Hβ

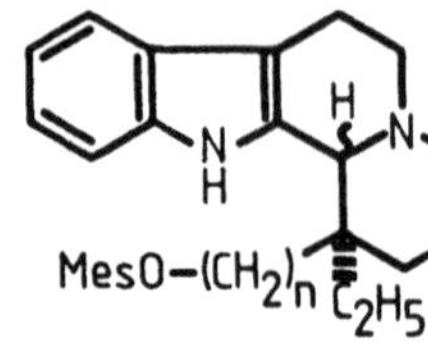

2

Let us study first the epimers 1. The *trans* isomer (4) was prepared by reacting the so called Wenkert enamine (3) with formaldehyde. This reaction is completely stereoselective. The *cis* compound (5) was prepared as follows. The OH was protected by acetylation. Oxidation followed by reduction and base catalyzed hydrolysis gave rise to the desired *cis* compound. Of course reduction of the C=N double bond is not completely stereo-selective, but the ester of the *trans* epimer hydrolizes much faster then the *cis* epimer in alkaline medium, thus the separation of the two isomers became facile.

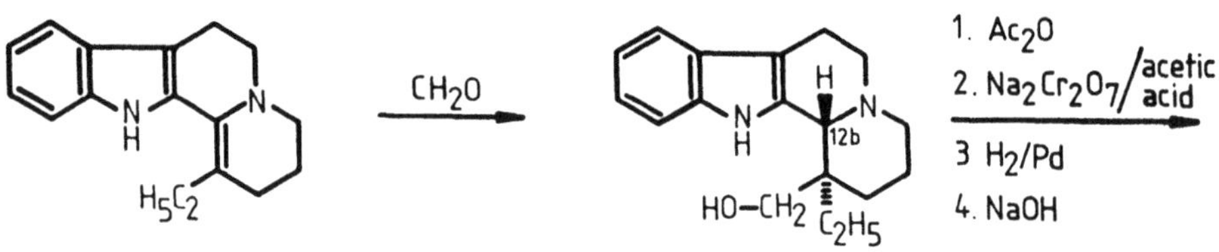

3 4

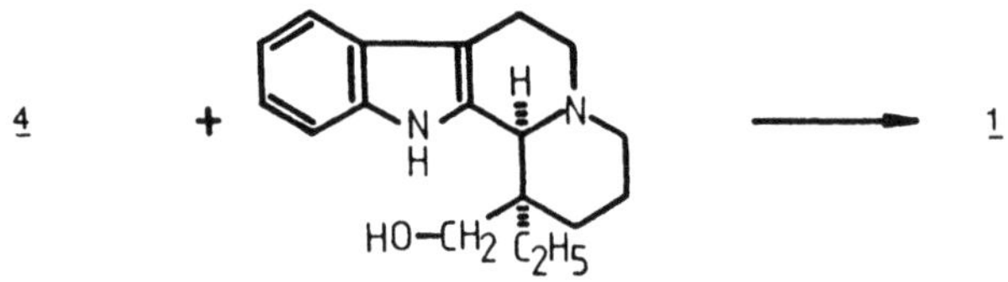

5

$\underline{4}:\underline{5} = 16:1$

Now let us see the behaviour of the mesylates of the isomers 4 and 5. Firstly the *cis* isomer (1a) was reacted with sodium cyanide, in boiling methanol. The expected nitrile (7) could be isolated only as a minor product, in about 1 % yield. The rearranged product 6 has been formed as a main product in 73 % yield. In addition to them a cyclic amidine (9), obviously formed from the nitrile *via* ring closure and a very small amount of a compound containing lo member ring (8) was isolated. The latter compound, however, was the only isolable derivative, if zinc cyanide was used instead of sodium cyanide. The hydrolysis of the amidine gave the alkaloid eburnamonine.

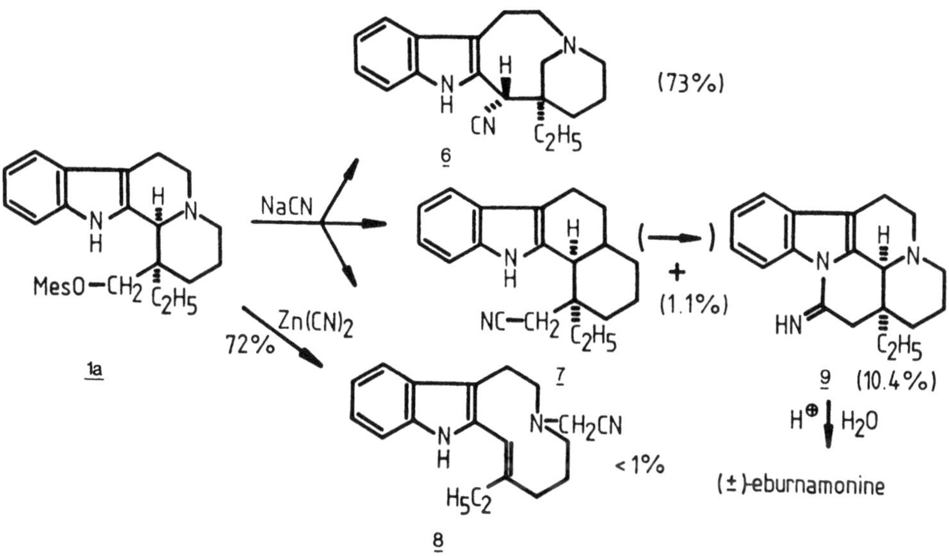

Now the intriguing question arises, how can we explain these reaction sequences?

It is very likely that the internal quaternary salt is the intermediate. This presumed intermediate was conveniently prepared by boiling the mesylate for 25 min. in methanol in almost quantitative yield, which on reacting with NaCN gave the same products in the same ratio.

A possible explanation of the different behaviour of sodium cyanide and zinc cyanide is the following. Sodium cyanide forms an ion-pair, the small anionic part of which readily attacks the most strongly electrophilic site of the molecule, namely C-12 (process A), and to lesser extent, the bridge carbon atom (B). On the other hand, owing to steric reasons, for zinc cyanide, which has considerably more covalent character, only the second strongest electrophilic site, the carbon atom of the bridge methylene is accessible; thus this is the site of the attack. However it is not quite clear, why process C predominates over process B when zinc cyanide is used.

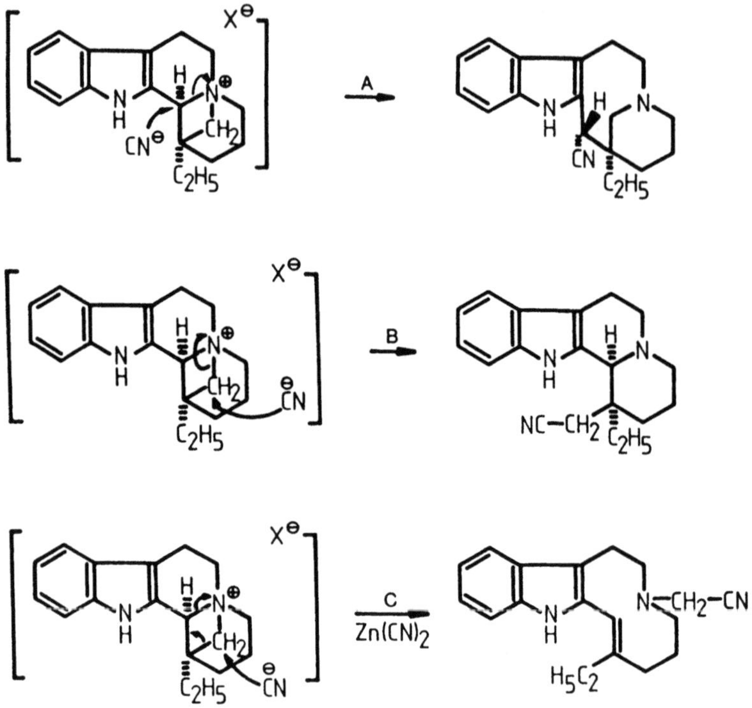

Now let us see the behaviour of the *trans* epimer (1b) which reacted quite differently with sodium cyanide. The reaction was considerably slower. Heating in DMSO at 100 °C for about four days was needed for disappearance of the starting material, and even then a well-defined derivative (10) could be isolated only in poor yield (25 %). Thus the behaviour of the *trans* epimer compared with the *cis* isomer is strikingly different.

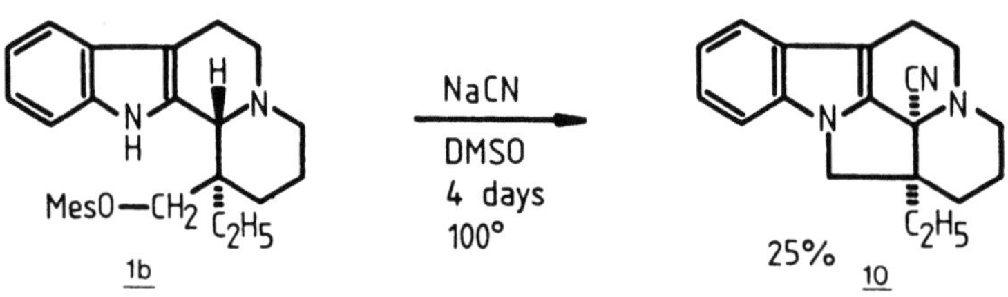

The probable way of formation of the substance 10 is that either before or after ring closure at the indole-nitrogen, the intermediate is oxydized during the long time heating in DMSO to iminium compound, which is converted by attack of cyanide ion into the pseudo-cyanide.

To investigate the two possibilities, first the iminium perchlorate (11) was treated in DMSO at 100 °C with excess NaCN. After two hours the cyanide (10) was isolated in 34.5 % yield. Second, the mesylate (1b) was converted with $KOBu^t$/DMSO to nor-eburnane (12). Oxidation of the latter gave the iminium salt (13), which rapidly reacted even at room temperature with sodium cyanide to give after 90 min the expected product (10) in 72 % yield.

We conclude that when the reaction is performed in a single operation, ring closure is the primary process followed by oxidation but that the alternative reaction sequence proceeds simultaneously can not be excluded.

The basic explanation of the different reaction route compared with the *cis* isomer is the difficult internal quaternerization of the *trans* compound.

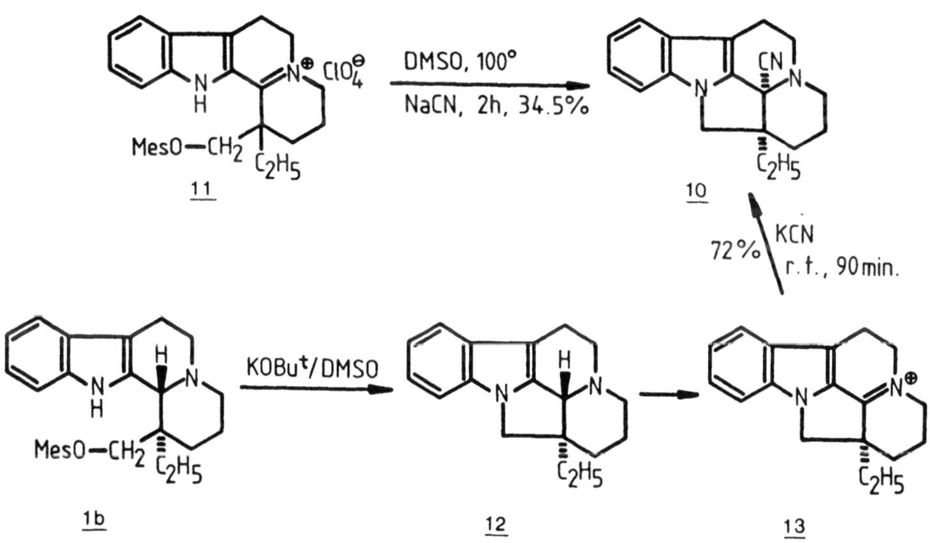

When *trans*-mesylate (1b) was reacted in methanol with sodium cyanide, the imino compound (14) was isolated in 51 % yield in addition to 8 % side product. Thus, under the given conditions, the nitrile presumably formed as an intermediate, very rapidly undergoes to ring closure. Treatment of the amidine (14) with aqueous hydrochloric acid afforded 3-epi-eburnamenine.

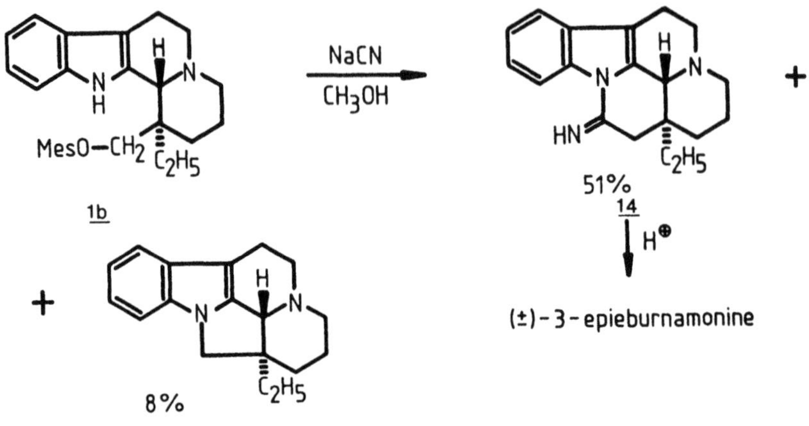

1b 14 12

In order to study the scope and limitations of the side chain length dependent reactions, the *cis* and *trans* homologous derivatives with a longer side chain were prepared and their reactions investigated. As starting compounds eburnamonine (*cis*) (15a) and 3-epi-eburnamonine (*trans*) (15b) were chosen, from which the corresponding carboxylic acids (16a and 16b) were prepared by hydrolysis.

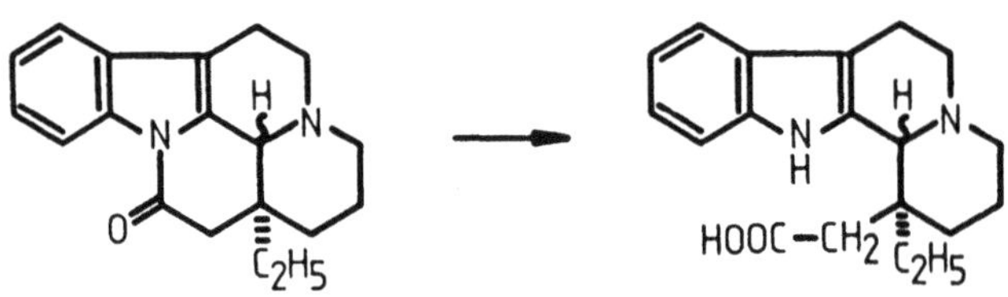

15 a Hα
15 b Hβ

16 a 12 b Hα
16 b 12 b Hβ

When the ester of the *trans* compound (17) was reduced by LAH 3-epi-eburnamine (18) and its isomer were obtained. It means, that either the intermediate aldehyde is trapped by the indole nitrogene, or the LAH catalyzes the ring closure and actually the eburnamonine is reduced afterwards.

However, when the free carboxylic acid (16b) was reduced with LAH, the desired alcohole (19b) was obtained.

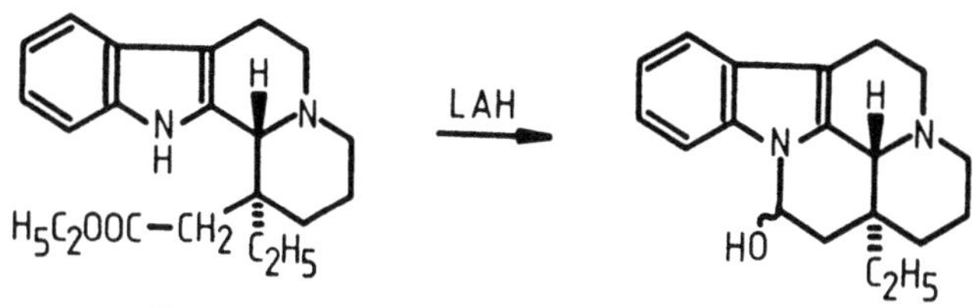

17 18

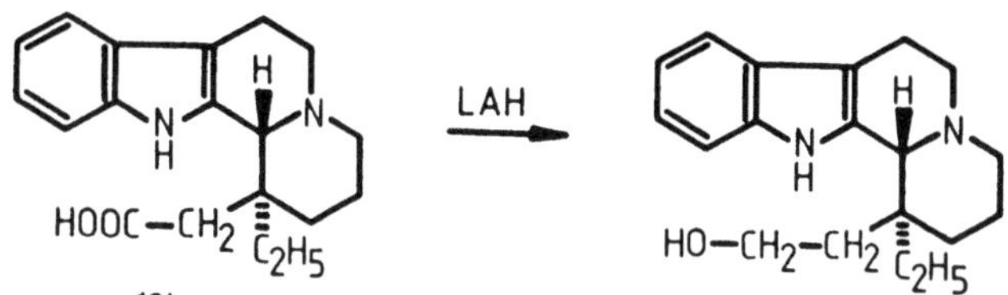

16b 19b

The *cis* compound (16a) behaved again quite differently. Reduction of the carboxylic acid with LAH or with other reducing agents failed to afford the hydroxyethyl compound (19a). The mixed

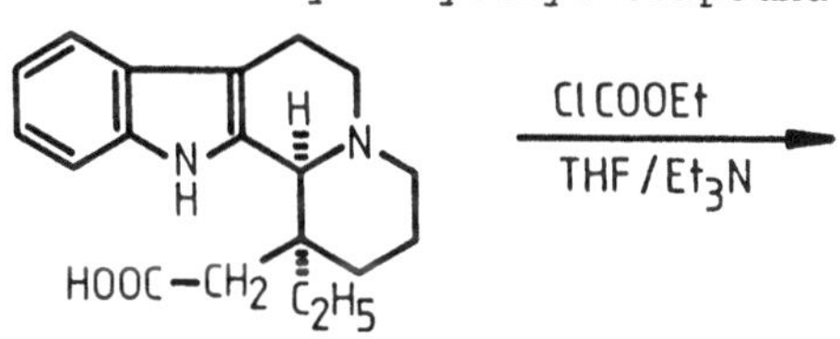

16a

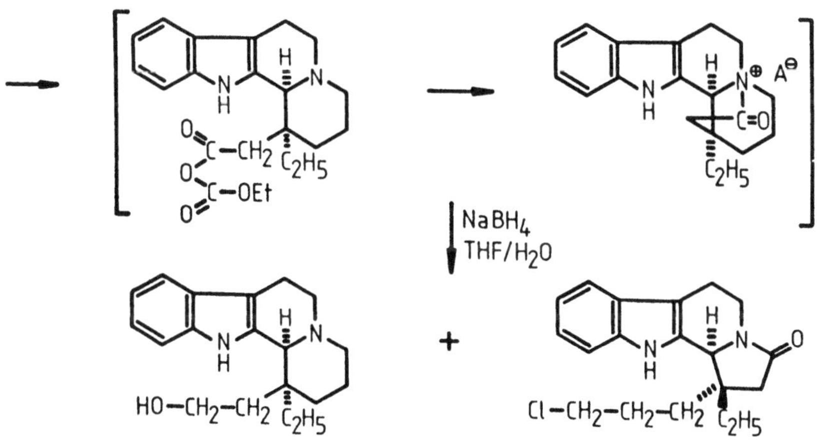

main - product
19a

20

anhydride, however, prepared by reacting the acid with ethyl chloroformate gave on reduction with $NaBH_4$ the expected compound. As a by-product the indolo-indolizine derivative (20) was isolated, which proves the existence of the intermediate acylium-ion. Of course, the alcohol can be deduced both from the acylium-ion and the mixed anhydride.

An alternative route to the *trans* compound involves acylation of the *cis* derivative with acetic anhydride followed by oxidation with $Na_2Cr_2O_7$. The isolated iminium salt (21) was reduced with $NaBH_4$ in ethanol to give the epimeric acetates. After separation and hydrolysis of the latter the desired *trans* epimer (19b) was obtained.

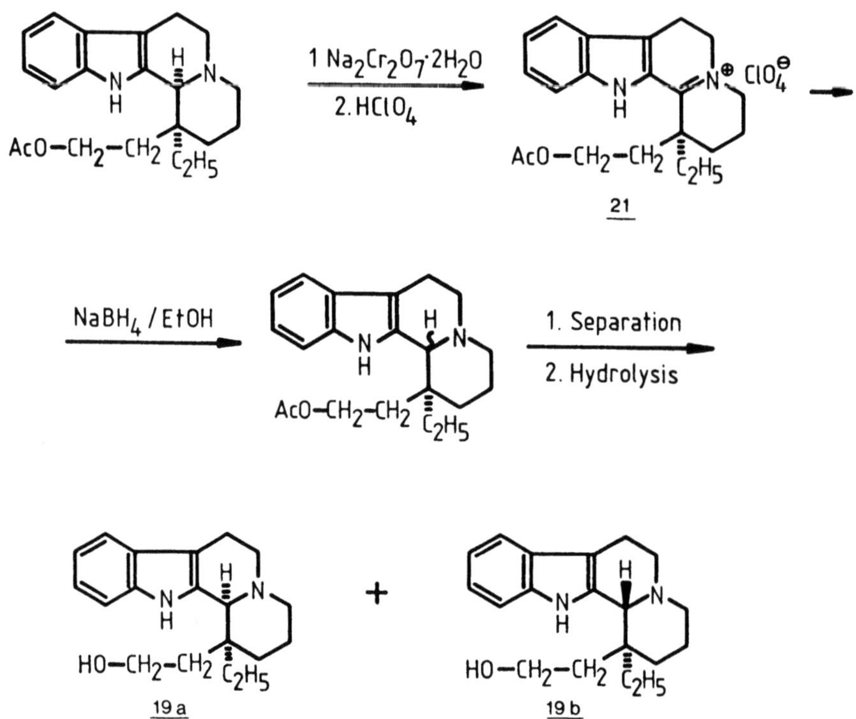

Mesylation of both epimers (19) afforded immediately the corresponding quaternary salts (22). Each of the stereoisomeric salts were reacted in DMF with NaCN. A mixture of epimers (23) was formed in both cases.

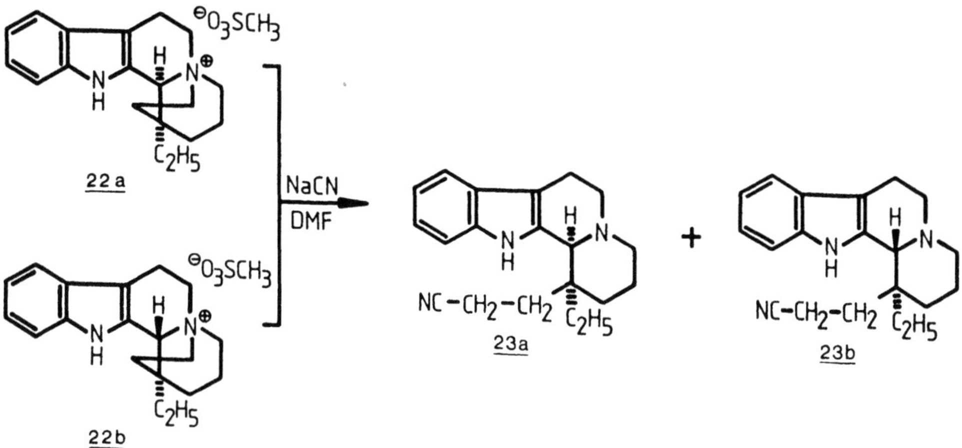

It seems that under the conditions applied the mesylate salts interconverted presumably *via* a seco intermediate.

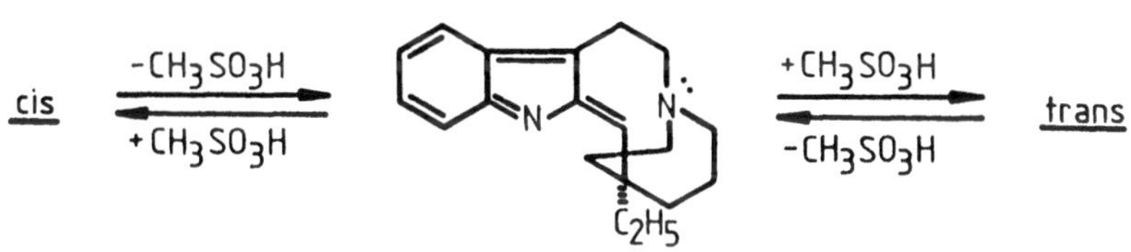

In order to support this assumption the $N_{ind.}$-methyl derivative and from that the mesylate salt (24) was prepared. Reaction of the latter in DMF with NaCN afforded - as expected - a single product (25), i.e. no epimerisation occured. The nitrile was also prepared by selective methylation of the corresponding indole derivative (26).

The quaternary salt in this case did not react with $Zn(CN)_2$.

We can conclude from the above investigations that the mesylates with 6,6,5 member ring system unlike their homologues give only the expected nitriles, but accompained by an unexpected epimerisation.

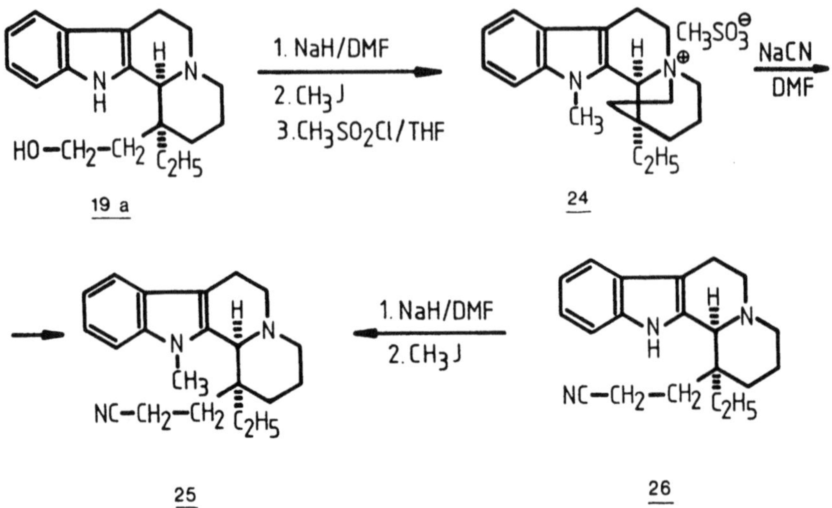

Now let us go one step further and investigate the homologue mesylate which containes one methylene group more in the side chain.

When the alcohol (27) was attempted to mesylate, the desired derivative could not be isolated. The formation of the internal quaternary derivative 28 is so fast, that this is the only product isolated in 71 % yield. This compound was very sluggish to react with sodium cyanide. The reaction has been performed in ethylene glycol at 200 °. Now no attack at an α-carbon with respect to the charged nitrogene occured, only an epimer-pair (29) was isolated in 68 % combined yield. Presumably a Hoffmann-type elimination is

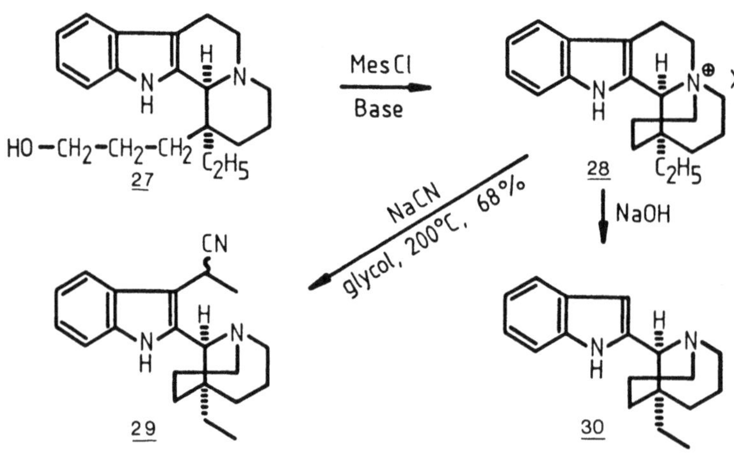

the first step followed by an addition of cyanide to the vinyl group. On the other hand, when the quaternary salt was treated with NaOH under the same conditions, the compound (30) with no side-chain at position 3 of the indole nucleus was isolated.

The quaternary salt in this case did not react with $Zn(CN)_2$.

In the course of the above studies[1-3] the interactions gave different products depending on the length of the side chain and the stereochemistry, i.e. our model can be attacked by the nucleophile (in this case by CN^-) at different points and the attack may lead to different products side by side.

At the same time there should be many attacks which are inconclusive, meaning that perturbation of encountering orbitals occurs without arriving at the transition state, consequently yielding no new product.

There is a third type of interaction. Removal of the acidic proton from the indole NH by base (CN^-) is easy and requires only a small activation energy, but because this is a reversible process, at the end no new products show up as a result of that particular attack. (However, as it was shown with compound 1b, if some steric requirements are fulfilled, the intermediate $N_{ind.}$ anion can be attacked by the internal mesylate affording new product (10).)

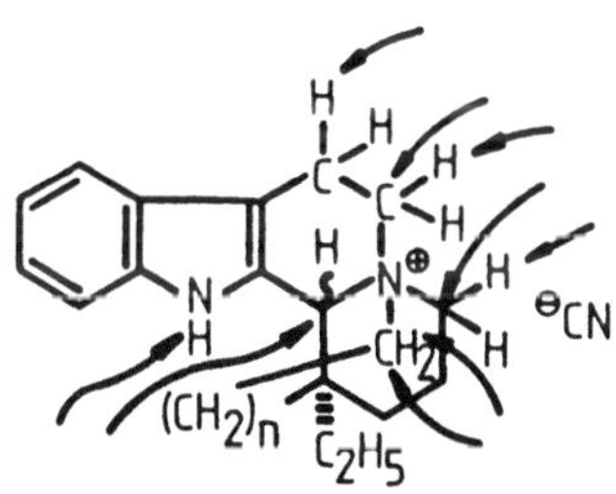

n= 0, 1, 2

Fig 1

One would think that the electrophilicity of the indicated centers in Fig.1 does not change much while varying n from 0 to 2. Nevertheless the "main attack", the most conclusive attack is shifted from the carbon in α-position with respect of the charged nitrogene to the hydrogene of the C-H bond being in β-position, resulting in Hoffmann-type elimination only because of a small change in the shape of the molecule.

We can draw now a simplified scheme (Fig.2), a graph with three types of curves, illustrating the different types of attacks:

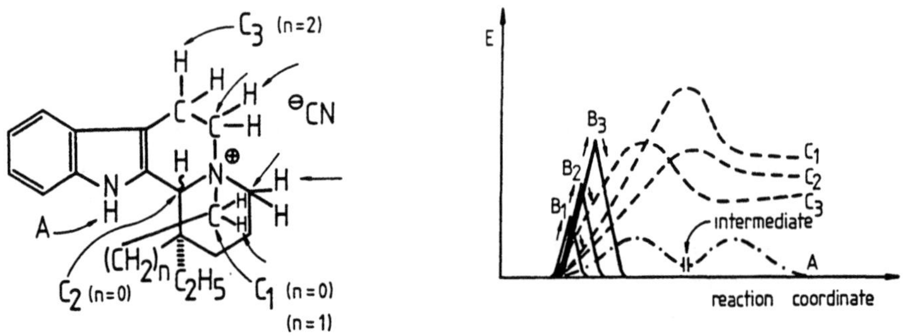

A -·-· Attack conclusive but product easily reconvertible

B —— Attack inconclusive (perturbation of orbitals without reaching transition state)

C --- Attack conclusive leading to new product

Fig. 2

A./ attack conclusive but product easily reconvertible, e.g. reversible removal of acidic proton; B./ attack inconclusive, i.e. perturbation of orbitals proceeds without reaching the transition state (dissociation of the "inconclusive complex", occurs before that point and molecules fall back from the hillside to the valley); C./ attack conclusive leading to new product.

It many cases it seems advantageous to view the chemical reaction as a *multicenter attack* regarding some of them as conclusive, the others as inconclusive ones, which can exchange their roles along with small modifications in the structure of the interacting molecules.

REFERENCES

1 Gy. Kalaus, N. Malkieh, I. Katona, M. Kajtár-Peredy, T. Koritsanszky, A. Kálmán, L. Szabó and Cs. Szántay, J. Org. Chem. 50 (1985) 3760.

2 M. Incze, F. Sóti, Zs. Kardos-Balogh and Cs. Szántay, Heterocycles 23 (1985) 2843.

3 Gy. Kalaus, N. Malkieh, M. Kajtár-Peredy, J. Brlik, L. Szabó and Cs. Szántay, J. Org. Chem. in press.

Recent Progress in Our Indole Alkaloid Synthesis

Yohimbines, Corynantheines, and Heteroyohimbines

Ichiya Ninomiya
Kobe Women's College of Pharmacy,
Motoyamakita, Higashinada, Kobe 658, Japan

Thank you very much for your kind introduction, Mr. Chairman. It is a great pleasure for me to be given this opportunity to visit Pakistan, to meet friends and to talk on what we have done during the last ten years in our laboratory in Kobe, particularly on our synthesis of indole alkaloids. I picked this title today because of the noted contribution in this field of chemistry by Professor Atta-ur-Rahman and his coworkers.

In the first slide I summarized what I am going to talk today, the list of target indole alkaloids. My talk consists of two parts, the first part is on the synthesis of yohimbine and its group and the second part on the synthesis of corynantheine and heteroyohimbine types of alkaloids.

Target Alkaloids in Today's Talk

1) Aromatized Alkaloids

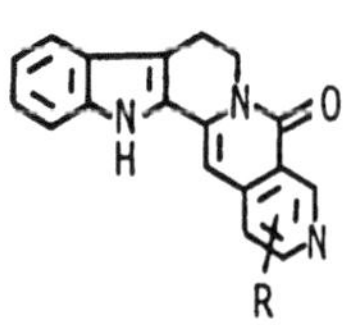

2) Yohimbines

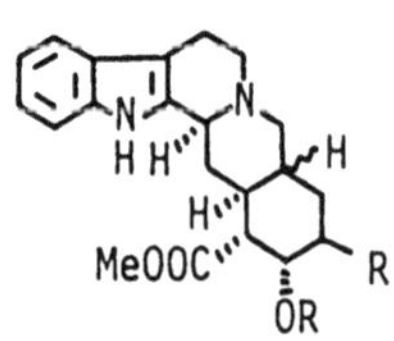

Yohimbine/Alloyohimbine
Deserpidine

3) Heteroyohimbines

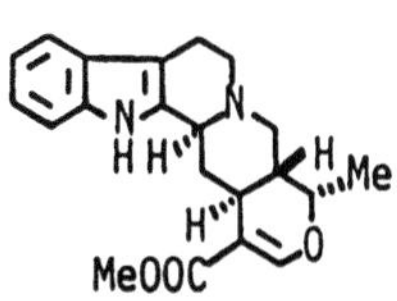

Ajmalicine

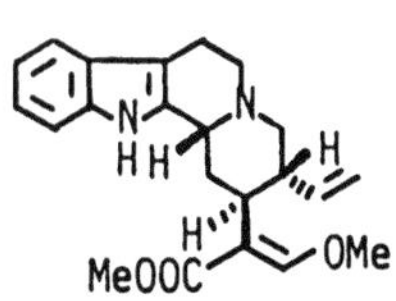

Hirsuteine

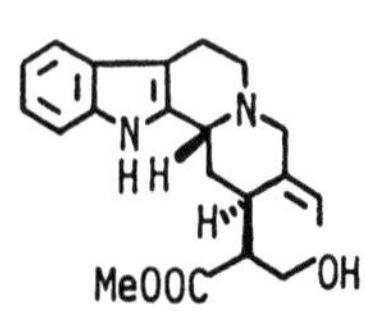

Rhazimanine

Our entry into indole alkaloid synthesis was triggered by the finding of a facile photochemical synthesis of the skeleton of yohimbine as shown in the following scheme. The fundamental reaction for the construction of the alkaloid skeleton was enamide photocyclization under reductive condition. The enamide was prepared from harmalane by benzoylation with benzoyl chloride. This enamide, which forms a conjugated six- π electron system, underwent smooth photocyclization under non-oxidative condition to afford the photocyclized product as a sole product, which had an unsaturated lactam structure which was regarded as an oxidative product although the irradiation was carried out non-oxidatively. This difference in the structures of photocyclized products would be caused by the instability of the intermediary structure of cyclization.

Yohimbine Skeleton (1976)

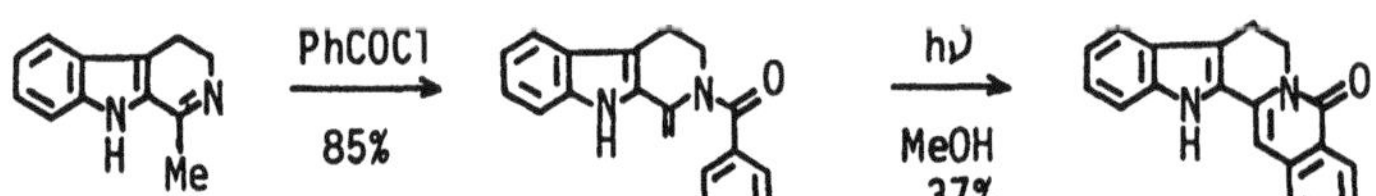

harmalane

However, the structure of this photocyclized lactam was the one which often appears in the alkaloids as shown in the table. These alkaloids belong to a relatively new class of indole alkaloids and often carry remarkable biological activity. Apparently, these alkaloids were very suited targets for the synthesis by enamide

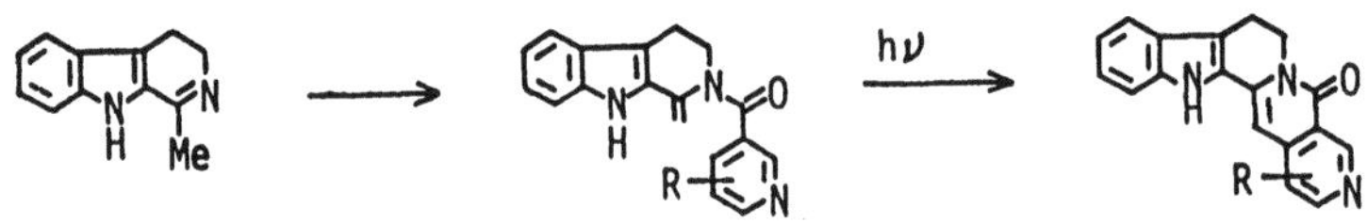

Yield of Photocyclization

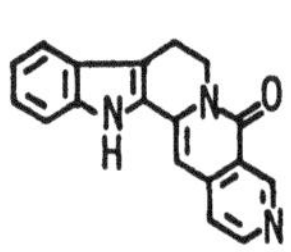

Nauclefine (48%)

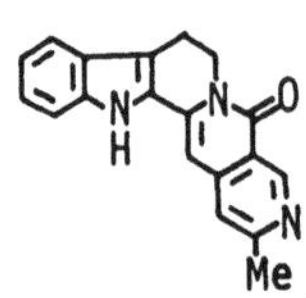

Angustidine (21%)

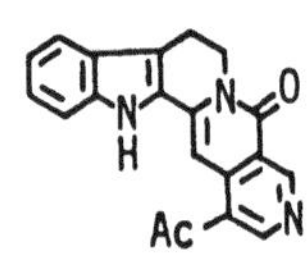

Naucletine (30%)

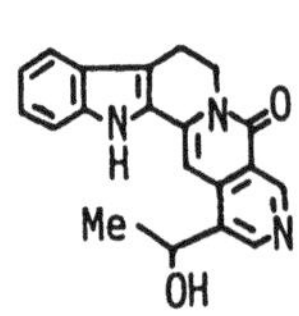

Angustoline (10%)

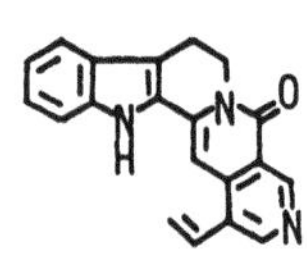

Angustine

photocyclization. Just by replacing benzoyl chloride to substituted pyridinecarbonyl chlorides, the enamides were prepared and irradiated. The products were obtained in the yields as shown in the table and they were the alkaloid themselves. Actually all these total syntheses were completed within a very short time, thus brought us to the brink of draught in the target alkaloids. Then new targets and new methodology were strongly needed for the continuation of our work in the field of indole alkaloids.
Just one example of total synthesis of new indole alkaloid nauclefcine was shown. Nauclefcine was isolated from the Chinese medicinal herbs recently and reported as having remarkable antineoplastic activity. In order to attain regioselective photocyclization, the enamide with two substitutents was prepared and irradiated. Contrary to our expectation, two photocyclized lactams were obtained as a result of poor regioselectivity. Introduction of a methoxy group into the ortho-position was aimed at the regioselective cyclization to the root of the methoxy group. Though the major product was the expected one, the presence of an additional methoxycarbonyl group caused only poor regioselective result. The alkaloid nauclefcine was obtained by the conversion of an ester group into the formyl group not by direct reduction but via reduction to the carbinol followed by back oxidation to a formyl group.

Synthesis of Nauclefcine (1986)

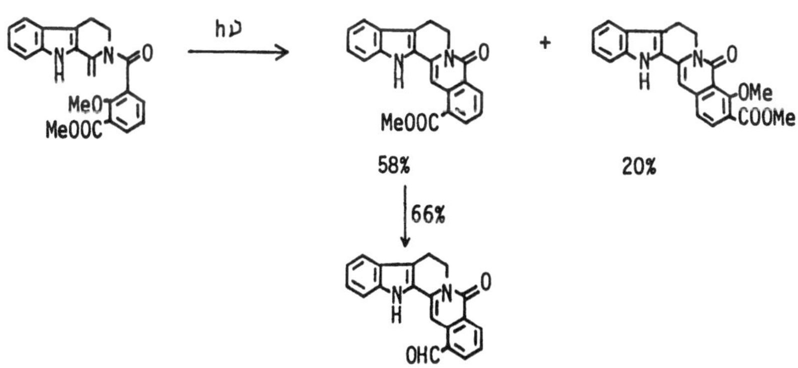

Nauclefcine

As an unavoidable situation we faced from smooth application of enamide photocyclization to alkaloid synthesis, we have investigated a new developement in enamide photocyclization and found a very efficient and most useful synthetic methodology in what we called " reductive photocyclization ".

As shown in the upper part of the scheme, enamide photocyclization would proceed via the cyclic intermediate, from which hydrogen would migrate suprafacially to give the lactam. When you look at

Non-oxidative Photocyclization of Enamide

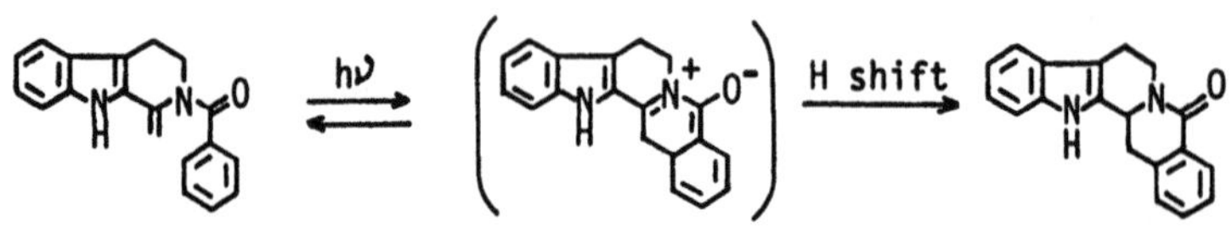

Reductive Photocyclization of Enamide

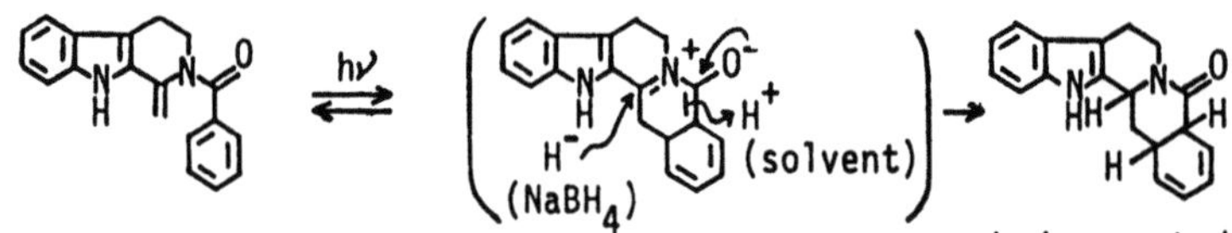

hydrogenated
lactam

the structure of this intermediate, which contains an imminium bond susceptible to reduction by hydride, it was suggested that when irradiation of enamide was carried out in the presence of hydride reagent, photocyclization would proceed to form a cyclic intermediate which would be then subjected to reduction by hydride present to afford the hydrogenated lactams as the final product. This assumption was confirmed and the reductive photocyclization was established by the experiments using deuterated reagents as shown in the scheme.

First, the same enamide was irradiated in the presence of a large excess of sodium borohydride in the solvents including 10% methanol which was added in order to dissolve hydride reagent. The expected reductive photocyclization proceeded smoothly to give a mixture of two hydrogenated lactams as shown. The one product had the conjugated diene structure and the other having an unconjugated diene moiety. In order to make this new methodology as a useful synthetic tool, the reaction condition for selective formation of single product was searched. As a result we found acetonitrile as the most suited solvent in this case, thus giving rise to the sole formation of the lactam with an unconjugated diene moiety in 98% yield. This result was just superb.

Choice of Solvent for Reductive Photocyclization

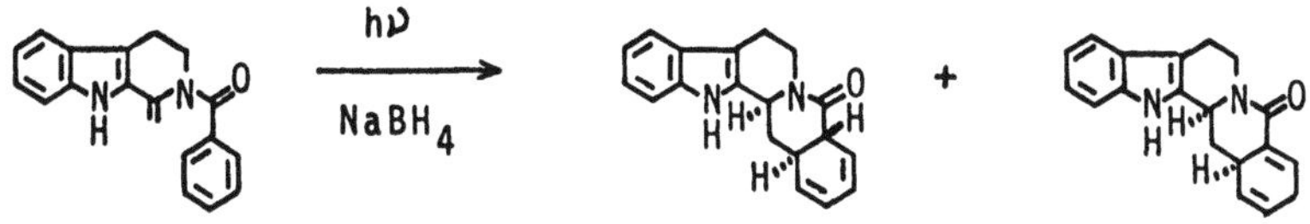

solvent		
C_6H_6-MeOH	7%	21%
Et_2O-MeOH	17%	32%
MeCN-MeOH	--	98%

The reaction mechanism for this reductive photocyclization was established as follows. When the enamide was irradiated in the presence of sodium borodeuteride in place of hydride, of course, in acetonitrile-methanol, the product was deuterated at the ring junction as expected, while the photocyclized lactam, when irradiated in the presence of hydride but in acetonitrile and deuterium methoxide, was deuterated at the benzene ring as shown. Thus, the reaction mechanism of this reductive photocyclization was firmly established by the structure in the parenthesis.

Deuterium Incorporation in Cyclization

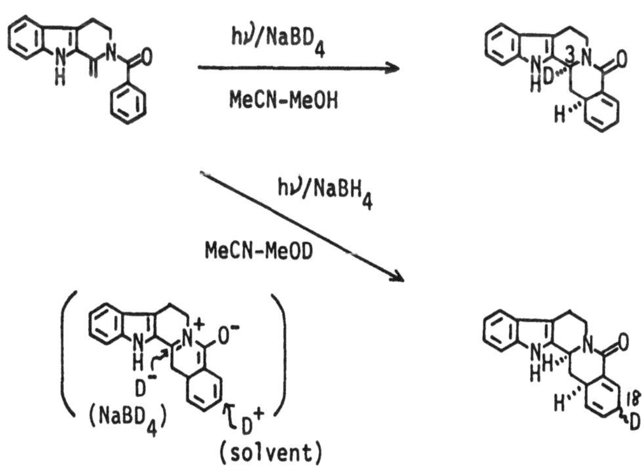

With this new development in synthetic methodology, we have picked our target in yohimbine group of alkaloids including yohimbine, alloyohimbine, deserpidine, and reserpine, which are well known as clinically useful medicinals.

From their importance as medicinals, there have been reported a number of synthetic achievements on these alkaloids as summarized in the scheme. Most of these previous syntheses had been achieved according to the strategy of starting from two building blocks which contained multi-chiral structural features at the start and construct the skeleton of the alkaloids at the final stage of total synthesis. Therefore these syntheses are good only for one

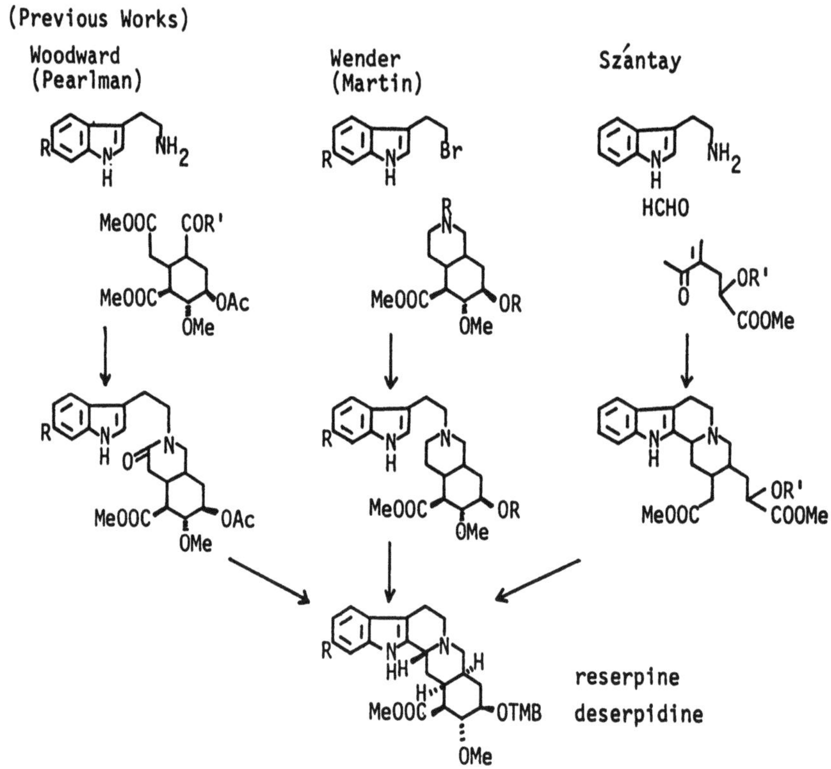

target alkaloid but not immediately applicable to other components having only slightly different structure in part. even though their structures, particularly their skeletal structure, are closely related. Quite contrarily, our synthesis for these alkaloids consists of the construction of the fundamental skeleton first and introduction of substituents at later stage, thus expecting a wide range of application. As a typical example, our synthetic strategy for the yohimbine group of alkaloids is depicted as follows. Starting from the enamide carrying an oxygen function, the skeletal structure could be formed at the first step of the synthesis and then introduction of a methoxycarbonyl group into the 16-position and a methoxy group into the 18-position would provide their total synthesis.

Synthetic Strategy for Yohimbine Group

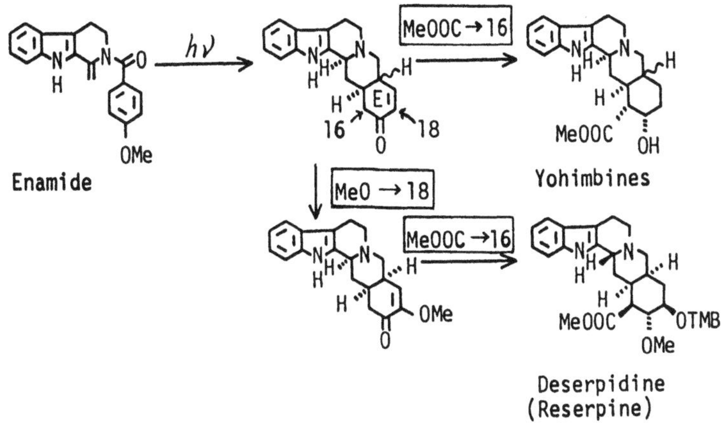

As the first example of the application of our methodology, we picked yohimbine and alloyohimbine. The enamide was prepared from harmalane by acylation with p-methoxybenzoyl chloride and irradiated according to the established reductive photocyclization condition to afford the lactam with a methoxy-substituted dihydrobenzene moiety, which then underwent a facile hydrolytic cleavage to give the corresponding ketone. After reduction of the lactam carbonyl, acid treatment converted the enol ether into the β,γ-unsaturated ketone in good yield.

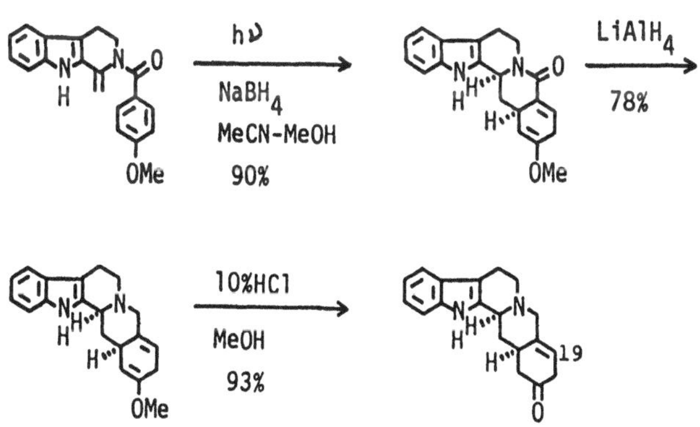

The next hurdles to overcome are the stereoselective formation of D/E-cis and trans ring systems and introduction of an ester group into the sterically hindered 16-position. Upon investigation of various conditions, we have succeeded in the stereoselective introduction of hydrogen into the 20-position with migration of a double bond. Acid treatment of this enone at an elevated temperature of 85°C furnished smooth migration of a double bond with stereoselctive introduction of hydrogen into the 20α-orientation while treatment of the same enone under ice-cooling temperature yielded the D/E-cis enone in an excellent yield. It was also found that the D/E-cis enone was smoothly isomerized to the thermodynamically stable trans isomer by heating at 85°.

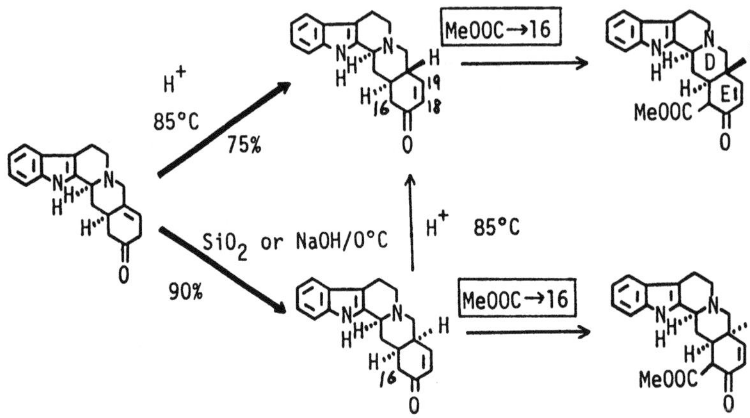

The next hurdle, introduction of an ester group into the 16-position, was overcome by taking up three factors into consideration as follows. Basically, the 16-position was sterically severely hindered, thus making introduction of any group very difficult. If the introduction of an acyl group into the 16-position were successful, the most convenient synthesis of this group of alkaloids be furnished. First, the reactivity of C-16 (sp3) can be differentiated from C-18 (sp2) by introducing a double bond into the 18,19-position.

Secondly, as Professor House discussed in his book and papers, the importance of the nature of intermediary metal enolate in the acylation reaction should be taken into consideration. Thus the procedure for the preferable formation of a contact ion pair type of metal enolate by adding magnesium salt to the lithiated solution should be preferred for the acylation on carbon over oxygen. Thirdly, recent introduction of cyanoformate as soft electrophile would be preferable because C-16 plays as soft nucleophile while oxygen of carbonyl group as hard nucleophile.

[Consideration on Sterically Unfavored C-16 Acylation]

1) Differentiating Reactivity of C_{16} over C_{18}

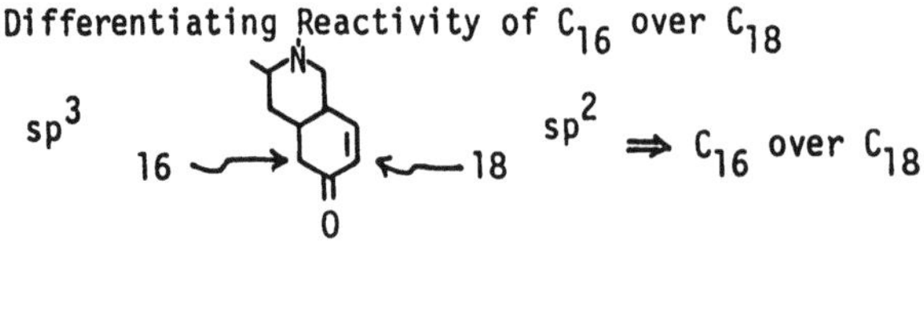

2) Type of Metal Enolates

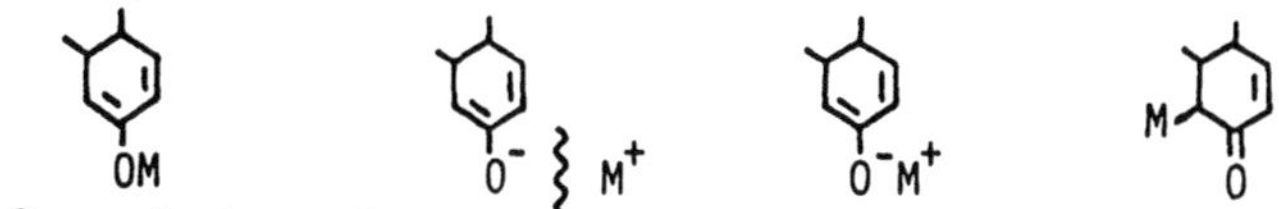

O-covalent (TMS) | solvent-separated ions (Li, Na, K) | contact ion pair (Mg, Zn) | C-covalent (Hg)

H. O. House, et al., J. Org. Chem., 38, 514 (1973).

3) Hard and Soft Acids and Bases Principle

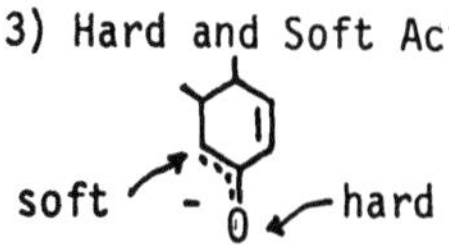

Cl–COOMe vs CN*–COOMe

hard — soft

as nucleophile — as electrophile

* L. N. Mander, et al., Tetrahedron Lett., 24, 5425 (1983).

By taking up the above factors into consideration on the reaction condition for acylation, introduction of a methoxycarbonyl group into the 16-position was investigated as summarized in the scheme. The D/E-trans enone was treated with LDA at -78°C for lithiation followed by methyl chloroformate. The desired product was obtained only in 14% yield as expected. When magnesium bromide was added to the lithiated solution, acylation with methyl chloroformate yielded the 16-acylated product in 68% yield. The same good result (70 % yield) was achieved by the use of methyl cyanoformate even without using magnesium salt. Quite similarly, selective acylation into the sterically hindered 16-position was achieved in the case of D/E-cis enone.

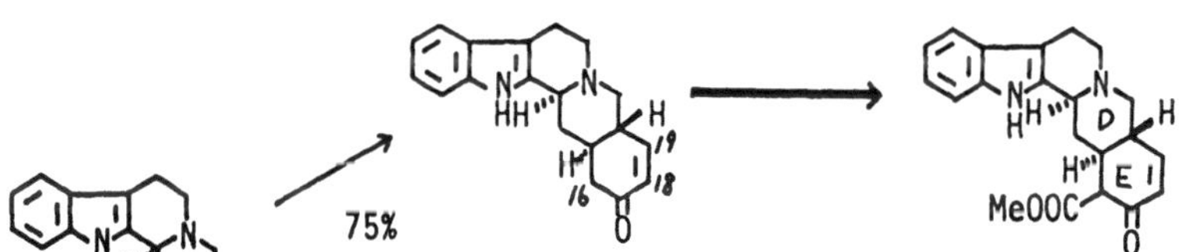

Base	E^+	Yield
LDA	ClCOOMe	14%
LDA-$MgBr_2$	ClCOOMe	68%
LDA	NCCOOMe	70%

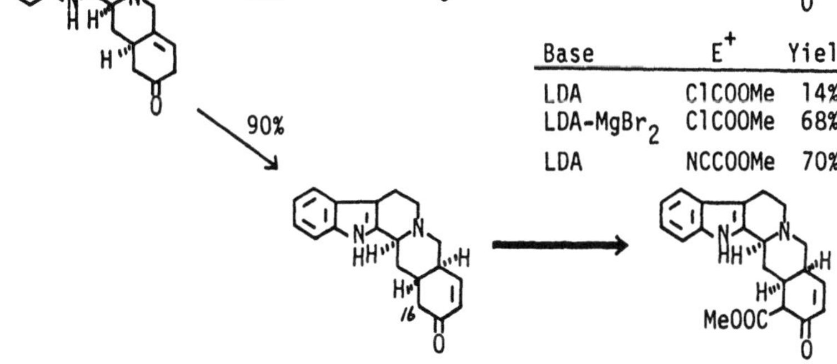

Base	E^+	Yield
LDA	ClCOOMe	26%
LDA-$MgBr_2$	ClCOOMe	69%
LDA	NCCOOMe	71%

Thus, the remaining steps to the alkaloids yohimbine and alloyohimbine gave us no problem, thus furnished a novel and practical synthesis of both alkaloids as summarized in the scheme.

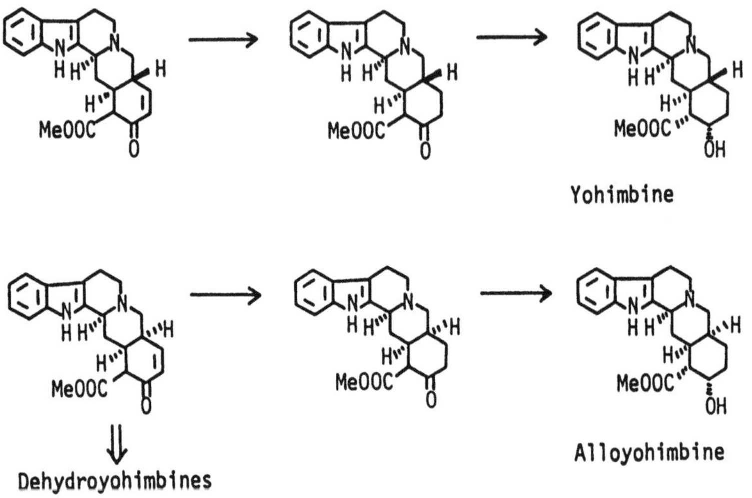

(Summary) Yohimbine Synthesis

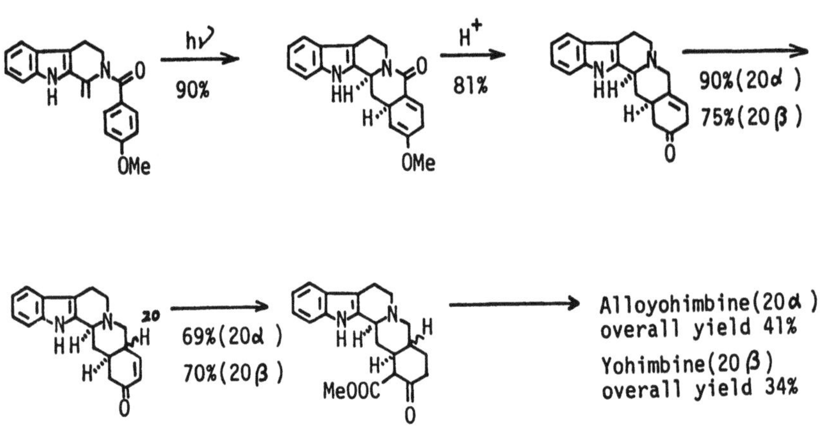

During the course of these synthesis, the synthesis of 19,20-dehydroyohimbines was also achieved by using the acylated D/E-cis enone, which has a folded ring conformation. On the other hand, the 19,20-unsaturated ketone has a rather planar structure. Therefore, it was expected that the back migration of a double bond at the 18,19-position into the 19,20-position would occur. Actually, acid treatment of the 18,19-enone yielded the 19,20-enone in an excellent yield. Subsequent reductions furnished the first total synthesis of three 19,20-dehydroyohimbines.

19,20-Dehydroyohimbines (1984)

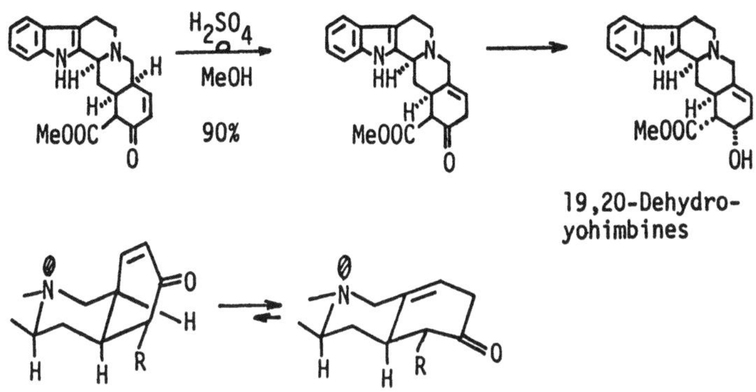

As the next target from this group of alkaloids, we picked deserpidine. For the synthesis of this alkaloid, only the introduction of a methoxy group into the 18-position was required. The D/E-cis enone, which was used for the synthesis of alloyohimbine, was oxidized with hydrogen peroxide in alkaline solution to give the 18,19-α-epoxide, which was then ring opened by the treatment of acid to furnish the 18-methoxylated D/E-cis enone. Subsequent conversion into the alkaloid deserpidine was carried out according to the procedure employed for the synthesis of yohimbine, thus completed another novel total synthesis of deserpidine. With this success in the synthesis of yohimbine, alloyohimbine and deserpidine, we are now in a good position to complete the synthesis of reserpine.

Synthesis of Deserpidine (1984)

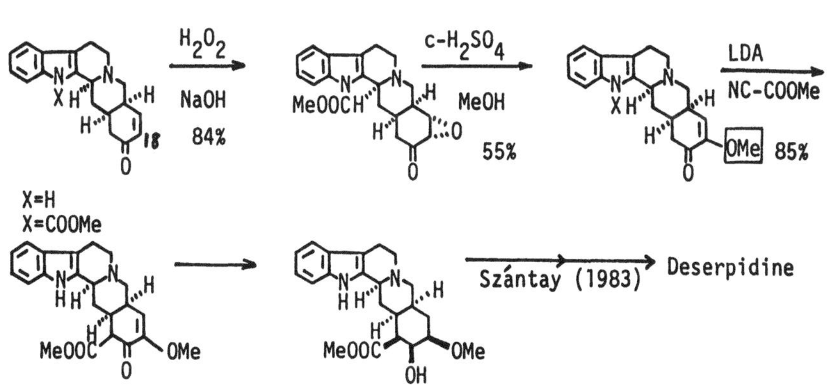

In summarizing of this part of talk, we depicted a tiny cascade of our synthesis of yohimbine group of alkaloids though not comparable to the arachidonate cascade. Thus, wide applicability of our synthetic methodology would be clearly demonstrated.

[Summary] Our Synthetic Cascade

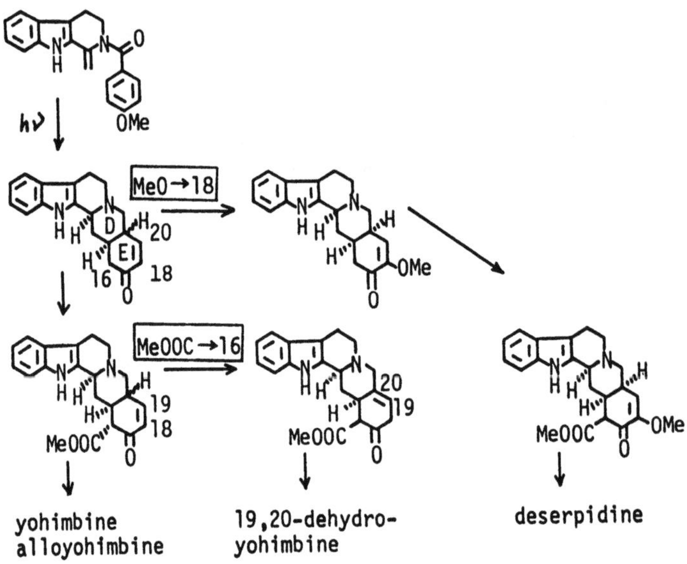

Now I am going to talk on the second part of our indole alkaloid synthesis.

In the first part of my talk, it is demonstrated that we have successfully applied reductive photocyclization of the enamide prepared from harmalane and p-methoxybenzoyl chloride to the synthesis of yohimbine group of alkaloids, which have a common benzoindoloquinolizidine skeleton.

After having shown the potentiality of our synthetic methodology to the synthesis of indole alkaloids of yohimbine group, our attention has then focused onto another representative group of indole alkaloids, that is, corynantheine and heteroyohimbine groups, as our next targets aiming at the synthesis of most of the members of these groups by one methodology. This is the feature of our synthesis. In fact, within a short span of time we have succeeded in the synthesis of a considerable number of alkaloids as summarized in the following scheme. As shown, these two groups of alkaloids are structurally closely related and chemically interrelated by facile conversions. The first group, heteroyohimbine group, consists of ajmalicine and akuamigine, of which ajmalicine has been regarded as the most noted because of its biological activity and clinical usefulness. And the second group, corynantheine group, consists of corynantheine, hirsuteine, geissoschizine, and many relatively untouched alkaloids.

Target Alkaloids (Corynantheine-Heteroyohimbine Group)

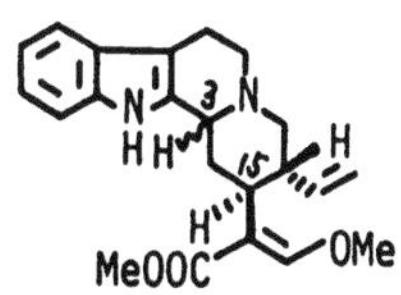

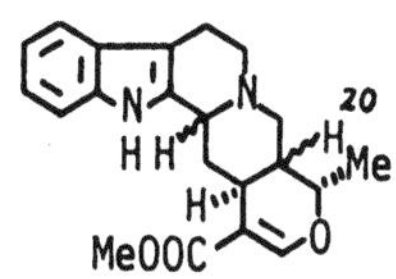

		20α	20β
3α	corynantheine	akuamigine	3-isoajmalicine
3β	hirsuteine	4H-alstonine	ajmalicine

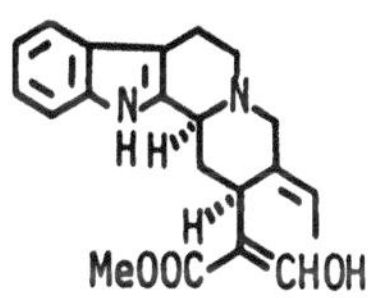

Geissoschizine

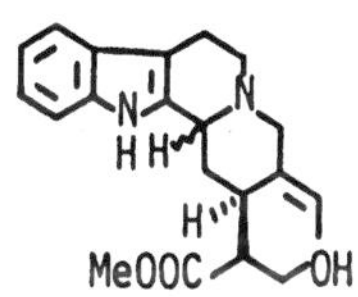

isositsirikine
rhazimanine
bhimberine

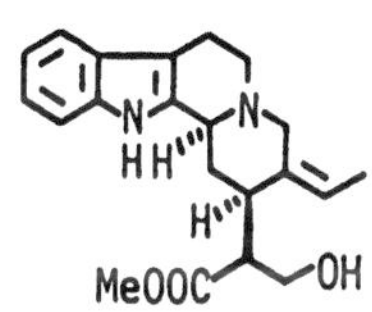

16-epi-Z-isositsirikine
Z-isositsirikine

The crucial point in our synthesis of these two groups of alkaloids was the synthesis of the oxygen-containing ring E in heteroyohimbines or its open-chained congeners in corynantheines. Previously, we had succeeded in the synthesis of ergoline group of alkaloids including lysergic acid by using a furan ring in the enamides. Photocyclization of the furoylenamine type of enamides proceeded smoothly to give the lactam with a cis-fused dihydrofuran moiety, which was then ring opened to yield the disubstituted lactams. This strategy on the use of a furan ring in the enamide was successfully applied to the synthesis of ipecac alkaloid emetine. The Bischler-Napieralski product, 3,4-dihydro-1-methylisoquinoline, was acylated with 2- or 3-furoyl chloride to give the corresponding enamides, N-(2- or 3-furoyl)-1-methylene-tetrahydroisoquinolines in good yield. These enamides were smoothly photocyclized under reductive condition to afford respective benzoquinolizidines with a dihydrofuran moiety. Ring opening of these dihydrofuran rings yielded the 2- and 3-substituted benzoquinolizidines respectively, from which the alkaloid emetine was prepared in two ways.

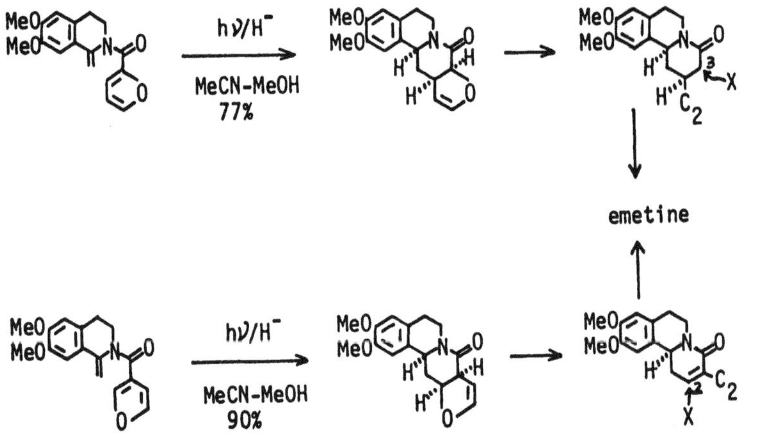

This success was now further extended to the indole derivatives by replacing a dimethoxybenzene ring to an indole ring. Thus it was suggested that photocyclization of enamides prepared from harmalane by acylation with 2- or 3-furoyl chloride would provide a novel synthesis of the biologically useful alkaloid ajmalicine and the hitherto untouched alkaloid hirsuteine. This synthetic strategy was just an another extension of the above emetine synthesis as shown in the scheme.

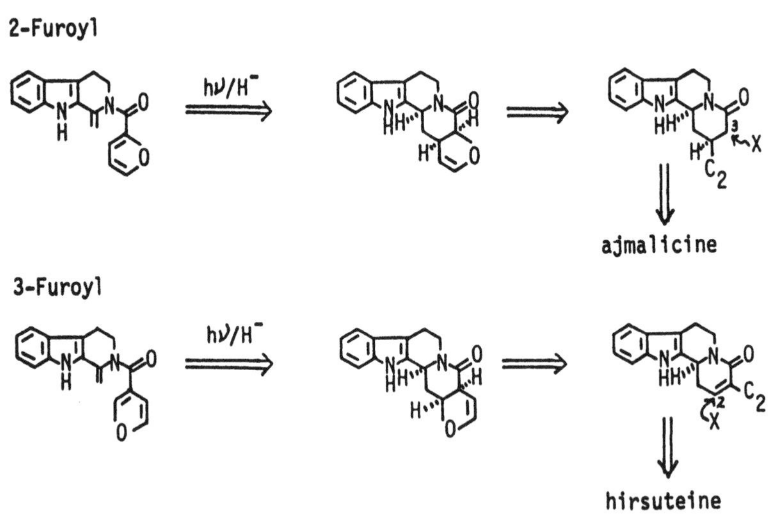

First, total synthesis of ajmalicine is shown as an example of the application of photocyclization of enamide prepared from harmalane with 2-furoyl chloride. The enamide was irradiated in the presence of sodium borohydride in acetonitrile and methanol solution. The lactam with a cis-fused dihydrofuran ring was obtained in an excellent yield. Its structure was established from n.m.r. data. This lactam has the structure equivalent to the desired quinolizidine with two-carbon substituent at the 3-position, therefore suggested as useful as the starting compound for the introduction of another two-carbon unit into the 2-position of the indoloquinolizidine by making good use of the reactivity of methylene group adjacent to the lactam carbonyl, though no such a reaction had ever been reported. The lactam was lithiated at -78°C followed by treatment with acetic anhydride to afford the N,3a-diacetylated lactam in 85% yield. The dihydrofuran ring was smoothly converted into the γ-lactone upon hydration with 15% sulfuric acid followed by oxidation of the resulting hemiacetal with dimethyl sulfoxide and acetic anhydride. The γ-lactone was then reductively ring-opened by treatment with zinc and acetic acid. Esterification with diazomethane gave the tricarbonyl compound in good yield which had been previously synthesized by Professor Massiot in his total synthesis of ajmalicine and was identified upon direct comparison with the authentic sample, thereby completed a formal total synthesis of this alkaloid. However, we have further gone to

[Synthesis of (±)-Ajmalicine and (±)-Corynantheine]

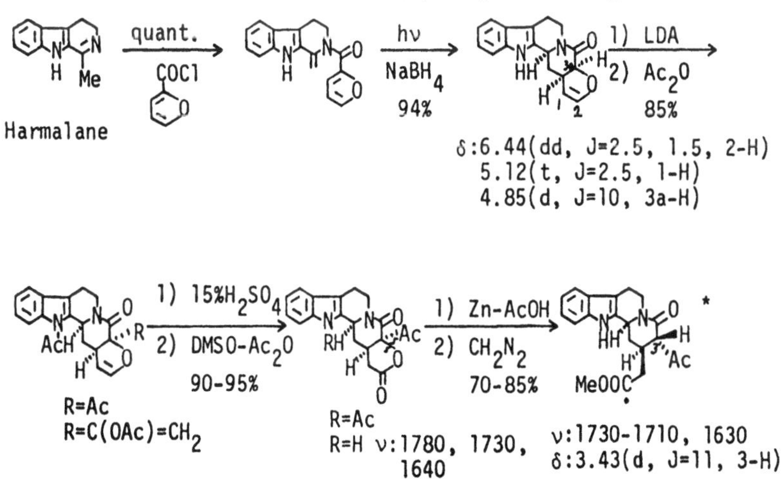

* G. Massiot, et al., Bull. Soc. Chim. Fr., 1982, II-241.

the δ-lactone via the route shown. By carrying out selective reduction of three carbonyl groups in this key compound, we have succeeded in the synthesis of the δ-lactone which was another key intermediate in the synthesis of ajmalicine by Professors van Tamelen, Winterfeldt, and Massiot.
First the lactam carbonyl group was reduced via the route involving conversion to thiocarbonyl with Lawesson's reagent followed by desulfurization with Raney-nickel and finally saturation of imine by sodium borocyanohydride. This acetyl ester was known as the key intermediate for the synthesis of corynantheine, thereby furnished a formal total synthesis of corynantheine. On the other hand, reduction of the acetyl carbonyl by sodium borohydride afforded the desired δ-lactone upon acid treatment, which was identical with the authentic sample prepared in the total synthesis of ajmalicine by Professor Winterfeldt. Thus this synthesis provided not only another total synthesis of this alkaloid but also an efficient route to a useful building block, δ-lactone, which is the structure with high potentiality of application to the synthesis of various monoterpenoid alkaloids as shown. Previously, Professor van Tamelen had completed total synthesis of ajmalicine by synthesizing the δ-lactone shown in the parenthesis, which was also the target compound by Professor Winterfeldt. Further Dr. Uskokovic's group had completed their synthesis by preparing the monoterpenoid δ-lactone as the key intermediate.

Usefulness of δ-Lactone as Key Intermediate for Ajmalicine Synthesis

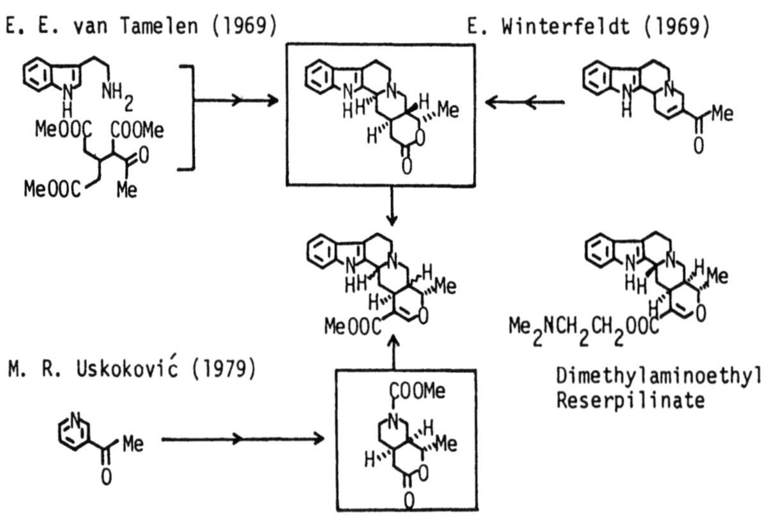

Further, the developement of alternative synthesis of the δ-lactone via the route shown in the following scheme in the parenthesis would give us a possibility of its application to various alkaloids summarized in this scheme.

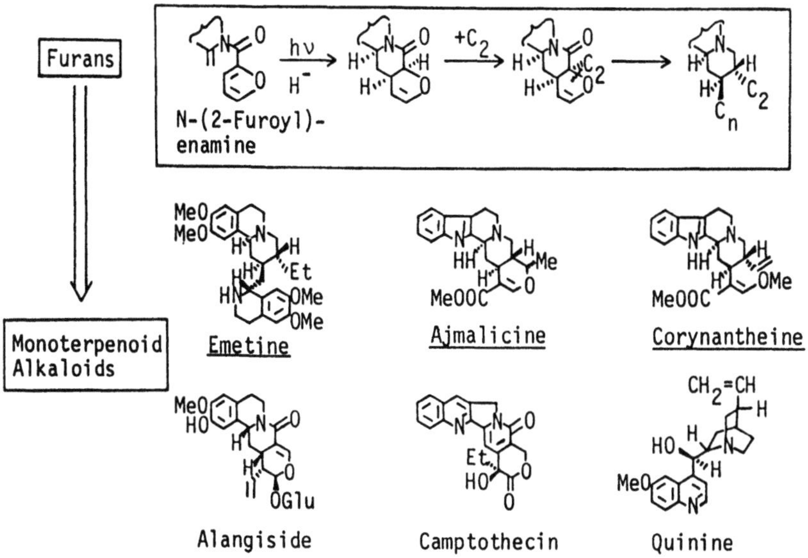

We have now proceeded to the use of enamide prepared from hamalane by acylation with 3-furoyl chloride. Upon irradiation, we could have expected to obtain the indoloquinolizidine fused also with a dihydrofuran ring but with different orientation. From

Synthetic Strategy for Corynantheine-Heteroyohimbine Alkaloids

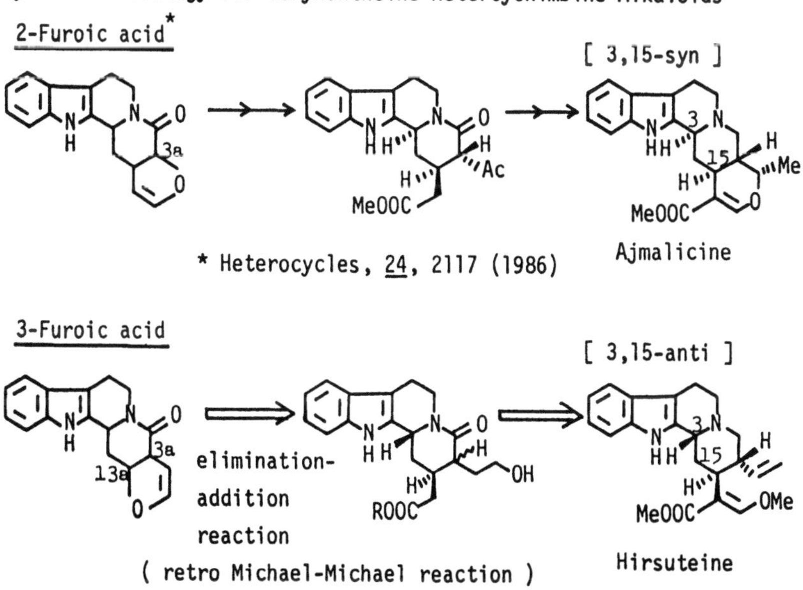

this photocyclized lactam, the ring opening would provide the quinolizidine with an enone moiety and two carbon side chain at the position adjacent to the lactam carbonyl. Therefore it is expected that Michael type of addition reaction would provide the disubstituted quinolizidine. As its consequence, corynantheine type of alkaloids such as hirsuteine, geissoschizine, isositsirikine, and related alkaloids could be synthesized and in addition, the problems pending on the structurally undetermined alkaloids rhazimanine and bhimberine could be solved.

Our synthetic strategy from the photocyclized lactam to alkaloids is not required specifically mentioned. The starting photocyclized lactam was prepared also from harmalane by acylation with 3-furoyl chloride followed by irradiation in the presence of sodium borohydride in good yield. The structure of this lactam was firmly established from 1Hn.m.r. data, particularly its NOE experiment which clearly proved stereochemical situation of two hydrogens at the ring junction as shown. Catalytic hydrogenation of the photocyclized lactam afforded the tetrahydrofuran derivative.

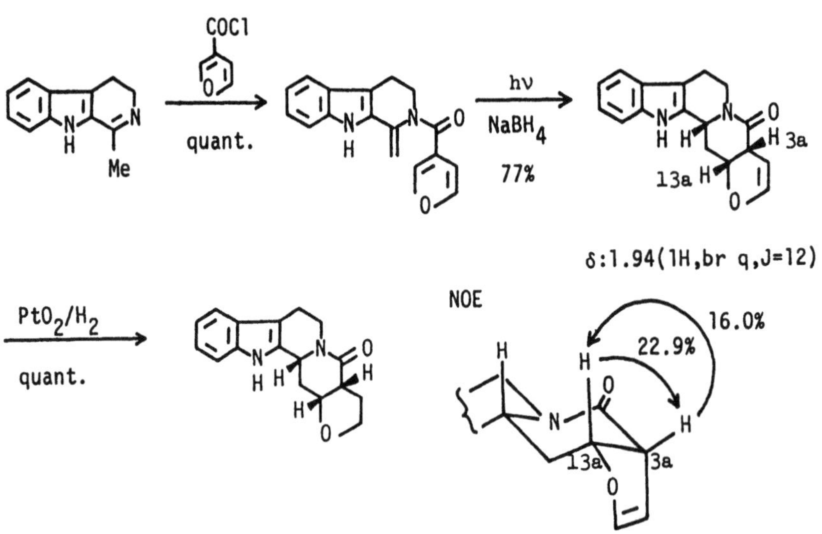

In order to introduce a carbon unit into the 2-position of the quinolizidine skeleton which would be obtained by ring cleavage of the dihydrofuran ring, we have investigated Michael type addition to the β-position of the enone moiety. When the tetrahydrofuran was treated with LDA at -78°C followed by treatment of methanol, the tetrafuran ring was nicely cleaved as expected to afford the enone with 2-hydroxyethyl group at the 3-position in good yield.

Michael Reaction using 2-Lithioacetamide as Donor

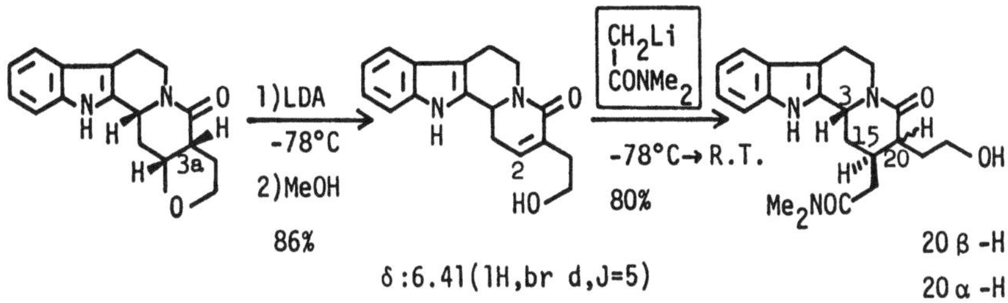

S. Sakai, et al., Chem. Pharm. Bull., 34, 77 (1986)

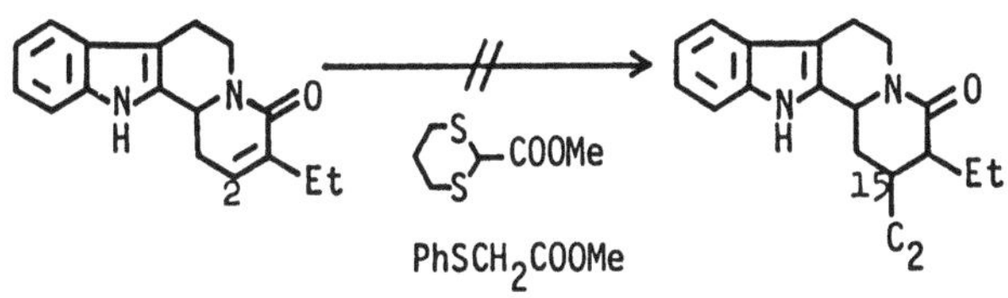

Then, we could carry on extensive study on the Michael reaction to the 2-position. According to the paper by Professor S. Sakai, Michael addition of the reactive esters as shown to the similar type of enone failed to give any desired C-alkylated product.

Michael Reaction using 2-Lithioacetate as Donor

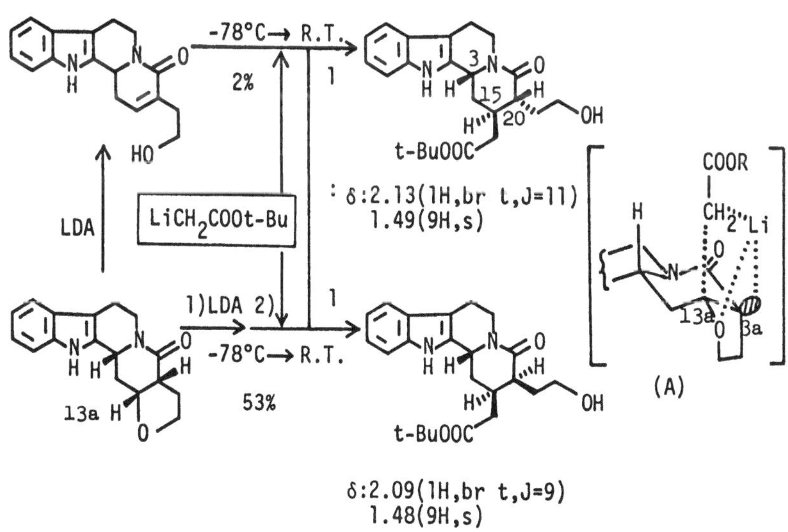

Therefore, we first investigated the reaction with the reactive dimethylacetamide. To the enone was added the lithiated acetamide at -78°C and the temperature was gradually brought up and finally to room temperature. As a result, we have succeeded in introducing

dimethylacetamide group into the 15-position, thus giving a mixture of two epimeric products in 80% combined yield. However, conversion of this amide group into the ester group did not occur smoothly. Then we had to switch back again to investigate the Michael reaction with the reactive acetate. The tetrahydrofuran was lithiated with LDA at -78°C. The lithioacetate which was prepared separately was added to the above lithiated solution at -78°C and the temperature was brought up to room temperature. The result was very poor giving only 1% combined yield of alkylated products. We thought that this poor result would be derived from the instability of the intermediary lithiated complex. Since this consists of ring-opening and the subsequent introduction of acetate group but under almost identical condition, we then carried out the Michael addition in one pot as follows.
The tetrahydrofuran was lithiated with excess of LDA and in situ added lithioacetate. The result was quite promising giving a mixture of two epimeric products in 53% combined yield. This success could be tentatively explained in terms of the formation of lithiated complex as shown by the structure in the parenthesis. The yield of this Michael addition was further improved by the adoption of the procedure for elimination-addition reaction as

[Elimination-Addition Reaction]
Heterocycles, 26, 1739 (1987).

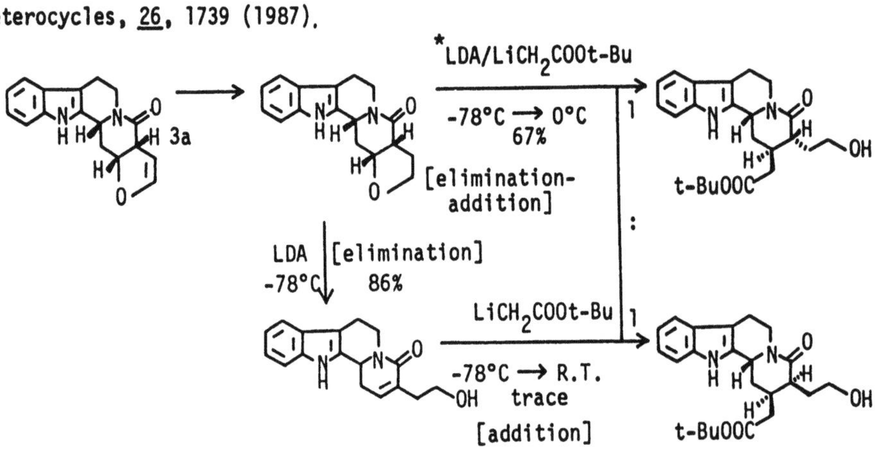

*Procedure
1) Addition of AcOt-Bu(6eq.) to LDA(11eq.) in THF at -78°C.
2) Stirred at -78°C for 30 min.
3) Addition of Furopyridone(1eq.).
4) Stirred at -78°C×10 min., -30°C×1-1.5h., 0°C×20 min..
5) Quenched by H_2O at 0°C.

follows. First tert-butyl acetate (6 equivalent amounts) was added to LDA (11 eq.) in tetrahydrofuran solution at -78°C and the solution was stirred well for 30 mins. Then to the above lithiated solution was added the furopyridone (1eq.) at -78°C and after stirring for a while, the temperature was gradually brought up to 0°C and finally the reaction was stopped by adding water at 0°C. According to this elimination addition procedure, we could have succeeded in improving the yield up to 67%. When this reaction was carried out stepwise, as first ring cleavage with LDA and addition of lithioacetate, only the trace amount of the product was obtained.

Now we could have prepared the quinolizidine with two side chains, which was then used for total synthesis of alkaloids.
One of the structural features of corynantheine type of alkaloids is to have a E- or Z-ethylidene group at the 20-position beside the vinyl group, as shown by the list of alkaloids as our targets.

Corynantheine type Alkaloids with Ethylidene Group

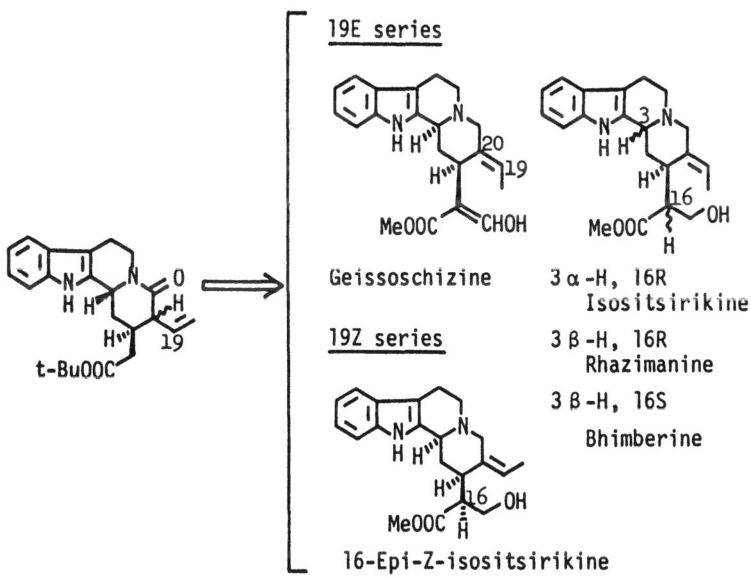

Therefore, the 2-hydroxyethyl group was converted into the vinyl group by selenation with o-nitrophenylcyanoselenide followed by oxidative cleavage of selenyl group with hydrogen peroxide in good yield. There are indole alkaloids with a vinyl group at the 20-

Conversion of Hydroxyethyl Group to Vinyl Group

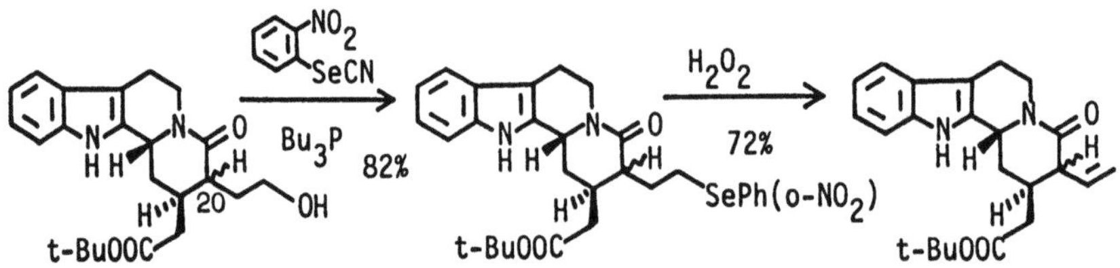

20β-H	20β-H	20β-H
20α-H	20α-H	20α-H
		; δ :5.75(1H,ddd,J=17.5, 10,7) 5.28-5.06(2H,m)

position as listed. As our first target we picked hirsuteine which had not been synthesized previously. The lactam with a vinyl group was reduced to the corresponding quinolizidine which was then formylated by the treatment with LDA followed by ethyl formate at -40°C. The tert. butyl ester was converted into the methyl ester which was then methylated, thus furnished the first total synthesis of hirsuteine which was identical with natural alkaloid. Further, the known conversion of this alkaloid to hirsutine also completed a formal total synthesis of the latter. Then we have gone further for the synthesis of other members of alkaloids as shown by the list.

First Total Synthesis of (±)-Hirsuteine

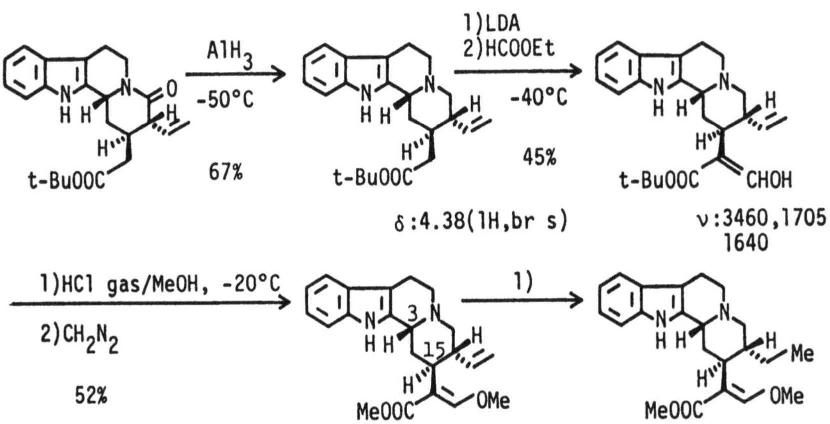

(±)-Hirsuteine[1),2)] (±)-Hirsutine

mp. 141-142°C

1) S. Sakai, et al., Yakugaku Zasshi, 93, 448 (1973)
2) E. J. Shellard, et al., Planta Medica, 21, 382 (1972)

There have been known a group of related alkaloids with 20-ethylidene group which could be obtained by the isomerization of a vinyl group. Geissoschizine is the central figure among them and most widely investigated. We picked this alkaloid as the next target, which had been synthesized by a couple of groups, van Tamelen (1974), Winterfeldt(1976 and 1979), Wenkert (1980), and Harley-Mason (1981). Another reason why we picked geissoschizine was that if we were successful, then our synthetic route could be also extended to closely related alkaloids such as isositsirikine, 16-epi-Z-isositsirikine, sitsirikine, rhazimanine, and bhimberine. Two of the last alkaloids mentioned are the alkaloids isolated and investigated by Pakistani researchers.

First the conversion of a vinyl group into ethylidene group was investigated. When the compound with a vinyl group was treated with sodium hydride in tetrahydrofuran, a facile isomerization was observed to give a single compound with an ethylidene group in

Stereoselective Isomerization of Vinyl Group to Ethylidene Group

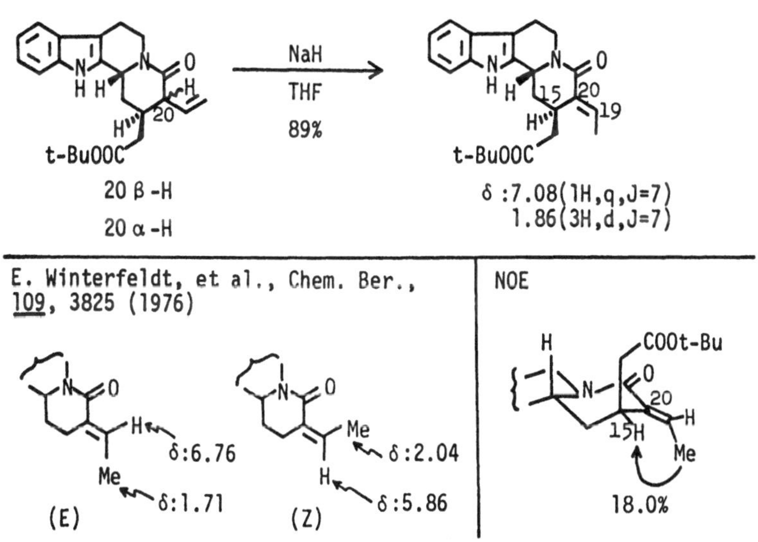

good yield. Its E-configuration was established from the NOE measurement as shown. The presence of a lactam carbonyl would be the major factor of restricting the migration of a double bond into one direction, that is, to the E-isomer. The strutural assignment of this stereochemistry, E- or Z-, could now be readily done by the comparison of the chemical shifts of hydrogen and methyl groups at the 19-position and NOE measurement between hydrogen at the 15-position and methyl group at the 19-position.

From this E-ethylidene compound, the alkaloid geissoschizine was synthesized in a similar way as in the synthesis of hirsuteine, beside the conversion of the configuration at the 3-position which was done first by dehydrogenation with cupric acetate and oxygen and the subsequent reduction with sodium borohydride. The products were a 1:1 mixture of 3α- and 3β-derivatives, which was then formylated as before to furnish total synthesis of geissoschizine.

Formal Total Synthesis of (±)-Geissoschizine

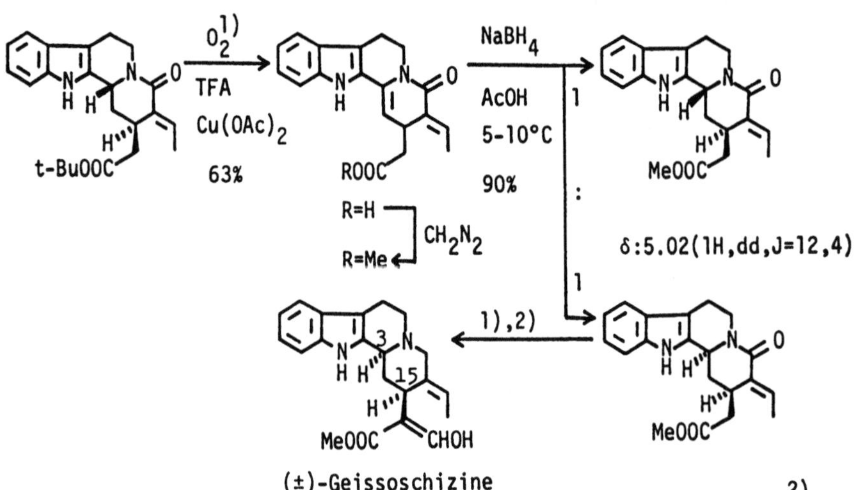

mp. 212-213°C(lit.[2] 213°C)
δ:4.85(1H,br t,J=5)

1) E. Winterfeldt, et al., Liebigs Ann. Chem., 1985, 1752
2) E. Winterfeldt, et al., Chem. Ber., 109, 3825 (1976)

Finally, we picked isositsirikine, rhazimanine and bhimberine as our next and final targets at the moment. Isositsirikine and 16-epi-E-isositsirikine are epimeric pair with respect to the stereochemistry at the 16-position among alkaloids with 3, 15-anti configuration, while rhazimanine and bhimberine are said to be the similar epimeric pair among the alkaloids with a 3,15-syn configuration though their structures are not fully established.

Rhazimanine and bhimberine are relatively new alkaloids isolated from Catharantus roseus and their structures are proposed by Professor Atta-ur-Rahman mainly from their analysis of 1Hn.m.r. spectra. Not only because of the similarity of their structures with geissoschizine and hirsuteine, but also from the reason that they are isolated by Pakistani researcher, we decided, first of all, to carry on their total synthesis.

The 20-E-ethylidene compound was reduced with aluminum hydride to give the quinolizidine with 20-E-ethylidene group and tert.butyl ester at the 15-position. After transesterification from the tert buty ester to the methyl ester, formylation of side chain adjacent to ester group was performed in 60% yield. Reduction of the formyl group was performed by the treatment with sodium borohydride to give a mixture of two epimeric compounds which was separated. The structures of separated isomers as of the proposed one by Professor Atta-ur-Rhaman were firmly assigned from 1Hn.m.r. spectra and further unambiguously confirmed by its X-ray analysis. However, the comparison of their 1Hn.m.r. spectra showed marked difference, therby suggesting the necessity of reinvestigation on the structures of these two alkaloids.

On the Synthesis of Bhimberine and Rhazimanine*

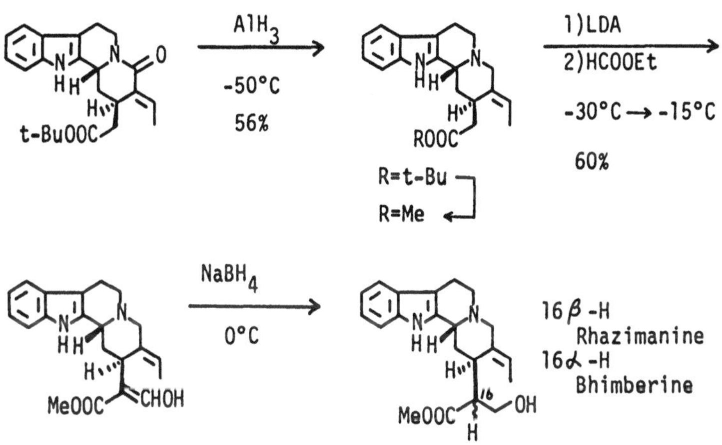

* Atta-ur-Rahman, et al, Heterocycles, 24, 703 (1986)
Phytochemistry, 25, 1731 (1986)

In conclusion of this part of synthesis, usefulness of enamide prepared from harmalane with 3-furoyl chloride was firmly demonstated by its application to the synthesis of relatively untouched alkaloids of corynantheine group of alkaloids. Actually, we have succeeded in the synthesis of geissoschizine and hirsuteine, thereby extended to the formal synthesis of corynantheine and hirsutine. Further, unexploited alkaloids rhazimanine and isositsirikine were also synthesized, thereby shed light on the chemistry of these groups of alkaloids. Full clarification of the problems on their structures will be achieved.

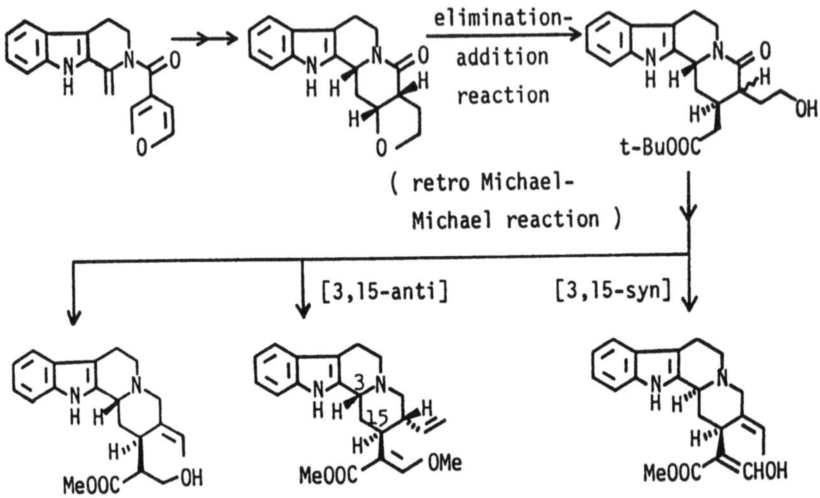

(±)-Hirsuteine

(±)-Geissoschizine

In concluding my talk today, I summarized what we have applied the enamide photocyclization to the synthesis of indole alkaloids as in the following scheme, which clearly demonstrates wide applicability of this methodology to the synthesis of indole alkaloids. All enamides are prepared from harmalane by acylation with either benzoyl, 2-furoyl or 3-furoyl chloride. Their photocyclizations proceed smoothly to give homogeneous lactams in excellent yields respectively. Subsequent modification of the ring E, which are either substituted dihydrobenzene or dihydrofuran ring, readily gives the substituents required for the construction of alkaloid themselves. One of the feature of this synthetic methodology is on ready construction of the ring skeleton of indole alkaloids, that is, indolobenzoquinolizidine or indolofuranoquinolizidine. Subsequent introductions of substituent for final touch of total synthesis are also achieved very smoothly thus providing one of the most versatile synthetic route to various indole alkaloids.

[Summary of Our Indole Alkaloid Synthesis]

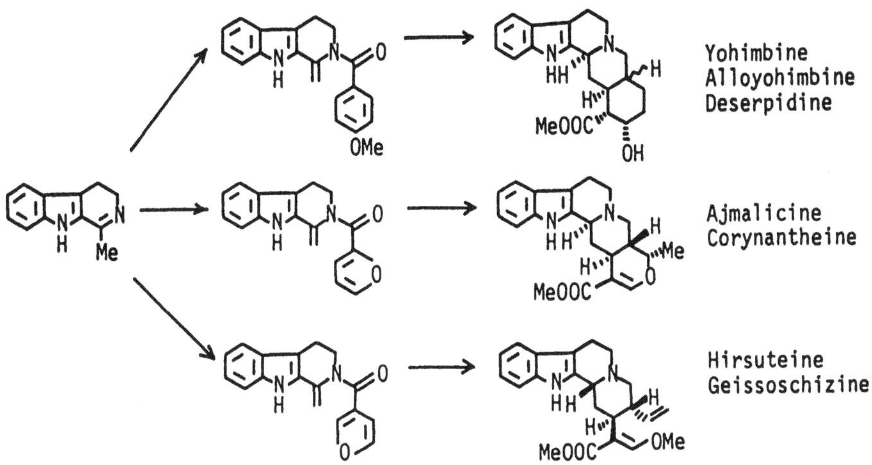

Finally by taking advantage of this chance I would like to express my sincere appreciation to my coworkers, especially to Dr. Okiko Miyata who has done almost all these works by her devotion to the works. And my gratitude should be extended to Professor Atta-ur-Rahman for his warm hospitality shown during my stay in this city and symposium, wishing his continuing success not only in his research but also in further development of his research institute. Thank you very much for your kind attention.

Synthetic Approaches to Carbazole Alkaloids

Jan Bergman and Benjamin Pelcman
Royal Institute of Technology
Department of Organic Chemistry
S-100 44 Stockholm, SWEDEN

Abstract: The occurrence, structure, physiological activities as well as the synthesis of alkaloids having a carbazole skeleton is briefly reviewed. Emphasis is made on the synthesis, in particular on approaches towards girinimbine and the carbazomycines. Some new synthetic approaches based on indoles and 2-vinylindoles, and the use of carbazoles in the synthesis of the cyclopent[b]indole alkaloid yuehchukene will also be discussed.

INTRODUCTION.

A large number of alkaloids containing an aromatic carbazole framework have been isolated from plants belonging to the *Rutaceae* family.[1-27] These alkaloids can be very simple, for example 3-methylcarbazole (**1**), but they frequently have terpenoid units of various length and complexity attached. A selection of *Rutaceae* carbazole alkaloids is given in Figure 1.

Taxonomically the *Rutaceae* carbazole alkaloids are of interest as they have been isolated exclusively from four closely related genera (*Glycosmis, Murraya, Clausena* and *Micromelum*).[6,9] In this respect the recent isolation[28] of ekeberginine (**2**) from *Ekebergia senegalensis* belonging to the *Meliaceae* family, with a structure similar to the *Rutaceae* carbazole alkaloids, merits attention.

Figure 1.

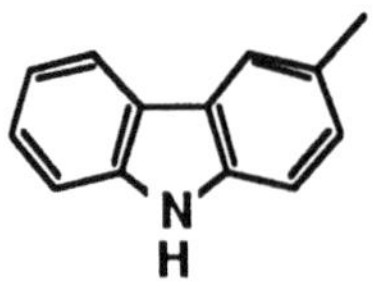

1

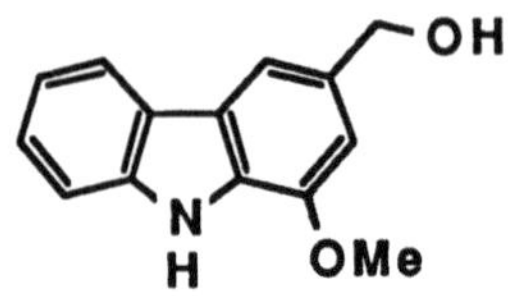

koenoline

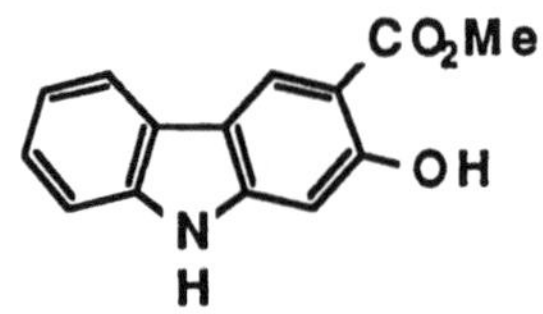

mukonidine

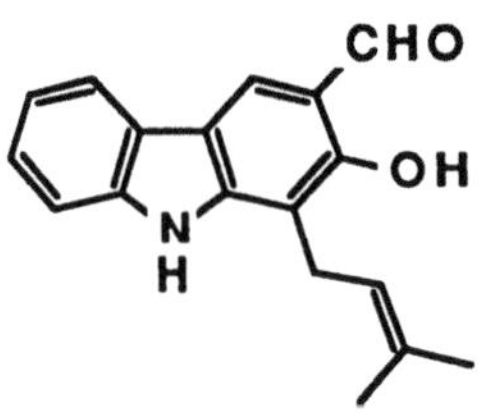

heptaphylline

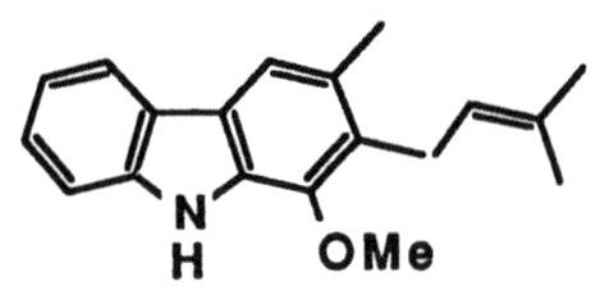

clausenapine

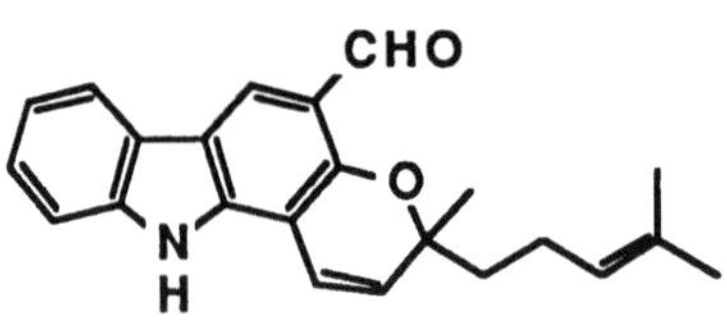

murrayacinine

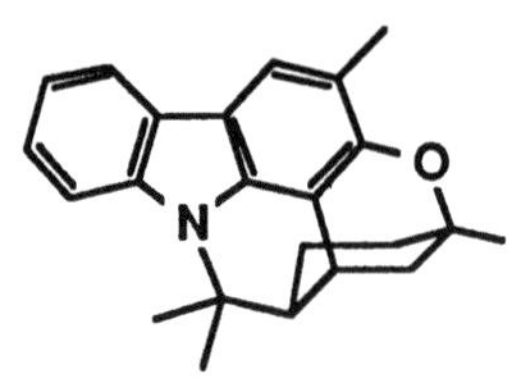

murrayazoline

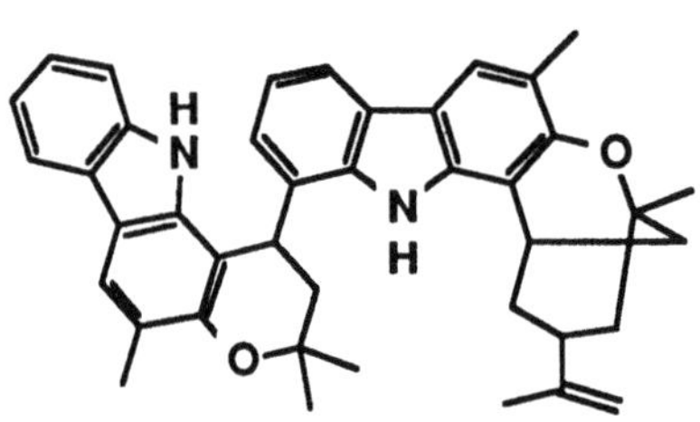

murrafoline

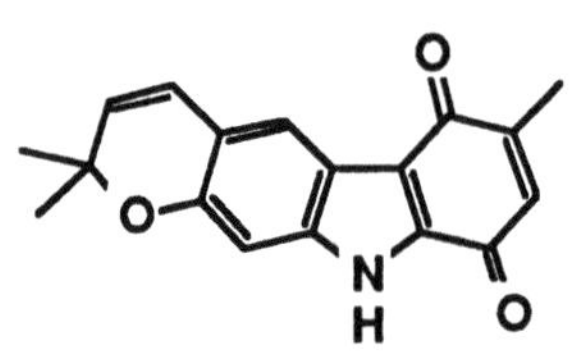

pyrayaquinone-A

Plants from the *Rutaceae* have been used in folk medicine in the treatment of eczema, rheumatism, stomachache, dysentery, herpes *etc.*[5,29] Presently, none of these effects have been correlated to any carbazole alkaloid. Some *Rutaceae* carbazole alkaloids have however been screened for antibiotic activity,[5] and cytotoxic activity has been found for koenoline (Figure 1).[1,20]

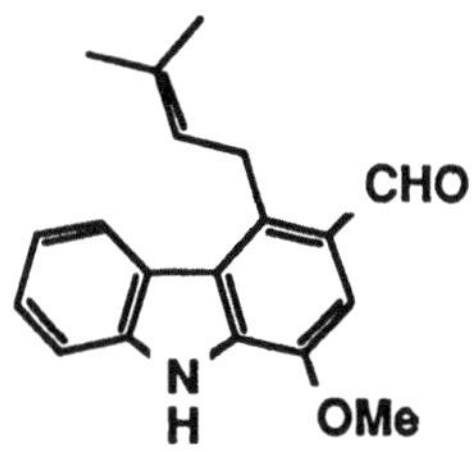

2

The hyellazoles (**3**) isolated from the blue green alga *Hyella caespitosa*[1,30], and the carbazomycins (**4**) produced by the actinomycete *Streptoverticillium ehimense*[1,4,31,32], show a substituent pattern different from the *Rutaceae* carbazole alkaloids. They all share the common features of carbon substituents in the 1- and 2-positions, and an oxygen functionality in the 3-position. Antibiotic activity is reported for carbazomycin B (**4b**).[32]

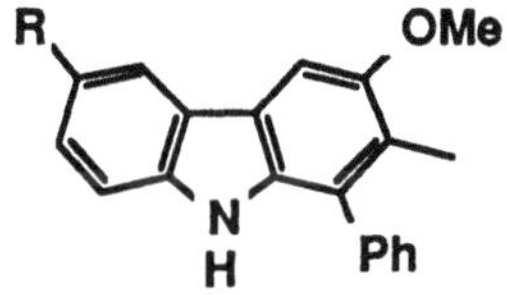

3a R = H hyellazole
b R = Cl chlorohyellazole

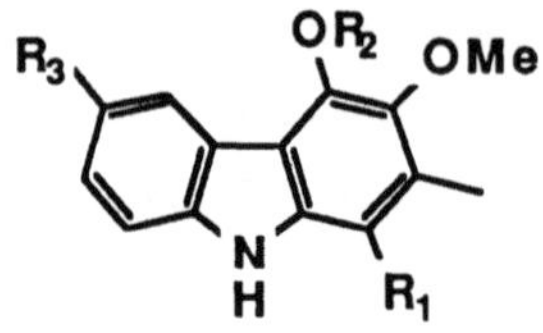

	R_1	R_2	R_3		
4a	Me	H	H	carbazomycin	A
b	Me	Me	H		B
c	Me	H	OMe		C
d	Me	Me	OMe		D
e	CHO	H	H		E (carbazomycinal)
f	CHO	H	OMe		F (6-methoxycarbazomycinal)

Alkaloids containing a pyrido[4,3-*b*]carbazole skeleton have been isolated from several genera of the *Apocynaceae* family (Figure 2). The interaction with DNA and the antitumoral activity of ellipticine (**5**) and olivacine have attracted enormous attention.[35-38] The chemistry of the pyrido[4,3-*b*]carbazoles has recently been reviewed[35-38] and will not be discussed in this treatise.

Figure 2.

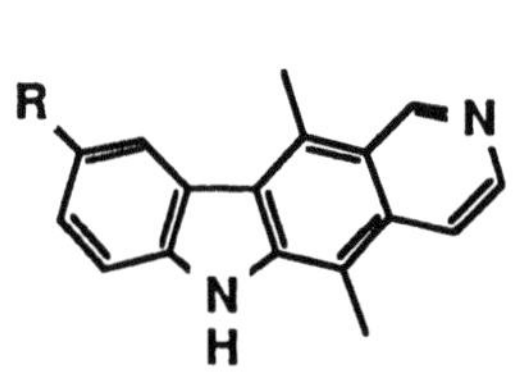

5 R = H ellipticine
R = OMe 9-methoxyellipticine

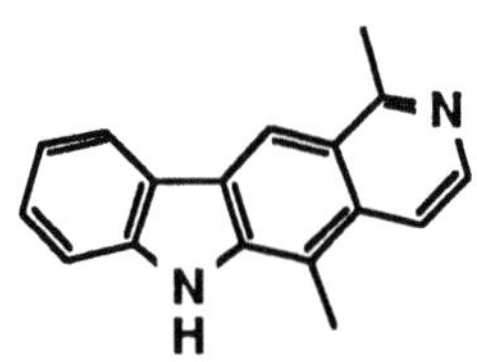

olivacine

The kinamycins (**6**), antibiotics produced by *Streptomyces murayamaensis*, are composed of a partially reduced carbazole nucleus.[39] A purple compound isolated in small quantities from the same species is believed to have the structure **7** on basis of spectroscopic data. Compounds similar to **7** are proposed as biosynthetic precursors of the kinamycins.[39,40]

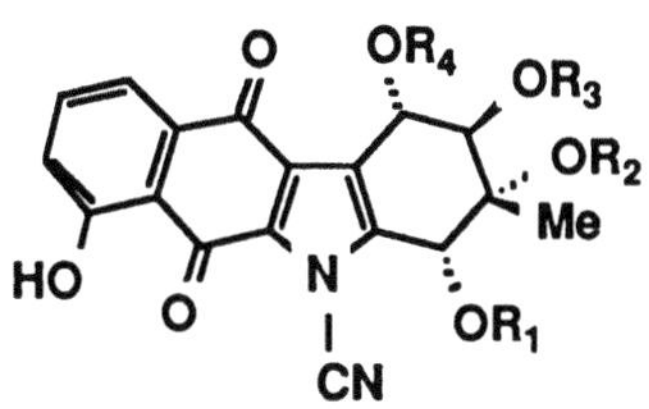

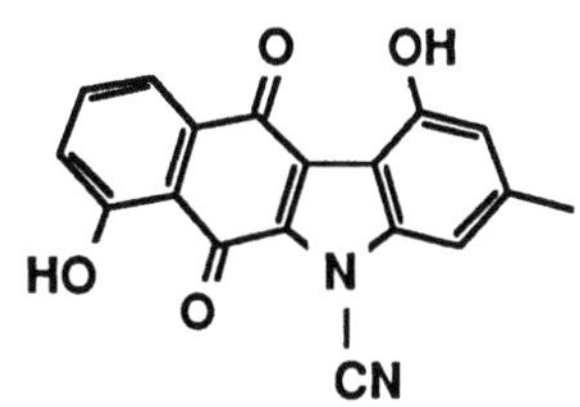

7

		R_1	R_2	R_3	R_4
6 kinamycin	A	Ac	Ac	Ac	H
	B	H	Ac	H	H
	C	Ac	H	Ac	Ac
	D	Ac	H	Ac	H

Recently still another class of compounds of fused carbazole alkaloids has been recognized. Two particulary interesting members of this class are staurosporine (**8a**) and its aglycon (**8b**), both potent inhibitors of protein kinase C. The chemistry of the indolocarbazole alkaloids has recently been thouroughly reviewed[41-42] and will therefore not be discussed in this treatise.

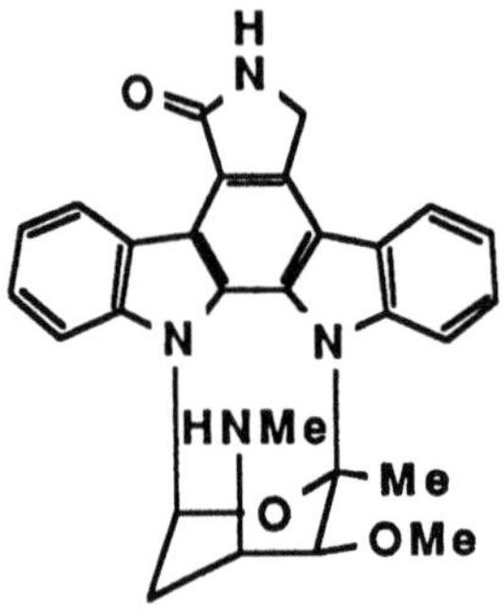

8a

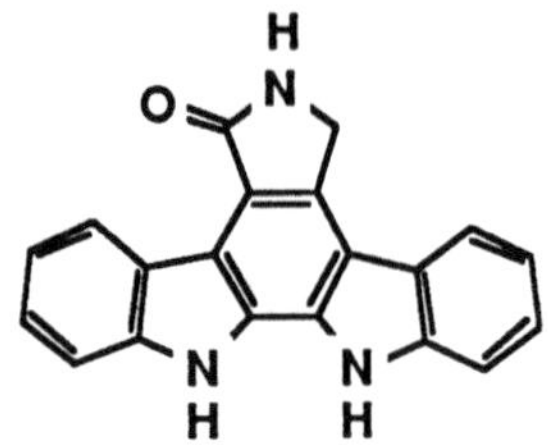

8b

SYNTHESIS OF CARBAZOLE ALKALOIDS.

The majority of the numerous synthetic approaches to carbazoles can be divided into four pathways, as depicted in Scheme 1.[43-48] All approaches have been employed in the synthesis of carbazole alkaloids.

The most extensively used approach is route A (Scheme 1). Some of the more commonly employed indole syntheses, such as the Fischer, Bischler, Nenitzescu and the Gassman procedures, also fall into this category.[49] Of these, the Fischer indole synthesis[50] (Scheme 2) has been most widely applied. However, the requirement of properly substituted starting materials, which can often be difficult to fulfil, and/or the lack of regio-selectivity in the cyclizative step, limit its utility. An example of the use of the Fischer indole synthesis in the synthesis of a carbazole alkaloid is given in Scheme 3.

In Scheme 4 an example of a synthesis of mahanimbine (**10**)[51] *via* the B route (Scheme 1) is given. Cyclization of an appropriately substituted *o*-nitrobiphenyl gave 2-methoxy-3-methylcarbazole (**9**), which on demethylation and subsequent con-

Scheme 1.

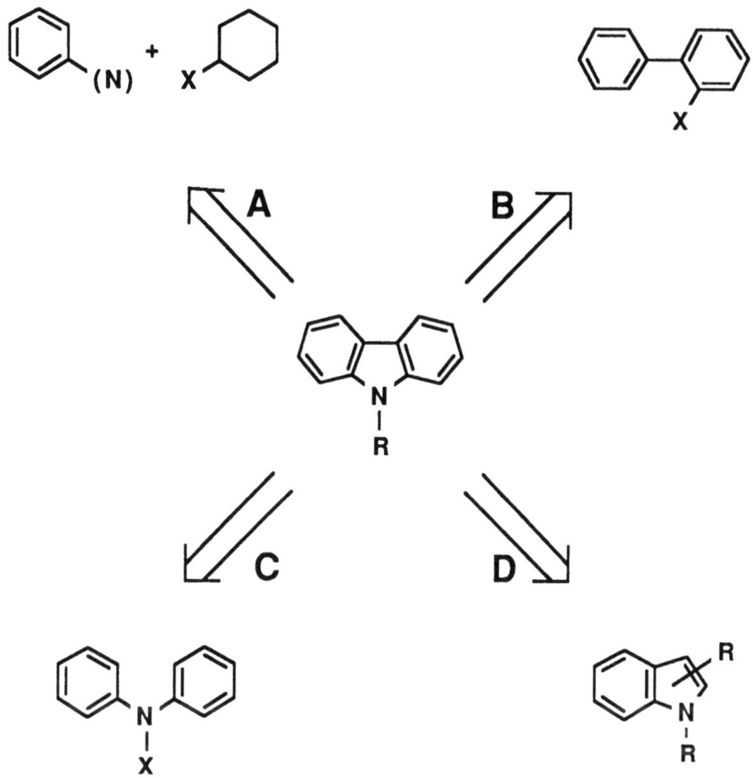

Scheme 2.

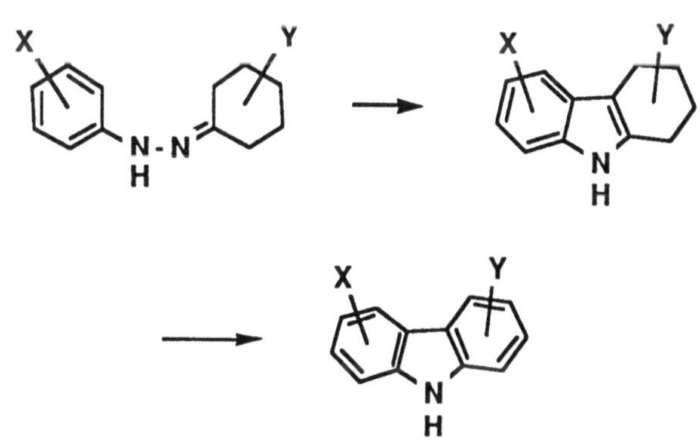

Scheme 3.

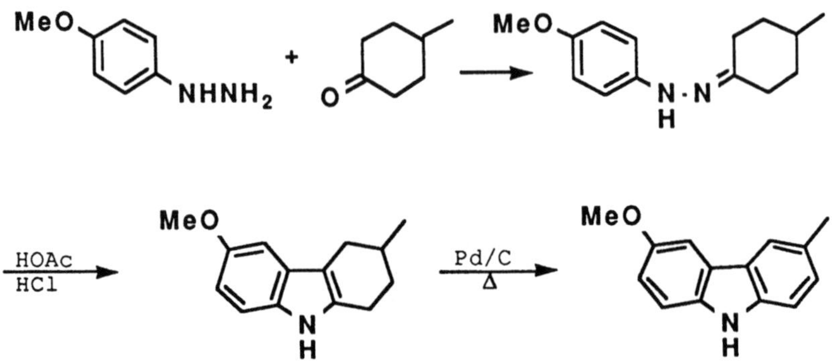

glycozoline

Scheme 4.

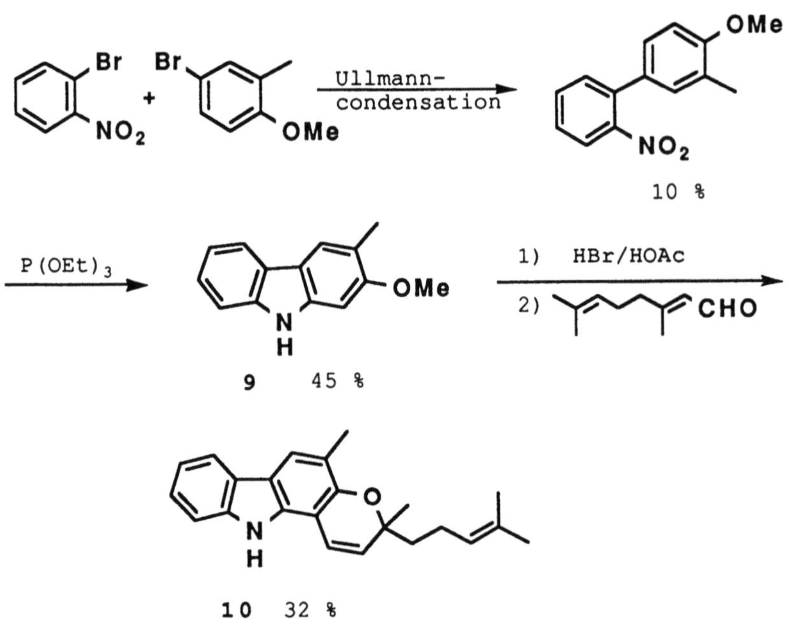

densation with citral afforded mahanimbine (**10**). In passing, it should be noted that both **9** and 2-hydroxy-3-methylcarbazole recently have been isolated[12,17] from the same species containing mahanimbine.

The main advantage using a preformed indole moiety[49] in the synthesis of carbazoles (Route D, Scheme 1) is that the problem encountered with regioselectivity in the cyclization steps of route A, B and C (Scheme 1) is overcome.

A simple synthesis of girinimbine (**11**) has been devised by Bergman[52-54] (Scheme 5). 2-Methylindole readily condensed with ethyl α-formylpropionate (commercially available) to give the 2:1 adduct **13**, which on thermolysis was split into **14** and 2-methylindole. The principle of thermolytic cleavage of bisindoles seems to be rather general for the *in situ* formation of indolenines and 3-vinylindoles, provided that the vinyl group is conjugated with an aromatic ring or a carbonyl group.

The ester **14** could be readily cyclized to 2-hydroxy-3-methylcarbazole (**15**), which was finally annelated to girinimbine(**11**)[55] by reaction with 3-chloro-3-methyl-1-butyne in the presence of $AlCl_3$ at 0 °C (Scheme 5). Compound **15** has also been prepared by Crombie[57] by a regiospecific (?) condensation of 2-hydroxycarbazole with $Zn(CN)_2$-HCl followed by a Wolff-Kishner reduction step. Mahanimbine (**10**) is readily available by condensation of **15** with citral (Scheme 4). Girinimbine (**11**) is also an interesting precursor to other carbazole alkaloids. Thus oxidation with DDQ in methanol gave murrayacine (**12**) and oxidation with *m*-chloroperbenzoic acid gave a mixture of *cis*- and *trans*- diols which could be separated. The *trans*-diol, 1,2-dihydroxygirinimbine, has recently been identified[16] as a natural product from *Murraya euchrestifolia*.

Base-catalyzed condensations on 2-alkyl functions in indoles seems to be rather general and the ellipticine analogue **16** is readily available by this route,[52] as outlined in Scheme 6.

Scheme 5.

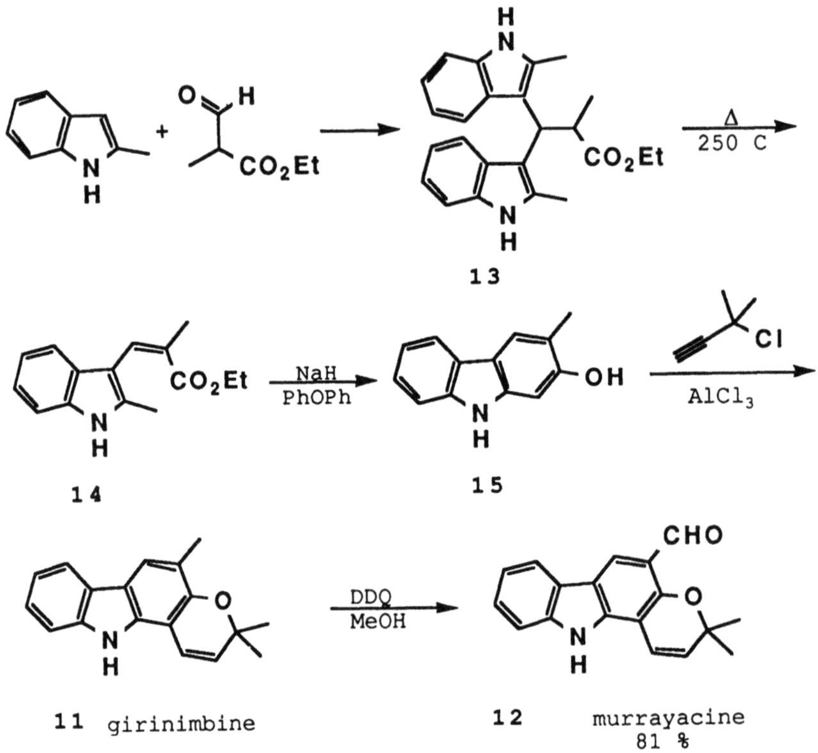

Scheme 6.

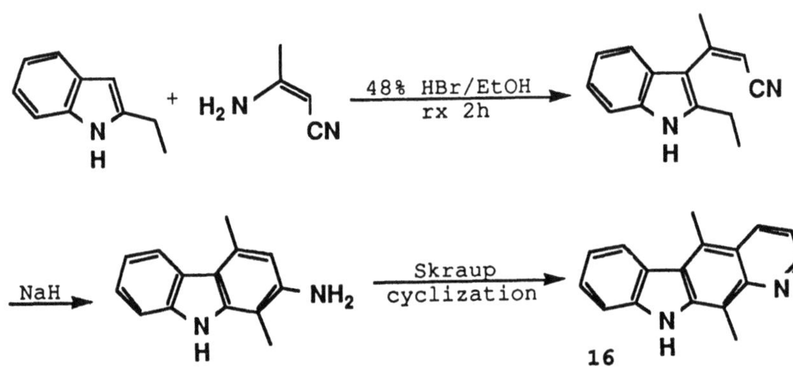

Recently, syntheses of several carbazolequinone alkaloids, isolated from *Murraya euchrestifolia*, have been reported.[58-61]

Hyellazole (**3a**) has been synthesized according to Scheme 7 [62-63]; regioselective lithiation of 1-benzenesulphonylindole in the 2-position, followed by addition of propiophenone and alkaline hydrolysis, yielded the 2-vinylindole **17**. A Vilsmeier reaction, followed by a Wittig reaction with methoxymethylenetriphenylphosphorane, gave the 2,3-divinylic indole **18**. Finally, heating in decaline in the presence of Pd-C cyclized **18** to produce hyellazole (**3a**). Chlorohyellazole (**3b**) was similarly obtained.[63]

Scheme 7.

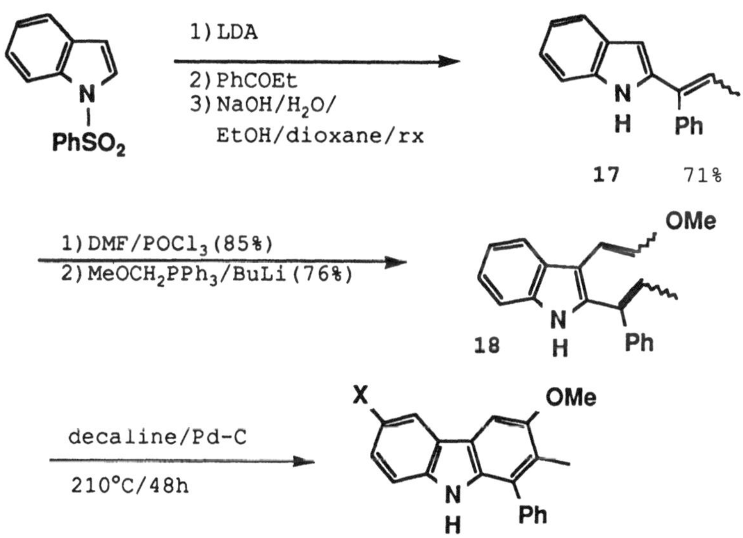

(41% from **17**)

3a X=H Hyellazole
b X=Cl Chlorohyellazole

According to Pindur[64,65] the Diels-Alder reaction of the 2-vinylindole **18** with dimethyl acetylenedicarboxylate (DMAD) gives a mixture of the cycloadducts **19** and **20** (Scheme 8). Transformation of the carbomethoxy groups of **19** to methyl groups and N-deprotection produces the carbazomycin analogue **21**.

Scheme 8.

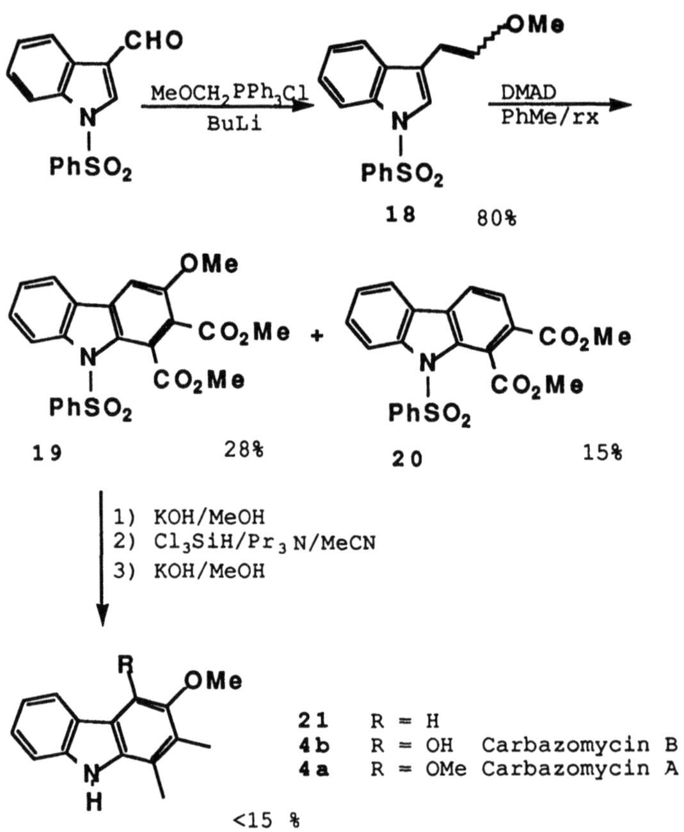

In an other protocol[66] towards alkaloids of the carbazomycin-type,the 2-vinylindoles **22a** and **22b** were, in a model study, treated with $POCl_3$/DMF followed by a period of reflux, which quickly and easily gave the carbazoles **25a** and **25b** as outlined in Scheme 9.

Scheme 9.

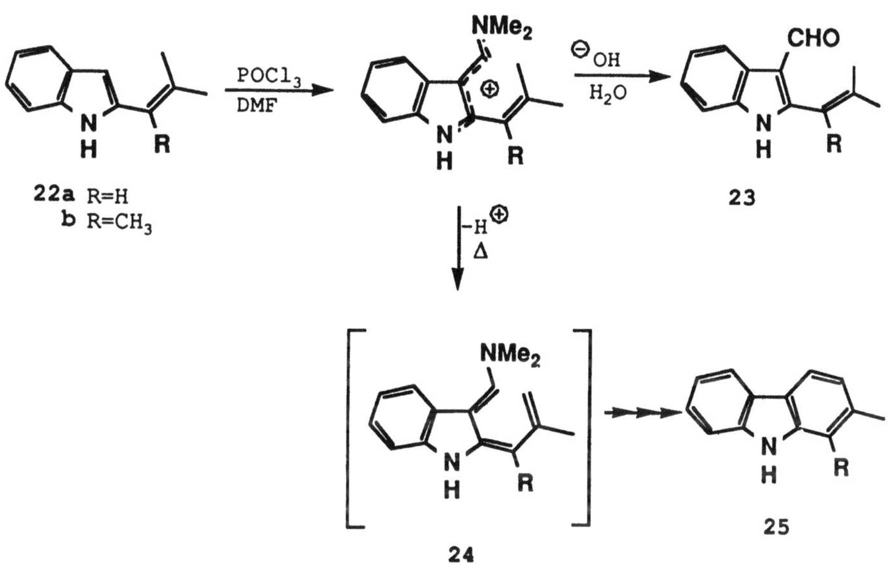

The vinylindoles **22a** and **22b** were quickly prepared in modest yields (~25%) from readily available starting materials as outlined in Scheme 10. It should be noted that α,β-unsaturated ketones previously[67-69] have been reported to be unsuccessful partners in attempted Fischer indolizations. The success in the present case probably critically depends on the stability of the cation formed by ring-opening of the intermediate 2-pyrazoline. The fact[70] that 2-styrylindole is formed when 1,5-diphenyl-3-

methylpyrazoline is heated in PPA is of interest in this connection.

Scheme 10.

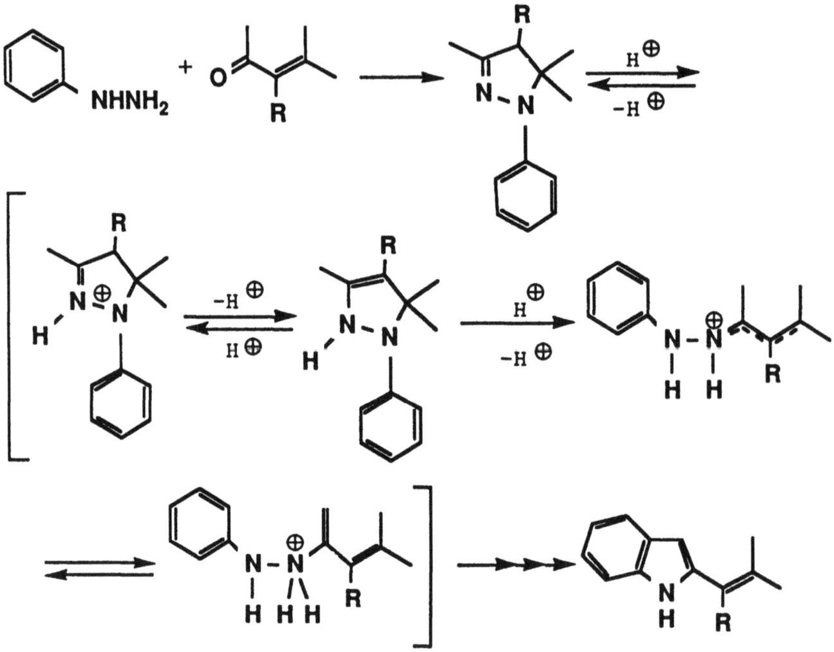

22a R=H 26 %
b R=Me 24 %

After having prepared carbazoles *via* the route in Scheme 9 the next exercise was to use a C_1-unit (like $COCl_2$ or $Me_2N{=}CCl_2^+\ Cl^-$) which would, hopefully, allow the introduction of a suitable substituent in the 4-position in the final carbazole. Unfortunately the reaction took a different course (Scheme 11).[71]

Parallel with these efforts the cyclizations outlined in Scheme 12 were also studied.[72] In the absence of Pd/C the yields dropped considerably.

Scheme 11.

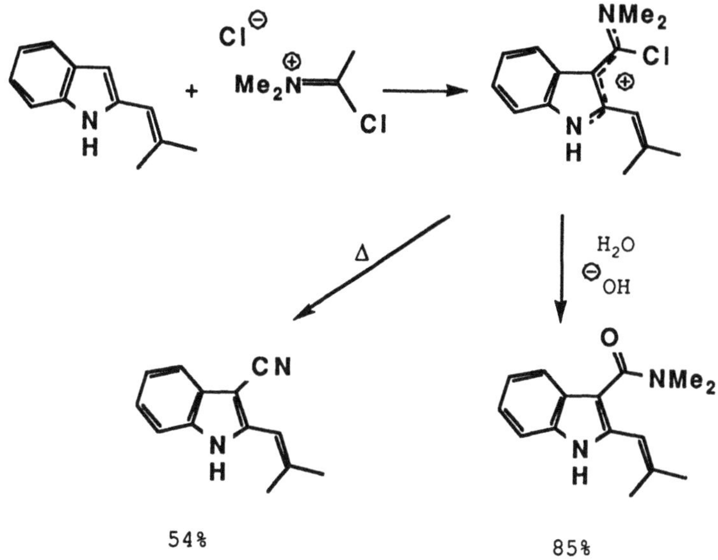

In another project[73] aimed at the synthesis of cyclopent[*b*]-indole alkaloids such as yuehchukene (**29**)[74-75], which exhibits anti-implantatory effects, it was found that carbazole derivatives like **27** could be readily prepared (Scheme 13). After dehydrogenation to **28**, a dienone rearrangement gave the desired carbazole **26b**.

The idea was now to apply the Barton-Ley reaction[76] (*i.e.* treatment with benzeneseleninic anhydride), which has been successfully used for *e.g.* conversion of 2-naphtol into 1,2-dihydroxynaphtalene (*via* the corresponding *o*-quinone). Unfortunately, this approach failed due to coupling reactions.

Scheme 12.

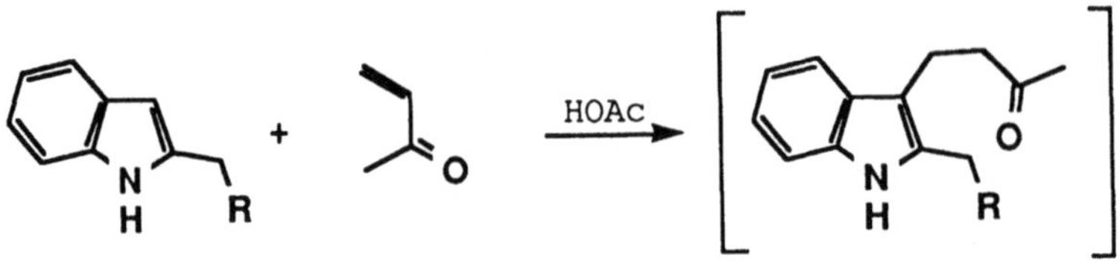

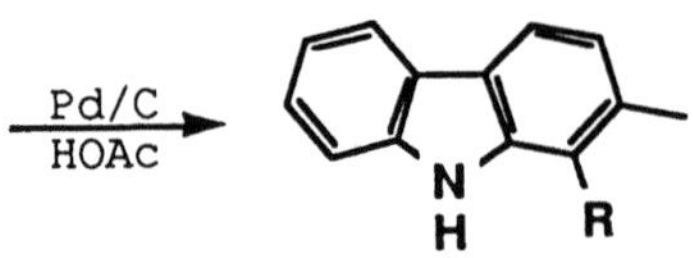

25b R=Me 78%
c R=Ph 80%

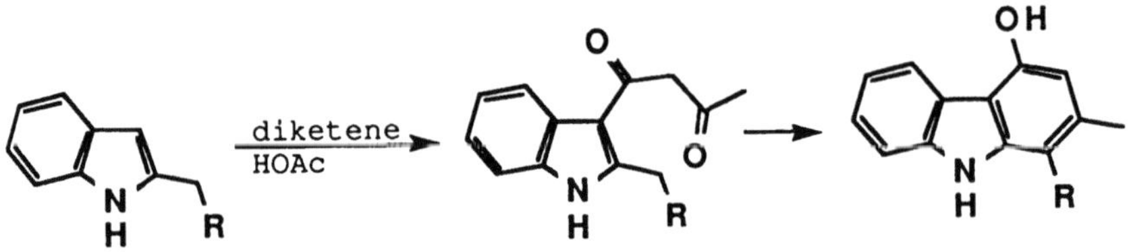

(low yield)

26a R=H
b R=CH_3

Scheme 13.

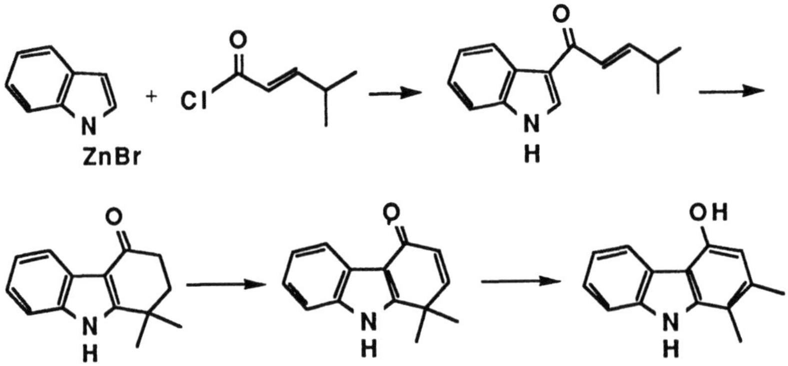

27 **28** **26b**

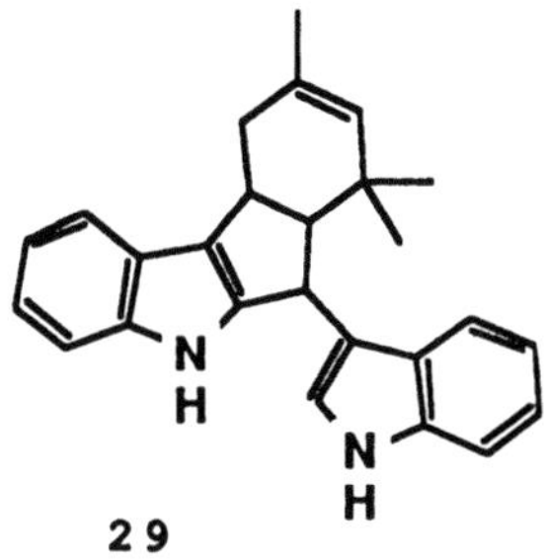

29

REFERENCES AND NOTES.

1. Several reviews on *Rutaceae* carbazole alkaloids have been published.[2-8] For more recent isolations of carbazoles from the *Rutaceae* see refs. 9-27.
2. P. Bhattacharyya, D.P. Chakraborty, *Fortschritte der Chemie organisher Naturstoffe,* **52**, 159 (1987).
3. H.-P. Husson, *The Alkaloids,* **26**, 1 (1985).
4. D.P. Chakraborty, *Trans. Bose Res. Inst.,* **47**, 49(1984).
5. P.G. Waterman, M.F. Grundon, "Chemistry and Chemical Taxonomy of the Rutales", Academic Press (1983).
6. D.P. Chakraborty, *Planta Medica,* **39**, 97 (1980).
7. D.P. Chakraborty, *Fortschritte der Chemie organisher Naturstoffe,* **34**, 299 (1977).
8. P.G. Waterman, *Biochem. Syst. Ecol.,* **3**, 149 (1975).
9. R.S. Kapil, *The Alkaloids,* **13**, 273 (1971).
10. C. Ito, T.-S. Wu, H. Furukawa, *Chem. Pharm. Bull.,* **35**, 450 (1987).
11. H. Furukawa, C. Ito, M. Yogo, T.-S. Wu, *Chem. Pharm. Bull.,* **34**, 2672 (1986).
12. P. Bhattacharyya, S.S. Josh, B.K. Chowdhury, *Chem. Ind. (London)* 246 (1986).
13. H. Furukawa, M. Yogo, C. Ito, T.-S. Wu, C.-S. Kuoh, *Chem. Pharm. Bull.,* **33**, 1320 (1985).
14. H. Furukawa, T.-S. Wu, C.-S. Kuoh, *Chem. Pharm. Bull.,* **33**, 2611 (1985).

15. H. Furukawa, T.-S. Wu, T.Ohta, C.-S. Kuoh, *Chem. Pharm. Bull.*, **33**, 4312 (1985).
16. H. Furukawa, T.-S. Wu, C.-S. Kuoh, *Heterocycles*, **23**, 1391 (1985).
17. P. Bhattacharyya, B.K. Chowdhury, *Indian J. Chem.*, **24B**, 452 (1985).
18. P. Bhattacharyya, B.K. Chowdhury, *J. Nat. Prod.*, **48**, 465 (1985).
19. P. Bhattacharyya, P.K. Chakrabartty, B.K. Chowdhury, *Phytochem.*, **24**, 882 (1985).
20. M. Fiebig, J.M. Pezzuto, D.D. Soejarto, A.D. Kinghorn, *Phytochem.*, **24**, 3041 (1985).
21. P. Bhattacharyya, B.K. Chowdhury, *Chem. Ind. (London)*, 301 (1984).
22. P. Bhattacharyya, T. Sarkar, A. Chakraborty, B.K. Chowdhury, *Indian J. Chem.*, **23B**, 49 (1984).
23. P. Bhattacharyya, A. Chakraborty, *Phytochem.*, **23**, 471 (1984).
24. P. Bhattacharyya, A. Chakraborty, B.K. Chowdhury, *Phytochem.*, **23**, 2409 (1984).
25. H. Furukawa, T.-S. Wu, T. Ohta, *Chem. Pharm. Bull.*, **31**, 4202 (1983).
26. M. Mukherjee, S. Mukherjee, A.K. Shaw, S.N. Ganguly, *Phytochem.*, **22**, 2328 (1983).
27. A.T. McPhail, T.-S. Wu, T. Ohta, H. Furukawa, *Tetrahedron Letters*, **24**, 5377, (1983).
28. D. Lontsi, J.F.Ayafor, B.L.Sondengam, J.D. Connolly, D.S. Rycroft, *Tetrahedron Letters*, **24**, 4249 (1985).
29. L. Perry, "Medicinal Plants of East and Southeast Asia: Attributed Properties and Uses", MIT Press, Mass.(1980).
30. J.H.Cardelina II, M.P. Kirkup, R.E. Moore, J.S. Mynderse, K. Seff, C.J. Simmons, *Tetrahedron Letters*, 4915 (1979).
31. For reviews on the carbazomycins, see refs.,3,32 and 33.
32. T. Naid, T. Kitahara, M. Kaneda, S. Nakamura, *J. Antibiotics.*, **40**, 157, (1987),and refs cited therein.
33. S. Nakamura, *Trans. Bose Res. Inst.*, **47**, 69 (1984).
34. M. Kaneda, K. Yamasaki, K. Sakano, T. Kitahara, T. Naid, K. Watanabe, Y. Ueki, K. Ishimura, S. Nakamura, *Tennen Yuki Kagobutsu Toronkai Koen Yoshishu*, 57 (1986); [Chem. Abstr., **106**, 64062m (1987)].

35. M. Suffness, A. Cordell, "The Alkaloids", **25**, 92 (1985).
36. G.W. Gribble, M.G. Saulnier, *Heterocycles*, **23**, 1277 (1985).
37. V.K. Kansal, P. Potier, *Tetrahedron*, **42**, 2389 (1986).
38. S. Archer, B.S. Ross, L. Pica-Mattocia, D. Cioli, *J. Med. Chem.*, **30**, 1204 (1987).
39. P.J. Seaton, S.J. Gould, *J. Am. Chem. Soc.*, **109**, 5282 (1987), and refs cited therein.
40. Synthetic procedures towards kinamycin are under development in the laboratory of K. Parker.
41. J. Bergman in "Studies in Natural Products Chemistry", Ed. A.-ur-Rahman, Elsevier, Amsterdam, to be published.
42. M. Gill, W. Steglich, *Fortschritte der Chemie organisher Naturstoffe*, **51**, 109 (1987).
43. J.A. Joule, *Adv. Het. Chem.*, **35**, 83 (1984).
44. M.J.E. Hewlins, A.H. Jackson, A. Long, A. Oliveira-Campos, P.V.R. Shannon, *J. Chem. Res. (M)*, 2645 (1986).
45. D.W.M. Benzies, P.M. Fresneda, R.A. Jones, H. McNab, *J. Chem. Soc., Perkin Trans. I*, 1651 (1986).
46. G.R. Stephenson, *J. Organomet. Chem.*, **286**, C41 (1985).
47. E. Differding, L. Ghosez, *Tetrahedron Letters* **26**, 1647 (1985).
48. 46. C. Kashima, S. Hibi, T. Maruyama, K. Harada, Y. Omote, *J. Heterocyclic Chem.*, **24**, 913 (1987).
49. Several reviews on the synthesis and the chemistry of indoles have been published: D.J. Chadwick, R.A. Jones, R.J. Sundberg, in "Comprehensive Heterocyclic Chemistry" Ed. A.R. Katritzky, C.W. Rees, Vol. 4, Pergamon Press, (1984), and refs. cited therein.
50. B. Robinson, "The Fischer Indole Synthesis", Wiley-Interscience (1982).
51. S.P. Kureel, R.S. Kapil, S.P. Popli, *J. Chem. Soc., Chem. Commun.*, 1120 (1969).
52. J. Bergman, R. Carlsson, *Tetrahedron Letters*, 4051 (1978).
53. J. Bergman, R. Carlsson, *Tetrahedron Letters*, 4055 (1978).
54. J. Bergman, unpublished results.
55. For a more complex synthesis of girinimbine, see ref.56.
56. Y. Oikawa, O. Yonemitsu, *Heterocycles*, **5**, 233 (1976).
57. W.M. Bandaranayake, M.J. Begley, B.O. Brown, D.G. Clarke, L. Crombie, D.A. Whiting, *J. Chem. Soc., Perkin I*, 988 (1974).

58. T. Martin, C.J. Moody, *J. Chem. Soc., Perkin I,* 235 (1988).
59. T. Martin, C.J. Moody, *J. Chem. Soc., Perkin I,* 241 (1988).
60. T. Martin, C.J. Moody, *J. Chem. Soc., Chem. Commun.,* 1391 (1985).
61. T. Martin, C.J. Moody, *Tetrahedron Letters,* **26**, 5841 (1988).
62. S. Kano, E. Sugino, S. Hibino, *J. Chem. Soc., Chem. Commun.,* 1241 (1980).
63. S. Kano, E. Sugino, S. Shibuya, S. Hibino, *J. Org. Chem.,* **46**, 3856 (1981).
64. U. Pindur, L. Pfeuffer, *Heterocycles,* **26**, 325 (1987).
65. A review on 2-vinylindoles has been published: E. Akgün, U. Pindur, *Chimia,* **39**, 264 (1985).
66. J. Bergman, B. Pelcman, *Tetrahedron Letters,* **26**, 6389 (1985).
67. W. Borsche, A. Witte, W. Bothe, *Ann,* **359**, 49 (1908).
68. J. Fränkel, *Diss,* Aachen, pp. 18-20, (1965).
69. Ref.49, pp. 764-766, and refs. cited therein.
70. F. Sannicolò, *Tetrahedron Letters,* **25**, 3101 (1984).
71. J. Bergman, B. Pelcman, *Tetrahedron Letters,* **27**, 1939 (1986).
72. J. Bergman, B. Pelcman, to be published.
73. J. Bergman, L. Venemalm, *Tetrahedron Letters,* **28**, 3741 (1987).
74. Y.-C. Kong, K.-F. Cheng, R.C. Cambie, P.G. Waterman, *J. Chem. Soc., Chem. Commun.,* 47 (1985).
75. E. Wenkert, E.C. Angell, V.F. Ferreira, E.L. Michelotti, S.R. Piettre, J.-H. Shen, C.S. Swindell, *J. Org. Chem.,* **51**, 2349 (1986).
76. S. Ley, in "Organoselenium Chemistry" Ed. D. Liotta, J. Wiley, New York, (1987), pp. 163-206.

Isoquinoline Alkaloid Synthesis via Arynes

L. Castedo and E. Guitián
Department of Organic Chemistry University of Santiago
15706 Santiago de Compostela Spain

A number of aporphine alkaloids have been isolated in this laboratory in such small quantities that their structure could at first only be stablished on the basis of their spectroscopic characteristics (Ref. 1). The need to confirm the structures of these new alkaloids and to obtain them in quantities sufficient to allow investigation of their pharmacological properties led us to attempt their synthesis by classical methods, but the problems encountered in some cases forced us to try modified or radically different approaches. As a result, we have developed a new synthetic procedure that is applicable to a variety of different kinds of isoquinoline alkaloids.

The classical methods pf preparing aporphinoids (1) involve the formation of bond A of the phenanthrene or dihydrophenanthrene system by cyclization of a stilbene or dihydrostilbene intermediate (2) by oxidative coupling (ferrocyanide, Cu_2Cl_2 / Py / O_2, vanadyl oxitrifluoride, etc), Pschorr cyclization or photocyclization (Ref. 2) (Scheme 1). The number of photochemical methods has increased considerably in recent years with the development of variants including, in particular, the cyclization of a halogenated derivative followed by dehydrohalogenation (Ref. 3). This reaction has two important advantages: firstly, the halogen atom determines the point at which cyclization takes place, thus preventing the formation of the other posible regioisomer when the substrate has asymmetric substitution; and secondly, the final dehydrohalogenation makes the process irreversible.

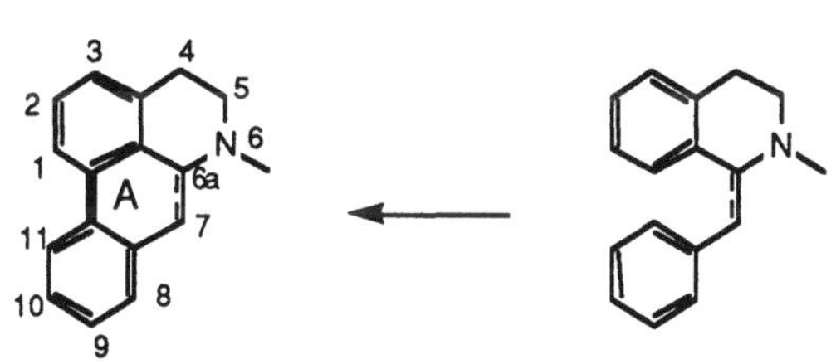

1 2

Scheme 1

The above procedure has succesfully been used for the synthesis of pontevedrine (**3a**), a dioxoaporphine isolated from Glaucium flavum var. vestitum (Ref. 1a) (Scheme 2). The benzylisoquinoline **4** acquires photochemically generated singlet oxygen to give (apparently via the endoperoxide **5**) the dioxocompound **6**; acid- or base-catalysed elimination of MeOH then affords **7**, which undergoes photocyclization followed by elimination of HBr to yield norpontevedrine (**3b**). Norpontevedrine can be transformed into the natural compound (**3a**) by means of methylation by methyl fluorsulphonate and NaH (Ref. 4).

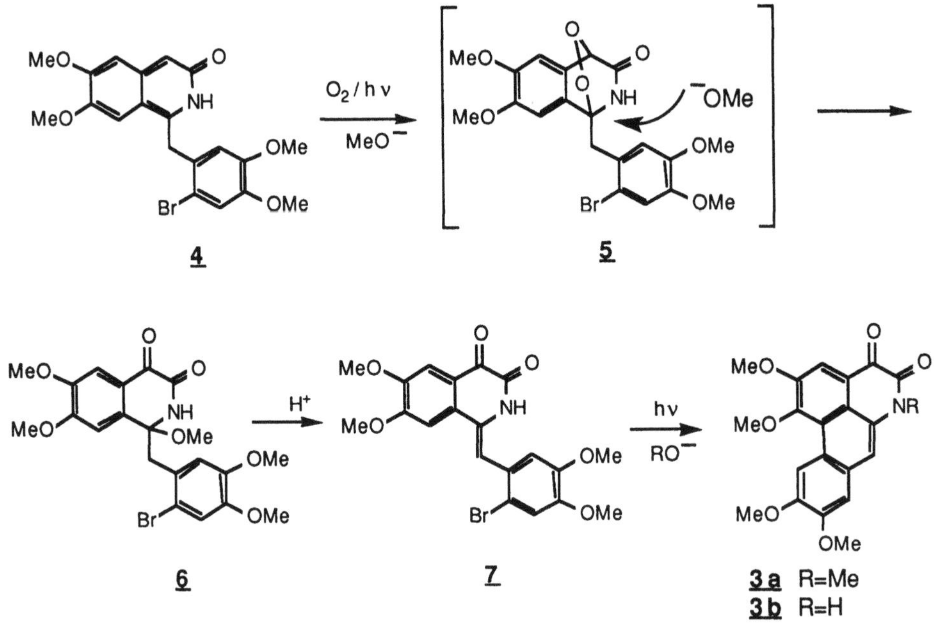

Scheme 2

One of the problems of this kind of reaction is low quantum efficiency, which generally results in long reaction times and mediocre chemical yields. Part of the cause of this problem is that most of the absorbed photons are often used up in E-Z isomerization of the stilbene double bond. In order to prevent this, we have fixed the stilbene configuration by forming a pyrrolinedione ring between the α carbon and the nitrogen of **8** by reaction with oxalyl chloride (Scheme 3). The product, compound **9**, undergoes rapid, efficient photochemical cyclization to afford a 60% yield of **10**, which can subsequently be transformed into the corresponding oxoaporphine (**11**) by oxidation with Fremy's salt (Ref. 5).

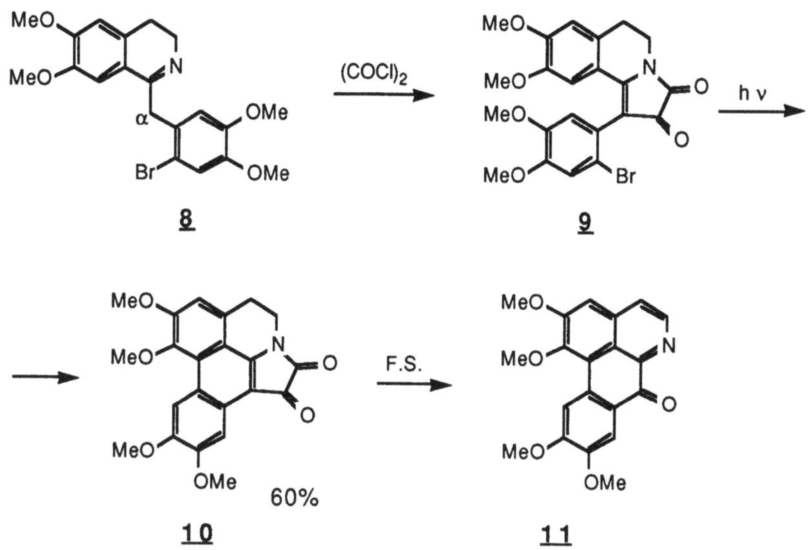

Scheme 3

The above procedure is fairly general, but gives unsatisfactory results when there are geometrical or sterical impediments to cyclization. The former problem arises, for example in the case of benzylidenephthalide (**12a**), which ought to cyclize to the aristolactam **13a** (Scheme 4). In spite of the latter's similarity to pontevedrine (**3a**), compound **12a** is not photocyclized under the same conditions, which we have attributed to there being a greater distance between the positions involved in cyclization. However, photocyclization of **12b** (probably via a radical mechanism) does afford a 25% yield of cepharanone B (**13b**) (Ref. 6).

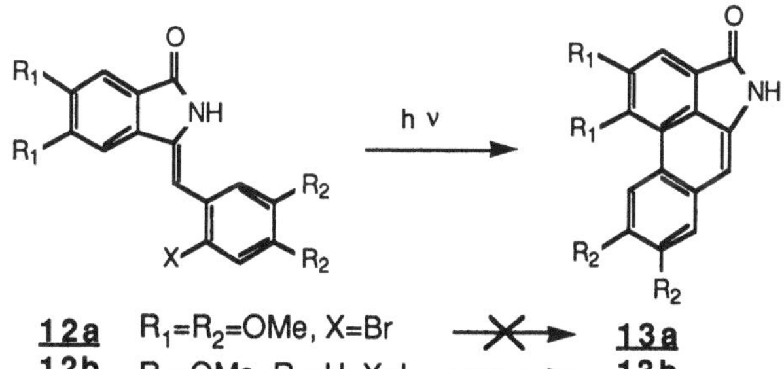

Scheme 4

Steric problems are presumably the cause of the difficulty of preparing 1,11-disubstituted aporphinoids by irradiation of the corresponding stilbenes. This hypothesis has led us to fix the geometry of the molecule in the optimal form for cyclization by introducing a bridge between positions 1 and 11 (aporphine numbering), which also avoids E-Z isomerization problems. The dialdehyde 14 constructed from two phenol units and α,α'-dichloro-o-xylene was converted into the stilbene 15 by reductive dicarbonylic coupling with low-valent titanium (Scheme 5); the phenanthrene carbamate 16 was then obtained in 85% yield by irradiation under oxidizing conditions (I_2, Cu(II)) and transformed by hydrolysis and cyclization (via the α-dicarbonyl derivative) into the oxoaporphine 17, from which corituberine (18) was obtained by reaction with MeI and hydrogenation/hydrogenolysis with Zn/HCl (Ref. 7).

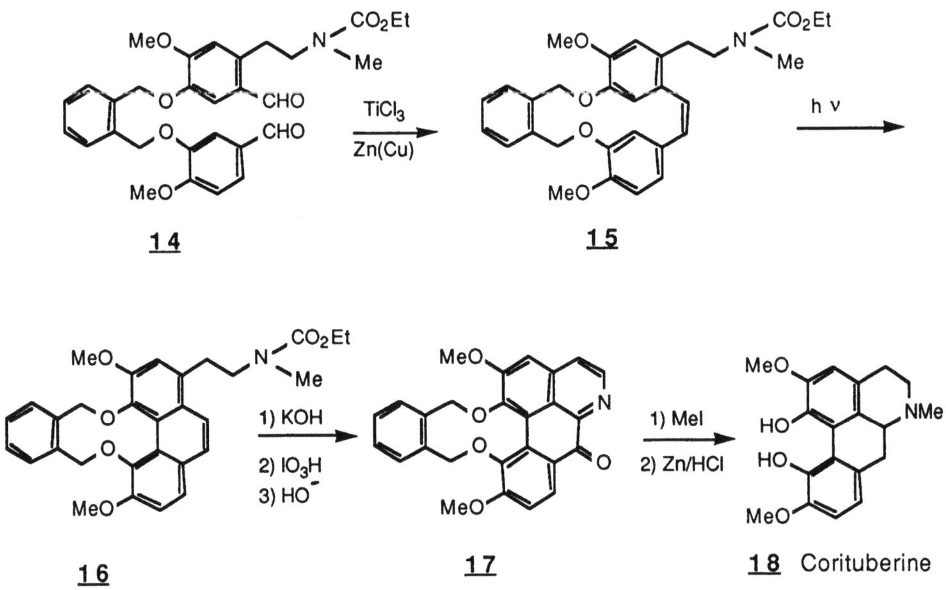

Scheme 5

At this point in our research we decided to investigate whether the need to devise *ad hoc* solutions for each separate aporphinoid might be done away with by developing a new method of aporphinoid synthesis that combined generality, convergence, simplicity and efficiency. Retrosynthetic analysis of the aporphinoids led us to examine the possibility of simultaneously constructing bonds A and B by means of a Diels-Alder reaction in the latter stages of the synthesis, i.e. by reaction as in Scheme 6 between a methyleneisoquinoline (**19**) and an aryne (**20**).

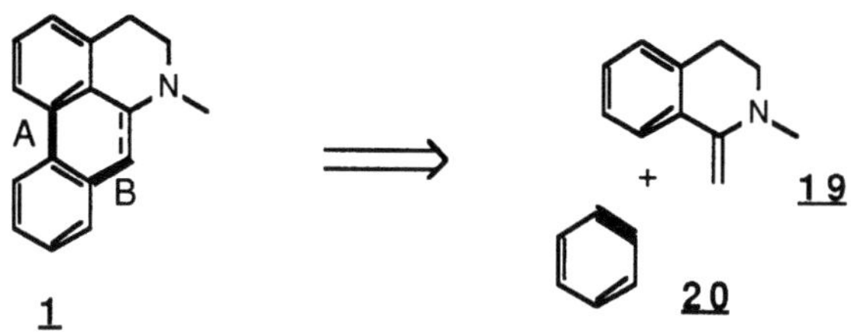

Scheme 6

Previous reactions of arynes with substituted styrenes had been discouraging, having produced mixtures of phenanthrenes and dihydrophenanthrenes, different regioisomers, and benzyne derivatives resulting from reaction with precursors, etc (Ref. 8). We nevertheless found that generation of the aryne by decomposition of the corresponding benzenediazonium 2-carboxylate in refluxing dimethoxyethane produced mainly phenanthrene compounds. In the case of benzyne, the formation of side products derived from precursors of the aryne was avoided by isolating the benzenediazonium 2-carboxylate as an internal salt that was then added to the refluxing reaction mixture (Method A) (Ref. 9). However, we have generated the substituted arynes *in situ* by aprotic diazotization (Method B) (Ref. 10).

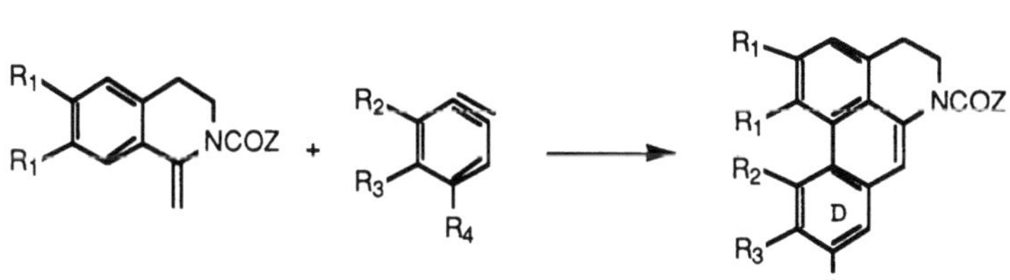

21a R_1=OMe
21b R_1+R_1=OCH_2O
21c R_1=H

22a R_2=R_3=R_4=H
22b R_2=H, R_3=R_4=OMe
22c R_2=OMe, R_3=R_4=H
22d R_2=R_3=OMe, R_4=H
22e R_2=R_4=H, R_3=OMe
22f R_2=H, R_3+R_4=OCH_2O

23a R_1=OMe, R_2=R_3=R_4=H
23b R_1+R_1=OCH_2O, R_2=R_3=R_4=H
23c R_1=R_3=R_4=OMe, R_2=H
23d R_1=R_2=OMe, R_3=R_4=H
23e R_1=R_2=R_3=OMe, R_4=H
23f R_1=R_3=OMe, R_2=R_4=H
23g R_1=R_4=OMe, R_2=R_3=H
23h R_1=R_4=H, R_2=R_3=OMe

Scheme 7

The first stage of our investigations consisted in determining the reactivity of benzyne with various kinds of methyleneisoquinoline . Compounds **21a** and **21b** reacted with unsubstituted benzyne (**22a**) to give the corresponding dehydroaporphines **23a** (Ref. 11) and **23b** (Ref. 12) (Scheme 7), and acetyl, trifluroacetyl and ethoxycarbonyl groups were found to be effective protecting groups for the nitrogen atom.

The 4,5-isoquinolinedione **24** which was obtained *in situ* by a process similar to that used for the synthesis of **7**, reacted with benzyne to afford norcepharadione B (**25a**) (Scheme 8) (Ref. 11).

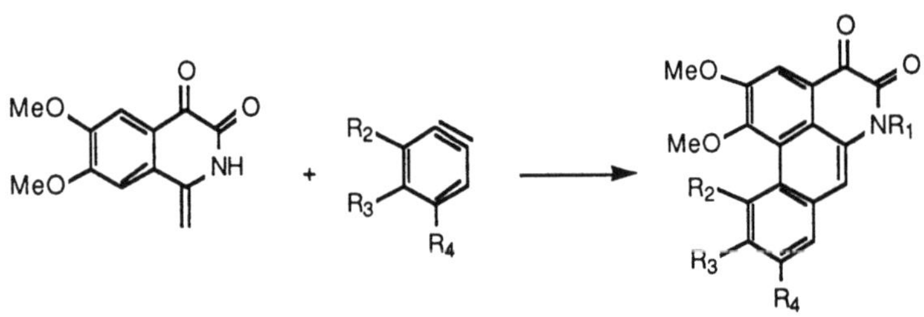

24

22a
22b

25a R_1=R_2=R_3=R_4=H , Norcepharadione B

3b R_1=R_2=H, R_3=R_4=OMe

3a R_1=Me, R_2=H, R_3=R_4=OMe, Pontevedrine

Scheme 8

Even the geometrically unfavourable molecule **26**, in which the distance between positions 1 and 4 of the diene is greater, reacted with benzyne to give the aristolactam **27** (though with a yield of only 20%) together with the product of unconcerted [2+2] addition (**28**) (Scheme 9) (Ref. 11).

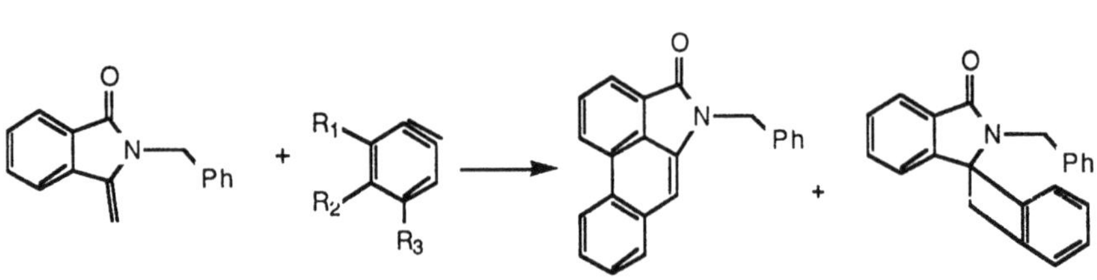

26 **22a** R_1=R_2=R_3=H **27** **28**

Scheme 9

The next step was to investigate the viability of the new method for preparing aporphinoids with substituents on ring D (see Fig. 23), i.e. the reactivity of methyleneisoquinolines with the substituted arynes 22. The simplest case is that of symmetrically substituted arynes such as 4,5-dimethoxybenzyne (22b), which reacts in the desired way with 24 to give norpontevedrine (3b), though with lower yields (20%) than in the case of unsubstituted benzyne (Ref. 12). When the aryne is asymmetrically substituted, regioselectivity is ensured by the presence of a methoxyl group *ortho* to the aryne triple bond (whereas mixtures of regioisomers are produced if the methoxyl group is *meta* to the triple bond). Thus 21a reacts with 3-methoxybenzyne (22c) and 3,4-dimethoxybenzyne (22d) to give 23d and 23e respectively (Ref. 13), but with 4-methoxybenzyne (22e) a mixture of the regioisomers 23f and 23g is produced (Ref. 14). Regioselectivity is also observed under these conditions if the aromatic ring of the styrene has no -OR substituents, as is shown by the production of 23h (apomorphine substitution) by reaction of 21c with 3,4-dimethoxybenzyne (22d) (Ref. 15).

As was mentioned above, the pyrrolinediones 10 can be transformed into oxoaporphines (11) by oxidation with Fremy's salt. In view of this, we proceeded to prepare the former by cycloaddition of an aryne to the pyrrolinedione 29a. Surprisingly, this reaction produced not only the desired compound 30 (in 11% yield; its oxidation with Fremy's salt led, as expected, to lysicamine (32) (Ref. 16), but also a 61% yield of the protoberberine 31, possibly via the mechanism indicated in Scheme 10.

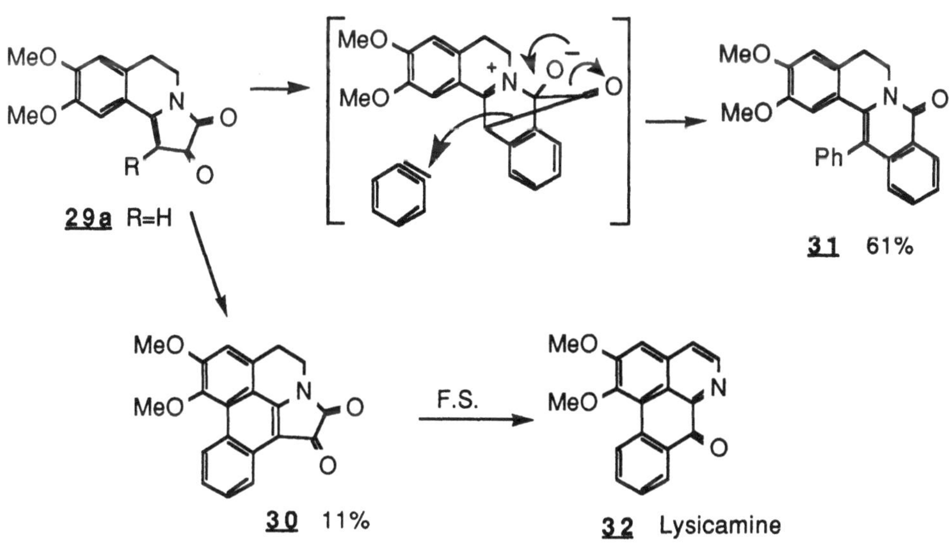

Scheme 10

Since in most natural protoberberines C_{13} has no substituent, or at most a methyl or methoxyl group (Refs. 2, 17), it is of interest to be able to avoid the arylation of C_{13} observed in **31**. Fortunately, this is achieved by the presence of a methyl group at the corresponding position in the pyrrolinedione, which has enabled us to carry out total synthesis of corydaline (**33**): 3,4-dimethoxybenzyne (**22d**) reacts regioselectively with the pyrrolinedione **29b** to give a 32% yield of the adduct **34a**, which reduction with lithium aluminium hydride and sodium borohydride converts into corydaline (Scheme 11)(Ref. 16).

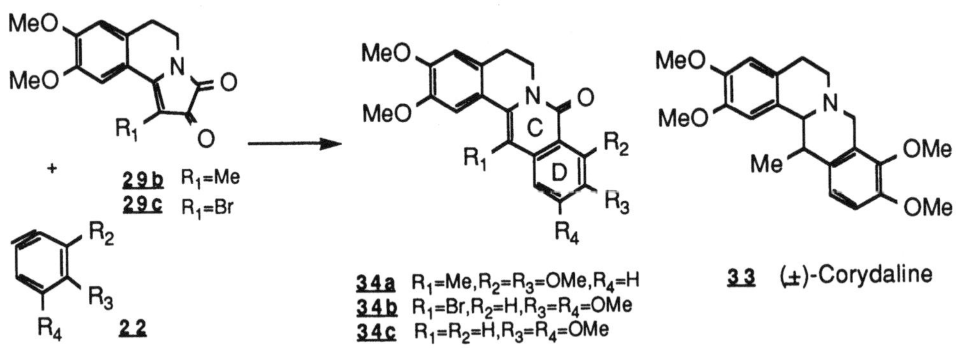

Scheme 11

Generalization of the above procedure requires that the group blocking phenylation of C_{13} be easy to remove once the protoberberine has been formed. Trials with various possible protecting groups have shown that bromine is effective in this role, as is illustrated by its use in the synthesis of the oxypseudopalmatine **34c**: the pyrrolinedione **29c**, which was obtained from **29a** by bromination with bromine, reacted with 4,5-dimethoxybenzyne (**22b**); in this case it was generated by Method A) to give **34b**, which was debrominated to 8-oxypseudopalmatine (**34c**) by hydrogenolysis with Pd/H_2 (Scheme 11) (Ref. 18).

The above method of synthesizing protoberberines involves the formation of an isoquinolone (ring C and D , fig. **34**) that can readily be reduced to the corresponding isoquinoline. The method should therefore prove to be of value for synthesizing some of the many isoquinoline alkaloids. These include, in particular, the benzophenanthridines (**35**), among which there are a number of compounds of great pharmacological interest (Refs. 2, 19). As the retrosynthetic analysis of Scheme 12 shows, it should be possible to obtain benzophenanthridine alkaloids by cycloaddition of an aryne (**19**) to the pyrrolinedione **37**.

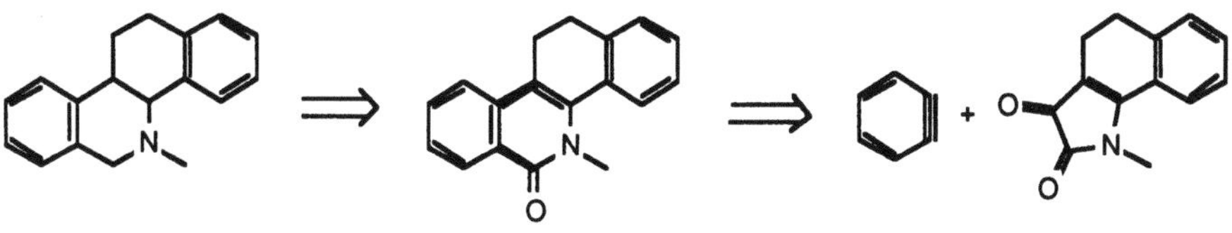

35 36 19 37

Scheme 12

In order to confirm this hypothesis, we synthesized the benzophenanthridines **40a** and **40b** by reaction of benzyne (**22a**) with the key intermediates **38a** and **38b**, which were obtained from the corresponding tetralones (**39a** and **39b**) by reaction with methylamine and titanium tetrachloride followed by addition of oxalyl chloride (Ref. 20). In view of our success, we then used the same procedure to prepare **40c** and **40d**, the precursors of avicine (**41a**) and nitidine (**41b**), the latter of which has considerable anti-tumour activity. Compound **40c** was obtained by reaction between **38b** and the aryne **22f**, and **40d** by reaction between **38b** and **22b**.

Finally, we have also found that the pyrrolinedione **38b** reacts regioselectively with the aryne **22d** to give the adduct **40e**, which is the precursor of chelerithrine (**41c**) (Ref. 15).

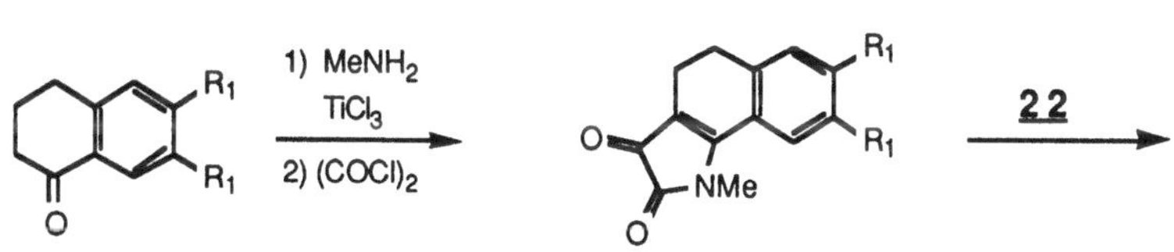

39a R_1=OMe
39b R_1+R_1=OCH_2O

38a R_1=OMe
38b R_1+R_1=OCH_2O

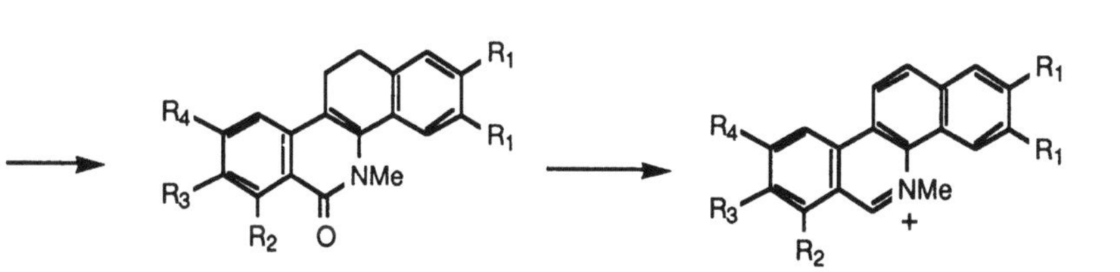

40a R_1=OMe, R_2=R_3=R_4=H
40b R_1=R_1=OCH_2O, R_2=R_3=R_4=H
40c R_1+R_1=R_3+R_4=OCH_2O, R_2=H
40d R_1+R_1=OCH_2O, R_2=H, R_3=R_4=OMe
40e R_1+R_1=OCH_2O, R_2=R_3=OMe, R_4=H

41a R_1+R_1=R_3+R_4=OCH_2O, R_2=H
41b R_1+R_1=OCH_2O, R_2=H, R_3=R_4=OMe
41c R_1+R_1=OCH_2O, R_2=R_3=OMe, R_4=H

Scheme 13

The above results show that in reactions with arynes, pyrrolinediones act as synthetic equivalents of azadienes. One therefore wonders how they behave in reactions with other dienophiles. Is it possible to generalize the method described above to obtain a new synthetic route to pyridones and pyridines? As a step in this direction, we have found that the pyrrolinedione **29b** reacts rapidly with dimethyl acetylene dicarboxylate (DMAD) to give the emetine **42** in 75% yield (Scheme 14)(Ref. 21).

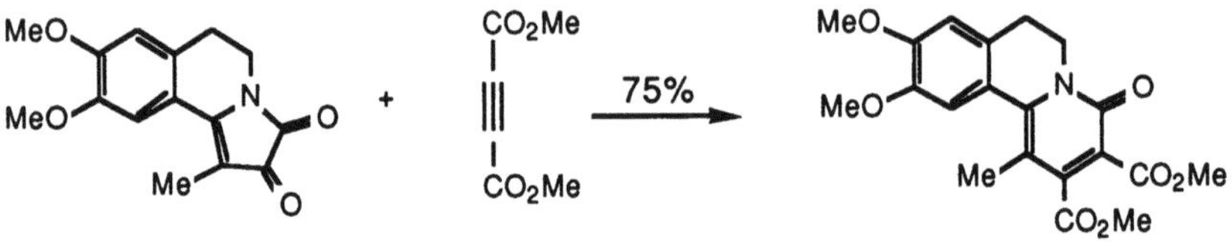

29b DMAD **42**

Scheme 14

In conclusion, the results described above prove the utility of aryne cycloaddition, and hint strongly at the synthetic promise of pyrrolinediones. More research will be needed to determine the extent of this potential.

REFERENCES

1.a.- I. Ribas, J. Sueiras and L. Castedo, Tetrahedron Letters (1971) 3093 and L. Castedo, A. Mouriño and R. Suau, Tetrahedron Letters (1976) 501.

b.- L. Castedo, D. Domínguez, J.M. Novo, A. Peralta, J.M. Saá and R. Suau, Heterocycles 24 (1986) 2781.

c.- L. Castedo, D. Domínguez, M. Pereira, J.M. Saá and R. Suau, Heterocycles 16 (1981) 533.

d.- J.M. Boente, D. Domínguez and L. Castedo , Heterocycles 24 (1986) 3359 , M.J. Campello, L. Castedo, D. Domínguez, A. Rodríguez de Lera, J.M. Saá and R. Suau, E. Tojo and M.C. Vidal, Tetrahedron Letters 25 (1984) 5933.

2.- M. Shamma, The Isoquinoline Alkaloids, Chemistry and Pharmacology, Academic Press, 1972 and M. Shamma and J.L. Moniot, Isoquinoline Alkaloid Research 1972-1977, Plenum Press, 1978.

3.- G. R. Lenz, Synthesis (1978) 489 . See also F.B. Mallory and C.W. Mallory, in Org. React. vol. 30 (1984).

4.- L. Castedo, J.R. Estévez, J.M. Saá and R. Suau, Tetrahedron Letters 19 (1978) 2179 and L. Castedo, J.R. Estévez, J.M. Saá and R. Suau, J. Heterocycl. Chem. 19 (1982) 1319. See also L. Castedo, J.M. Saá, R. Suau and J.R. Estévez, An. Quím. 79C (1983) 329.

5.- L. Castedo, C. Saá, J.M. Saá and R. Suau, J. Org. Chem., 47 (1982) 513.

6.- L. Castedo, E. Guitián, J.M. Saá and R. Suau, Heterocycles, 19 (1982) 279.

7.- J.A. Seijas, M.C. Villaverde and L. Castedo, in preparation. See also I. Ben, L. Castedo, J.M. Saá, J.A. Seijas, R. Suau and G. Tojo, J. Org. Chem., 50 (1985) 2236 and J.A. Seijas, A. Rodríguez de Lera, M.C. Villaverde and L. Castedo, Heterocycles 23 (1985) 3079.

8.- S.F. Dyke, A.J. Floyd and S.E. Ward, Tetrahedron 26 (1970) 4005.

9.- L. Friedman and F.M. Logullo, J. Am. Chem. Soc., 85 (1963) 1549.

10.- F.M. Logullo, A.H. Seitz and L. Friedman, Org. Synth. 48 (1968) 12.

11.- L. Castedo, E. Guitián, J.M. Saá and R. Suau, Tetrahedron Letters 23 (1982) 457.

12.- E. Guitián, L. Castedo, J.M. Saá and R. Suau, in preparation.

13.- L. Castedo, E. Guitián, C. Saá, R. Suau and J.M. Saá, Tetrahedron Letters 24 (1983) 2107.

14.- C. Saá, E. Guitián, L. Castedo and J.M. Saá, in preparation.

15.- G. Martín, E. Guitián, L. Castedo and J.M. Saá, in preparation. In these cases arynes were generated from the corresponding benzenediazonium 2-carboxilate hydrochloride, see ref. 8.

16.- C. Saá, E. Guitián, L. Castedo, R. Suau and J.M. Saá, J. Org. Chem., 51 (1986) 2781.

17.- D.S. Bhakuni and S. Jain, in The Alkaloids vol XXVIII, Ed. A. Brossi, Academic Press, 1986.

18.- A. Cobas, E. Guitián, L. Castedo and J.M. Saá, submitted for publication.

19.- V. Simánek, in The Alkaloids vol. XXVI , Ed. A. Brossi, Academic Press, 1985.

20.- G. Martín, E. Guitián, L. Castedo and J.M. Saá, Tetrahedron Letters, 28 (1987) 2407.

21.- N. Atanes, E. Guitián, L. Castedo and J.M. Saá, in preparation.

Biosynthetic Studies of Protoberberine and Related Alkaloids Using Plant Cell Cultures

MARTINA RUEFFER

Institut für Pharmazeutische Biologie
Universität München
Karlstr. 29, D-8000 München 2, F.R.G.

INTRODUCTION

The benzylisoquinolines comprise the largest group of alkaloids with about 2000 structures some of which possess valuable medicinal importance (such as codeine or morphine). During the past years the biosynthetic pathway leading to one group of these alkaloids, namely the protoberberines, was elucidated at the enzyme level (refs. 1-2). This progress was only possible by using cell suspension cultures as plant material. These plant cell cultures have the advantage of growing independently of climatical and geographical conditions and in a defined media. In some cases they are able to produce natural products in much higher amounts over shorter time periods than the intact plant (refs. 3-4). The cell cultures used in the investigations of protoberberine biosynthesis had been optimized for a high production of these alkaloids (ref. 5).

THE BIOSYNTHESIS OF RETICULINE

Recently the first steps in the enzymatic sequence leading to (S)-reticuline had to be revised (Fig. 1). Previously norlaudanosoline, the condensation product of dopamine and 3,4-dihydroxyphenylacetaldehyde had been assumed to be the universal precursor for the benzylisoquinolines (refs. 6-7), a hypothesis, which was supported by the discovery of an enzyme catalyzing the coupling of these two units to form (S)-norlaudanosoline (ref. 8).

Feeding experiments on the other hand showed that the two building blocks have different biogenetic origins. While labelled tyrosine is equally incorporated into both the amine and the aldehyde part, the other putative precursors 3,4-dihydroxyphenylalanine (DOPA), tyramine and dopamine nearly exclusively

FIG. 1 Traditional and revised pathway leading to (S)-reticuline.

1 = (S)-Norcoclaurine synthase (= (S)-Norlaudanosoline synthase)
2 = (S)-Norcoclaurine-6-O-methyltransferase (= (S)-Norlaudanosoline-6-O-methyltransferase)
3 = (S)-Coclaurine-N-methyltransferase
4 = Phenolase
5 = (S)-6-O-Methyllaudanosoline-4'-O-methyltransferase (= (S)-6-O-Methylnorlaudanosoline-4'-O-methyltransferase).

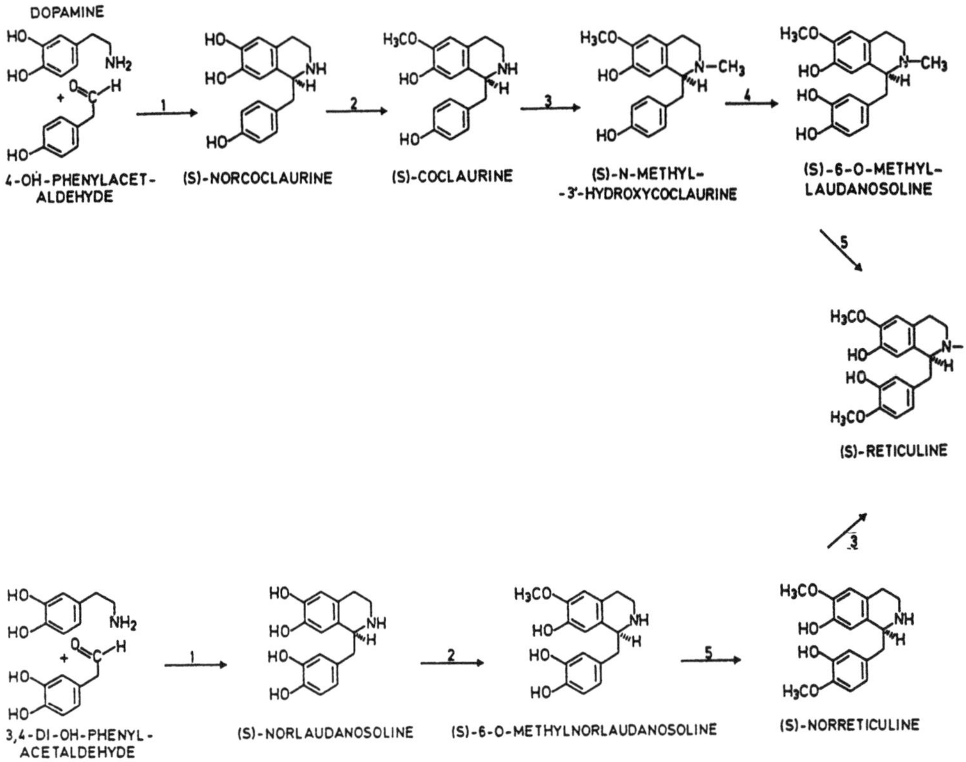

labelled the isoquinoline moiety of the benzylisoquinoline molecules (refs. 9-12). These results can be explained by assuming that 4-hydroxyphenylacetaldehyde formed by transamination and subsequent decarboxylation of tyrosine (ref. 13) could be the immediate precursor for the benzylic portion of the benzylisoquinoline nucleus. This assumption could be verified by the results obtained with feeding experiments in which (S)-6-O-$^{14}CH_3$- or (S)-6-O-$^{13}CH_3$-coclaurine were incorporated into (S)-reticuline in **Annona reticulata**, jatrorrhizine and berbamunine in **Berberis stolonifera**, macarpine in **Eschscholtzia californica** and thebaine in **Papaver somniferum** (refs. 12, 14). In addition, 1-^{13}C-norcoclaurine was well incorporated into morphine and thebaine in **Papaver** seedlings (ref. 12). Another argument for propagating norcoclaurine as the first condensation product is the fact that though it could be shown that norlaudanosoline was incorporated into the morphinan alkaloids (ref. 15), this proported intermediate never could be detected as natural product in plants, whereas there are numerous reports of the isolation of norcoclaurine (also called higenamine or lotusine) (refs. 16-18) and coclaurine (refs. 19-20).

In conclusion it is now proposed that dopamine and 4-hydroxyphenylacetaldehyde condense to form (S)-norcoclaurine as the first benzylisoquinoline alkaloid in the pathway leading to (S)-reticuline.

The decision which of the two aldehydes is the condensation partner of dopamine could not be made using the isolated enzyme, for the substrate specificity of this first enzyme in the pathway is relatively low. The previously so called (S)-norlaudanosoline synthase has approximately the same K_M-value for 4-hydroxyphenylacetaldehyde (0.9 mM) as for 3,4-dihydroxyphenylacetaldehyde (0.7 mM) and the V_{max}-values are also quite similar (27.9 pmol/s and 34 pmol/s) (ref. 8). The next enzyme in the pathway, which introduces the first methyl group into the 6-position to form coclaurine also showed a very broad substrate specificity and accepted (R)- and (S)-norcoclaurine as well as (R)- and (S)-norlaudanosoline as substrates (ref. 21). By all the data gained so far in the genus **Berberis** it is now assumed that coclaurine is then methylated again to N-methylcoclaurine (ref. 22). In the next step, a hydroxylation reaction is also catalyzed by an enzyme with a broad substrate specificity to yield N-methyl-3'-hydroxycoclaurine (= 6-O-methyllaudanosoline) (ref. 21). Finally, the third methyl-group is introduced in the 4'-position by an O-methyltransferase with a higher substrate specificity than the enzyme described above (ref. 22) resulting in (S)-reticuline, an important intermediate in the biosynthesis of the protoberberines.

ENZYMATIC FORMATION OF PROTOBERBERINES

The first reaction in the direction of the protoberberines is the formation of the berberine bridge (Fig. 2), catalyzed by the berberine bridge enzyme (abbreviated: BBE), which specifically only accepts (S)-reticuline and not the (R)-enantiomer as substrate (ref. 23). As a next step a methylation at C-9 follows to form (S)-tetrahydrocolumbamine catalyzed by a methyltransferase with a much higher substrate specificity than the methyltransferases described before (ref. 24). In the genus **Berberis** (S)-tetrahydrocolumbamine is then oxidized by an enzyme called (S)-tetrahydroprotoberberine oxidase (abbreviated: STOX) (ref. 25) and finally the methylene dioxy bridge is formed by the so called berberine synthase (ref. 26).

Nevertheless in 1985 an oxidase was reported from crude extracts of **Coptis japonica** cell cultures with different properties as compared to the enzyme from **Berberis**. This led to a systematic comparison of the two cell cultures concerning the last steps of the berberine biosynthesis. The purified oxidase from

FIG. 2 Biosynthetic pathway leading from (S)-reticuline to the protoberberine alkaloids in **Berberis** and **Coptis**.

1 = Berberine bridge enzyme (BBE)
2 = (S)-Scoulerine-9-O-methyltransferase
3 = (S)-Tetrahydroprotoberberine oxidase (STOX)
4 = Berberine synthase
5 = Columbamine-O-methyltransferase
6 = (S)-Canadine synthase
7 = (S)-Canadine oxidase.

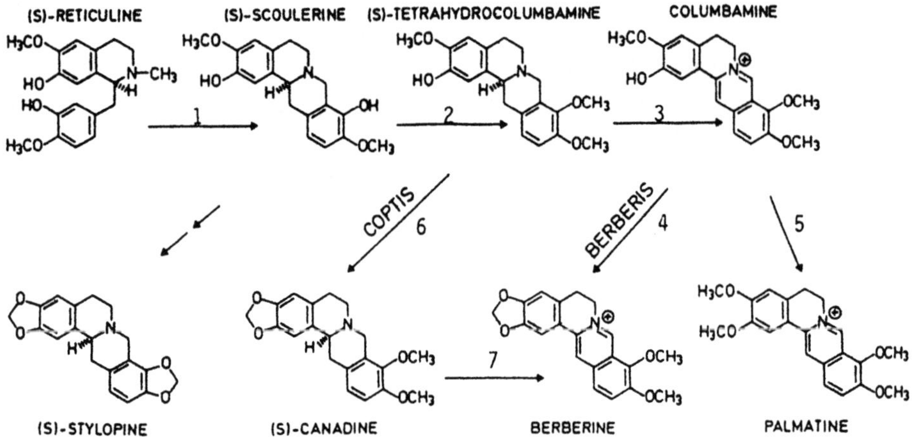

Coptis indeed is different from the **Berberis** enzyme (ref. 27). Most strikingly diverse are the substrate specificities. The enzyme from **Berberis** accepts a great variety of tetrahydroprotoberberines as substrates, among them (S)-canadine, (S)-scoulerine and the simple benzylisoquinoline (S)-norreticuline, whereas in **Coptis** (S)-canadine is the major substrate and (S)-tetrahydropalmatine and (S)-stylopine are transformed in much minor amounts.

In addition, the methylene dioxy bridge in **Coptis** is formed at an earlier stage of the biosynthetic sequence (Fig. 2). In **Berberis** this bridge is formed as the last step of the biosynthesis with columbamine as substrate while in **Coptis** first the formation of the methylene dioxy bridge and then the oxidation takes place. The **Coptis** pathway is also realized in the four **Thalictrum** species so far investigated. This led to the conclusion that the **Ranunculaceae** and **Berberidaceae** though having close taxonomical relationship (both belong to the order **Ranunculales**) have developed different biosynthetic strategies to form berberine.

In **Berberis** columbamine is a late branch point intermediate: on the one hand by formation of the methylene dioxy bridge berberine is formed and on the other hand columbamine serves as substrate for another methylation reaction catalyzed by a highly specific methyltransferase to form palmatine, another naturally occurring protoberberine (ref. 28).

SUBCELLULAR LOCALIZATION OF PROTOBERBERINE BIOSYNTHESIS

The enzymes involved in the late steps of protoberberine biosynthesis, beginning with the berberine bridge enzyme all have a high pH-optimum of 8.9. This fact led to an investigation of the subcellular localization of these enzymes. After fractionation of the organelles by sucrose density gradient centrifugation the enzymes could be located in a fraction with the density $\rho = 1.14\ g/cm^2$ (ref. 29). Electron microscopy of this fraction showed membranous vesicles with diameters between 0.1 to 1.0 µm.

These vesicles could be detected in **Berberis** as well as in **Coptis** (ref. 27) and seem to be the general principle for the biosynthesis and the transport of protoberberines. Once reticuline has entered this vesicle its fate to be metabolized to protoberberine alkaloids is determined. Since the quaternary end products are not able to pass membranes these vesicles also seem to be the transport vehicles to the final storage compartment, the vacuole. Electron microscopic pictures show that they seem to fuse with the vacuole and release their contents into it (ref. 29) which strongly supports this assumption.

BIOSYNTHESIS OF PROTOPINE ALKALOIDS

Protopines are a widely distributed group of alkaloids occurring in the **Berberidaceae, Fumariaceae, Papaveraceae, Ranunculaceae, Rutaceae**, and **Sapindaceae** (ref. 30). Feeding experiments (refs. 31-32) have shown that (S)-tetrahydroprotoberberines possessing a **cis**-N-methylated configuration are the precursors for these alkaloids. As a first enzyme leading in this direction an N-methyltransferase has been isolated and characterized that methylates (S)-tetrahydroprotoberberines in the **cis**-position (ref. 33). This N-methyltransferase was detected only in those tissues lacking protoberberine alkaloids, suggesting that the STOX enzyme responsible for the formation of protoberberines was not present in high amounts (Table 1).

With the help of this N-methyltransferase the substrates for testing the hydroxylation reaction to the protopines were prepared (Fig. 3) from tetrahydroprotoberberines that had been labelled by reduction of the corresponding protoberberines with NaB^3H_4.

TABLE 1 Distribution of (S)-tetrahydroprotoberberine-**cis**-N-methyltransferase in plant cell cultures using different substrates

Cell culture	Substrate: Canadine	Stylopine (pkat/g dry weight)	Sinactine	Tetrahydro-palmatine	Nandinine
Berberidaceae					
Berberis beaniana	0.9	0.9	1.7	0	0
Berberis wilsoniae	0	0	0	0	0
Nandina domestica	0	0.9	2	1	0
Fumariaceae					
Corydalis vaginans	52	80	29	31	93
Dicentra spectabilis	28	40	12	50	12
Fumaria officinalis	44	42	n.d.	52	7
Menispermaceae					
Chondodendron tomentosum	2	2	n.d.	4	3
Papaveraceae					
Argemone platyceras	41	41	57	49	40
Bocconia cordata	20	18	12	20	23
Eschscholtzia caespitosa	77	107	30	74	96
Papaver somniferum	4	3	2	4	4
Ranunculaceae					
Thalictrum glaucum	0.4	0.4	0.8	1.1	0.5
Thalictrum bautrinii	0	4	0	0	0

FIG. 3 Test system for the formation of protopine alkaloids.
Protoberberines were reduced with NaB^3H_4. The tritiated (S)-tetrahydroprotoberberines were methylated by the (S)-tetrahydroprotoberberine-**cis**-N-methyltransferase. If the (S)-**cis**-N-methyltetrahydroprotoberberines are hydroxylated a ^3H-release into water can be measured.

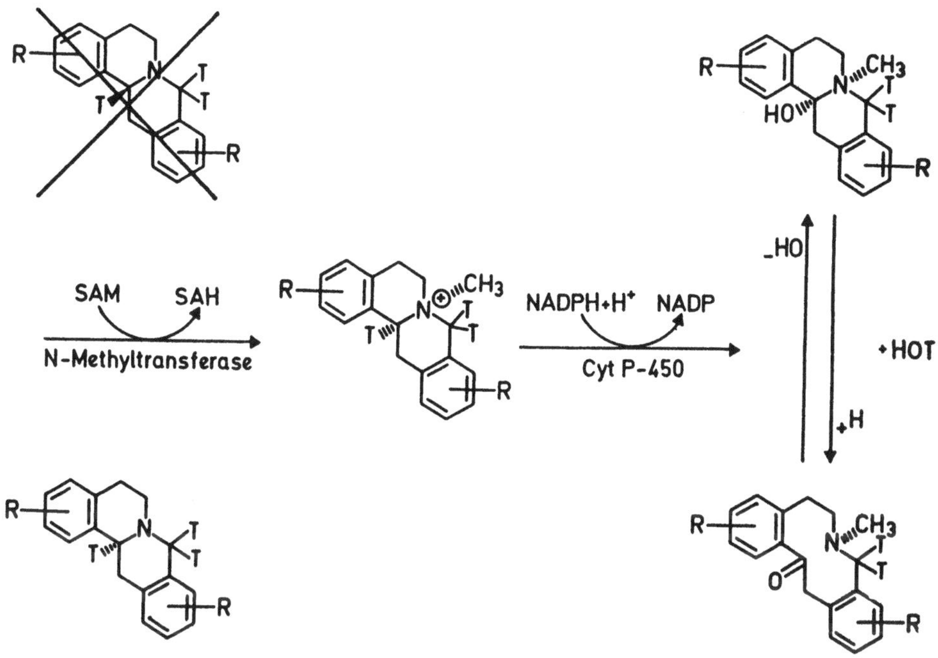

Cell-free extracts prepared as usual did not show any hydroxylase activity. Only when microsomal preparations (ref. 34) were incubated in the presence of NADPH could a ^3H-release be measured (ref. 37).

The enzyme appears to be a cytochrome P-450, for it showed the typical UV-spectrum when reduced with sodium dithionite or oxidized with carbon monoxide. In addition, it can be inhibited by specific cytochrome P-450 inhibitors (Table 2).

It was specifically detected in cell cultures of the **Papaveraceae** and **Fumariaceae** (ref. 35) but not in non-alkaloid producing cultures like **Rosa canina** or **Quercus robur**.

This is the first monoxygenase reported for the metabolism of benzylisoquinolines.

TABLE 2 Inhibtion of N-methylstylopine hydroxylase by cytochrome-P-450 inhibitors. The concentrations shown are those for 50% inhibition of the enzyme.

Inhibitor	Concentration (μM)
Ancymidol	12
Cytochrome c	40
Ketoconazole	1.5
Metyrapone	2
Tetcyclacis	12
Propiconazole	4
Prochloraz	6

In addition to the pathway leading to the protoberberines, a second one leading to the protopine alkaloids has now been elucidated at the enzyme level utilizing plant cell cultures as the enzyme source. Since the protopines are regarded as the biosynthetic precursors of the benzazepine and benzophenanthridine alkaloids (refs. 36-39) these now can be the target of further enzymatic investigations.

ACKNOWLEDGEMENTS

This work was supported by the Deutsche Forschungsgemeinschaft, Bonn (SFB 145). Linguistic help by Dr. T.M. Kutchan is greatfully acknowledged.

REFERENCES

1 M.H. Zenk, Enzymology of benzylisoquinoline formation, in: J.D. Phillipson, M.F. Roberts and M.H. Zenk (eds.), The Chemistry and Biology of Isoquinoline Alkaloids, Springer, Berlin, Heidelberg, New York, Tokyo (1985), pp. 240-256.

2 M.H. Zenk, M. Rueffer, M. Amann, and B. Deus-Neumann, Benzylisoquinoline biosynthesis by cultivated plant cells and isolated enzymes, J.Nat.Prod. **48** (1985), 725-738.

3 Y. Fujita and M. Tabata, Secondary metabolites from plant cells - pharm. applications and progress in commercial production, in: C.E. Green, D.A. Somers, W.P. Hackett, and D.D. Biesboer (eds.), Plant Tissue and Cell Culture, Alan R. Liss Inc., New York (1987), pp. 169-186.

4 A. Fontanel and M. Tabata, Production of secondary metabolites by plant tissue and cell cultures, present aspects and prospects, in: C. Boella, H. Schelling and C. Schlienger (eds.), Nestlé Research News (1987), pp. 93-103.

5 H. Hinz and M.H. Zenk, Production of protoberberine alkaloids by cell suspension cultures of **Berberis** species, Nat.Wiss. **67** (1981), 620.
6 E. Winterstein and G. Trier, in: Die Alkaloide, Gebrüder Bornträger, Berlin (1910), p. 307.
7 R. Robinson, The structural relations of natural products, Oxford University Press (1955).
8 H.M. Schumacher, M. Rueffer, N. Nagakura, and M.H. Zenk, Partial purification and properties of (S)-norlaudanosoline-synthase from **Eschscholtzia tenuifolia** cell cultures, Planta Med. **48** (1983), 212-220.
9 I.D. Spenser, Biosynthesis of isoquinoline alkaloids, in: M. Florkin and H. Stotz (eds.), Comprehensive Biochemistry XX, Elsevier, Amsterdam (1968), p. 231.
10 H.L. Holland, P.W. Jeffs, T.M. Capps, and D.B. MacLean, The biosynthesis of protoberberine and related isoquinoline alkaloids, Can.J.Chem. **57** (1979), 1588-1597.
11 A.R. Battersby, R.C.F. Jones, R. Kazlauskas, C. Poupat, C.W. Thornber, S. Ruchiwarat, and J. Staunton, The biosynthesis of hasubanonine and protostephanine, J.Chem.Soc. Chem.Commun. (1974), 773-775.
12 S. Loeffler, R. Stadler, N. Nagakura, and M.H. Zenk, Norcoclaurine as biosynthetic precursor of thebaine and morphine, J.Chem.Soc. Chem.Commun. (1987), 1160-1162.
13 M. Rueffer and M.H. Zenk, Distant precursors of benzylisoquinoline alkaloids and their enzymatic formation, Z.Naturforsch. **42c** (1987), 319-332.
14 R. Stadler, T.M. Kutchan, S. Loeffler, N. Nagakura, B. Cassels, and M.H. Zenk, Revision of the early steps of reticuline biosynthesis, Tetrahedron Lett. **28** (1987), 1251-1254.
15 A.R. Battersby, R. Binks, R.J. Francis, D.J. McCaldin, and H. Ramuz, Alkaloid Biosynthesis Part IV, J.Chem.Soc. (1964), 3600-3610.
16 A. Furukawa, Yakugaku Zasshi **85** (1965), 335.
17 T. Kosuge and M. Yokota, Studies on cardiac principles of aconite root, Chem.Pharm.Bull. **24** (1976), 176-178.
18 H. Wagner, M. Reiter and W. Ferstl, Neue herzwirksame Drogen I, Planta Med. **40** (1980), 77-85.
19 R.R. Arndt (-) N-Methylcoclaurine from **Phylica rogersii**, J.Chem.Soc. 2 (1963), 2547-2549.
20 H. Otsuka, Y. Ogihara, and S. Shibata, Isolation of coclaurine from **Zizyphus jujuba** by droplet counter current chromatography, Phytochem. **13** (1974), 2016.
21 S. Loeffler, unpublished results.
22 T. Frenzel, Ph.D Thesis (1987).
23 P. Steffens, N. Nagakura and M.H. Zenk, The berberine bridge forming enzyme in tetrahydroberberine biosynthesis, Tetrahedron Lett. **25** (1984), 951-952.
24 S. Muemmler, M. Rueffer and M.H. Zenk, S-Adenosylmethionine: (S)-Scoulerine-9-O-methyltransferase, a highly stereo- and regiospecific enzyme in tetrahydroberberine bioynthesis, Plant Cell Reports **4** (1985), 36-39.
25 M. Amann, N. Nagakura and M.H. Zenk, (S)-Tetrahydroprotoberberine oxidase, the final enzyme in protoberberine biosynthesis, Tetrahedron Lett. **25** (1984), 953-954.
26 M. Rueffer and M.H. Zenk, Berberine synthase, the methylenedioxy group forming enzyme in berberine synthesis, Tetrahedron Lett. **26** (1985), 201-202.
27 E. Galneder, M. Rueffer, G. Wanner, M. Tabata, and M.H. Zenk, Alternative final steps in berberine biosynthesis in **Coptis japonica** cell cultures, Plant Cell Reports, in press.
28 M. Rueffer, M. Amann, and M.H. Zenk, S-Adenosyl-L-methionine: Columbamine-O-methyl transferase, a compartmentalized enzyme in protoberberine biosynthesis, Plant Cell Reports **3** (1986), 182-185.
29 M. Amann, G. Wanner and M.H. Zenk, Intracellular compartmentation of two enzymes of berberine biosynthesis in plant cell cultures, Planta **167** (1986), 310-320.
30 H. Guinaudeau and M. Shamma, The protopine alkaloidss, J.Nat.Prod. **45** (1982), 237-246.

31 A.R. Battersby, J. Staunton, H.R. Wiltshire, R.J. Francis, and R. Southgate, Biosynthesis XXII, J.Chem.Soc. Perkin I (1975), 1147-1156.
32 N. Takao, K. Iwasa, M. Kamigauchi and M. Sugiura, Studies on the Alkaloids of Papaveraceous Plants XXV, Chem.Pharm.Bull. **24** (1976), 2859-2868.
33 M. Rueffer and M.H. Zenk, S-Adenosyl-L-methionine: (S)-7,8,13,14-Tetrahydroberberine-**cis**-N-methyltransferase, a branch point enzyme in the biosynthesis of benzophenanthridine and protopine alkaloids, Tetrahedron Lett. **27** (1986), 5603-5604.
34 H. Diesperger, C.R. Mueller, H. Sandermann, Rapid isolation of a plant microsomal fraction by Mg2+ - precipitation, FEBS Letters **43** (1974), 155-158.
35 M. Rueffer and M.H. Zenk, Enzymatic formation of protopines by a microsomal cytochrome P-450 system of **Corydalis vaginans**, Tetrahedron Lett. **28** (1987), 5307-5311.
36 H. Roensch, Zur Biosynthese der Alkaloide vom Rhoeadin-Typ, Eur.J.Biochem. **28** (1972), 123-126.
37 H. Roensch, Biosynthesis of alpinigenine by way of tetrahydroprotoberberine and protopine intermediates, Phytochem. **16** (1977), 691-698.
38 C. Tani, K. Tagahara, Studies on berberine derivatives and related alkaloids VII, Chem.Pharm.Bull. **22** (1974), 2457-2459.
39 N. Takao, M. Kamigauchi and M. Okada, Biosynthesis of benzo[**c**]phenanthridine alkaloids sanguinarine, chelirubine and macarpine, Helv.Chim.Acta **66** (1983), 473-484.

Secondary Metabolism in Cell Cultures of Some Terpenoid-Indole Alkaloid Producing Plants

R. VERPOORTE*, R. van der Heijden, J. Schripsema, T. van der Leer, E.J.M. Pennings, P.A.A. Harkes, S.S. Hoekstra, J.J. Meijer and H.J.G. ten Hoopen

* Division of Pharmacognosy, Center for Bio-Pharmaceutical Sciences, Leiden University, P.O. Box 9502, 2300 RA Leiden, The Netherlands

1. INTRODUCTION

In the past years it has been proven that large scale culturing of plant cells is feasible. The production of shikonin (1) and of ginseng biomass (2) are examples of a commercial scale production with plant cell cultures.

Being technically feasible, how are the economics of a plant cell biotechnological production method? We have made a design for a large scale plant cell culture system using airlift fermentors, in order to be able to calculate the costs (Fig. 1)(3).

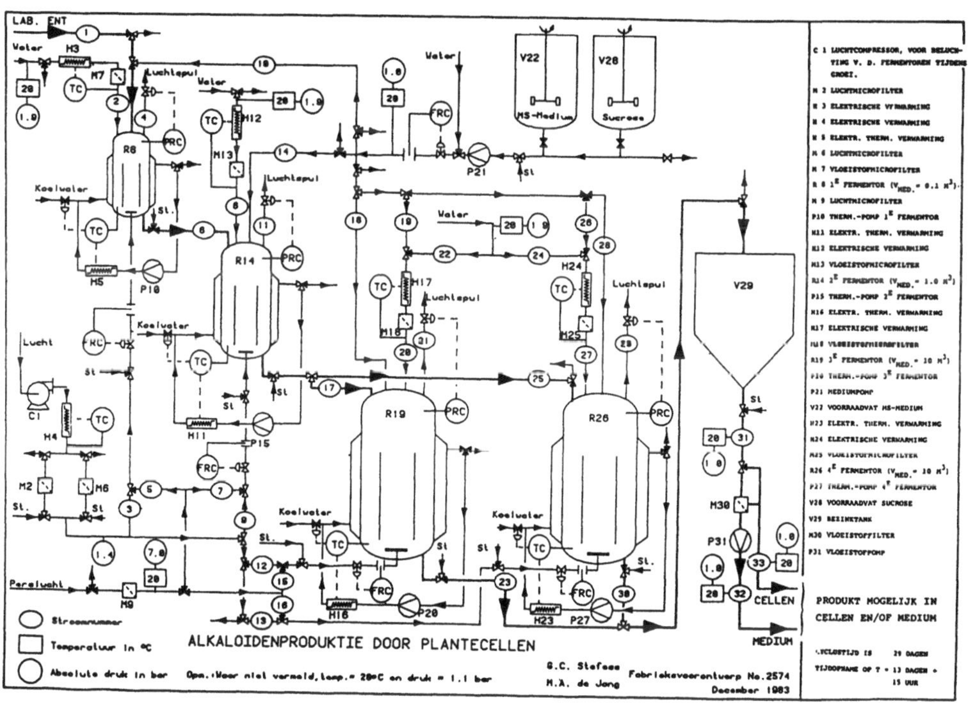

Fig. 1. Design for industrial scale plant cell production unit.

In this design a 10.000 l airlift fermentor is the final step. Two of these fermentors (Fig. 1: R19 and R26)are operated parallel, but out of phase, one being in the growth phase, the other in production phase, each lasting one week.

The inoculum for these bioreactors is in a series of airlift fermentors, increasing a factor 10 in size grown from laboratory scale (see Fig. 1 bioreactors R8 of 100 l and R14 of 1000 l). With the parameters as summarized in Table 1 production costs of about fl 5000,- (ca. $ 2000,-) per kg was calculated, based on the investment in the equipment and the energy and media costs, but not including downstream processing and labor costs. Similar prices were reported by Goldstein (4) and Rhodes (5). These prices are quite high, although several of the plant derived chemicals have a price in this order of magnitude.

TABLE 1
Parameters used to calculate production costs of a secondary metabolite with large scale plant cell cultures in a process design as in Fig. 1.

Growth rate u	0.164 day^{-1}
Biomass density	20 g/l (dry weight)
Product yield	1%
Production/year	50 kg
Size bioreactor	10.000 l airlift

Cost price fl 5000 (ca. $ 2500) exclusive downstream processing and labor costs.

In this example airlift bioreactors were used as it was generally assumed that plant cells are difficult to grow in stirred fermentors, due to their sensitivity for shear forces. However recent studies in our laboratories have shown that this is not a general characteristic of plant cells. Efforts to measure the shear sensitivity of Catharanthus roseus cell cultures were not succesful, as even at stirrer speeds as high as 1000 rpm (normal speed 100 rpm) in a 3 l vessel, the cells were still viable after one month. Similar results were found for tobacco cell cultures (6). This means that stirred fermentors can also be considered for large scale culture, which makes the economy more favourable, as such fermentors do already exist in fermentation industry, whereas airlift bioreactors are scarcely used.

As the cost calculation is made using rather conservative parameters for growth and production, compounds with higher yields than 0.2 g/l show a more favourable picture. Some examples of high producing cell cultures are presented in Table 2. It is obvious that some of these have commercial interest (shikonin, berberine). However not for all products the situation is equally favourable, in fact quite a few of the products with pharmaceutical interest are not produced in cell suspension cultures, or only in very small amounts.

TABLE 2
Some high producing plant cell cultures

Product	Plant species	% of dry weight	g/l medium	reference
rosmarinic acid	Coleus blumei	21.5	5.6	7
	Anchusa officinalis	12	1.7	8
shikonine	Lithospermum erythrorhizon	23.2	3	9
anthraquinones	Morinda citrifolia	18	2.5	10
	Rubia fruticosa	20		11
verbascoside	Syringa vulgaris	16	1.4	12
berberine	Coptis japonica	13.2	2.5	13
	Thalictrum minus	12.1	0.8	14
jatrorrhizine	Berberis stolonifera	10	1.7	15
raucrafficine	Ruawolfia serpentina	1.6	0.5	16
serpentine	Catharanthus roseus		0.6	17

In Table 3 some examples are presented. Why certain compounds are produced in cell cultures and others are not is not clear yet, although some ideas exist.

TABLE 3
Some examples of cell suspension cultures of wellknown medicinal plants in which the desired secondary products are not being produced or in very small amounts only.

plant species	desired product
Papaver somniferum	morphine, codeine
Cinchona species	quinine, quinidine
Atropa belladonna	atropine, scopolamine
Hyoscyamus niger	atropine, scopolamine
Datura stramonium	atropine, scoplamine
Catharanthus roseus	vinblastine, vincristine
Digitalis lanata	digoxin, digitoxin

Secondary product formation is a form of differentiation, induction of the required type of differentiation does lead to secondary product formation, e.g. phytoalexins will readily be formed upon stress conditions (additions of fungal elicitors) and compounds usually found in the roots will also be found in root cultures or hairy-root (cells transformed with Agrobacterium rhizogenes) cultures. In Table 4 some examples are summarized of the influence of differentiation on secondary metabolite production.

TABLE 4
Some examples of organ cultures and their secondary product content

Product	Plant species	% g/dw in cultures				plant	references
		unorganized	root	hairy root	shoot		
atropine	Duboisia myoporides	n.d.	0.005			0.01	
scopolamine		n.d.	0.004			0.006	18
atropine	Scopolia japonica			1.3		0.15	
scopolamine				0.5		0.16	19
atropine	Hyoscyamus muticus	n.d.		0.6		0.6	20
atropine	Duboisia leichhardtii		0.53				
scopolamine			1.16				21
atropine	Atropa belladonna	traces	0.3			0.5	22
thebaine	Papaver bracteatum	traces	0.03				23
codeine and thebaine	Papaver somniferum				0.5	1.1	24
vinblastine	Catharanthus roseus				0.01	0.21	25

In order to be able to control the production of a certain secondary metabolite one has to know more about the regulation of its biosynthesis, preferably at the level of the genes. This could eventually open the way to genetic engineering of plant cells with the aim to improve the yields. A better understanding of the biosynthesis may furthermore be useful for identifying biotechnologically interesting bioconversions, which also can be considered for genetic engineering of microorganisms.

In the subsequent paragraphs some of the work in our laboratories in this field will be reviewed. It particularly involves the secondary metabolism in cell cultures of plants capable of producing terpenoid-indole and related alkaloids, i.e. Cinchona and Tabernaemontana.

2. CINCHONA CELL CULTURES

With a year production of 300-500 tons (26), the Cinchona alkaloids (quinine 1 and quinidine 2) probably form one of the largest markets of fine chemicals derived from higher plants. They are extracted from stem and rootbark of Cinchona trees, containing 5-18% of alkaloid, with an average of about 8% (27). Because of the high demand a number of attempts have been made to develop a commercial synthesis (28 and references cited therein) of the quinoline alkaloids. Although successful syntheses have been reported they could not be commercialized.

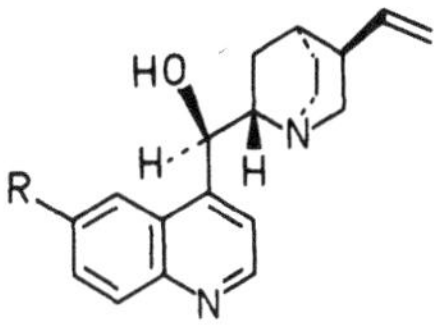

1 R = OCH_3 quinine

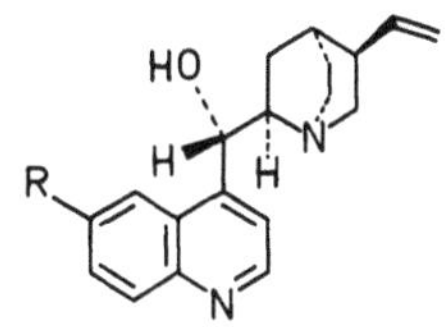

2 R = OCH_3 quinidine

Only the stereospecific conversion of quinine in quinidine through an oxidation yielding quininone, which can be reduced stereospecifically to quinidine, is applied industrially to meet the higher demands of quinidine (28 and the references cited therein). The commercial interest has also initiated studies into biotechnological production methods. The complex biosynthesis of the alkaloids (Scheme 1), includes a number of enzyme catalyzed steps, each controled by a different gene. The introduction of a number of genes in a microorganism will be too difficult for the present state of the art. Furthermore the plant genes responsible for the biosynthesis are still unknown. Only certain steps of the biosynthesis, e.g. a stereospecific reduction of quinidinone, could be of interest as goal for the genetic engineering of microorganisms. In fact for such a complex biosynthesis one should rather use the genetic information as present in the plant cell itself, i.e. the use of plant cell or tissue cultures.

Among others in our laboratories cell and tissue cultures of Cinchona species have been initiated. For a review of this work is referred to Wijnsma and Verpoorte (29). Summarizing the results: cell suspension cultures of Cinchona are very poor producers of alkaloids. With an increase of organisation of the culture an increase of the alkaloid production can be observed (Fig. 2).

For the analysis of the alkaloids in screening programs of high producing cell cultures we have investigated several methods. First of all commercially available immunoassays for quinidine were tested for their sensitivity and selectivity in the analysis of plant material. It was found that the sensitivity was in the same range as HPLC, furthermore the FIA showed considerable cross reactivity for other quinoline alkaloids (Table 5). As only one alkaloid is measured and information on other alkaloidal constituents is missed by this method we prefered HPLC as method of analysis. As the cell cultures produced considerable amounts of anthraquinones (see below) an extensive clean-up procedure was necessary for the quantitative analysis of the alkaloids (30). Therefore a semiquantitative TLC method was developed for the rapid screening of large numbers of cell extracts (31). Using this method we were able to select cell lines that had special production characteristics.

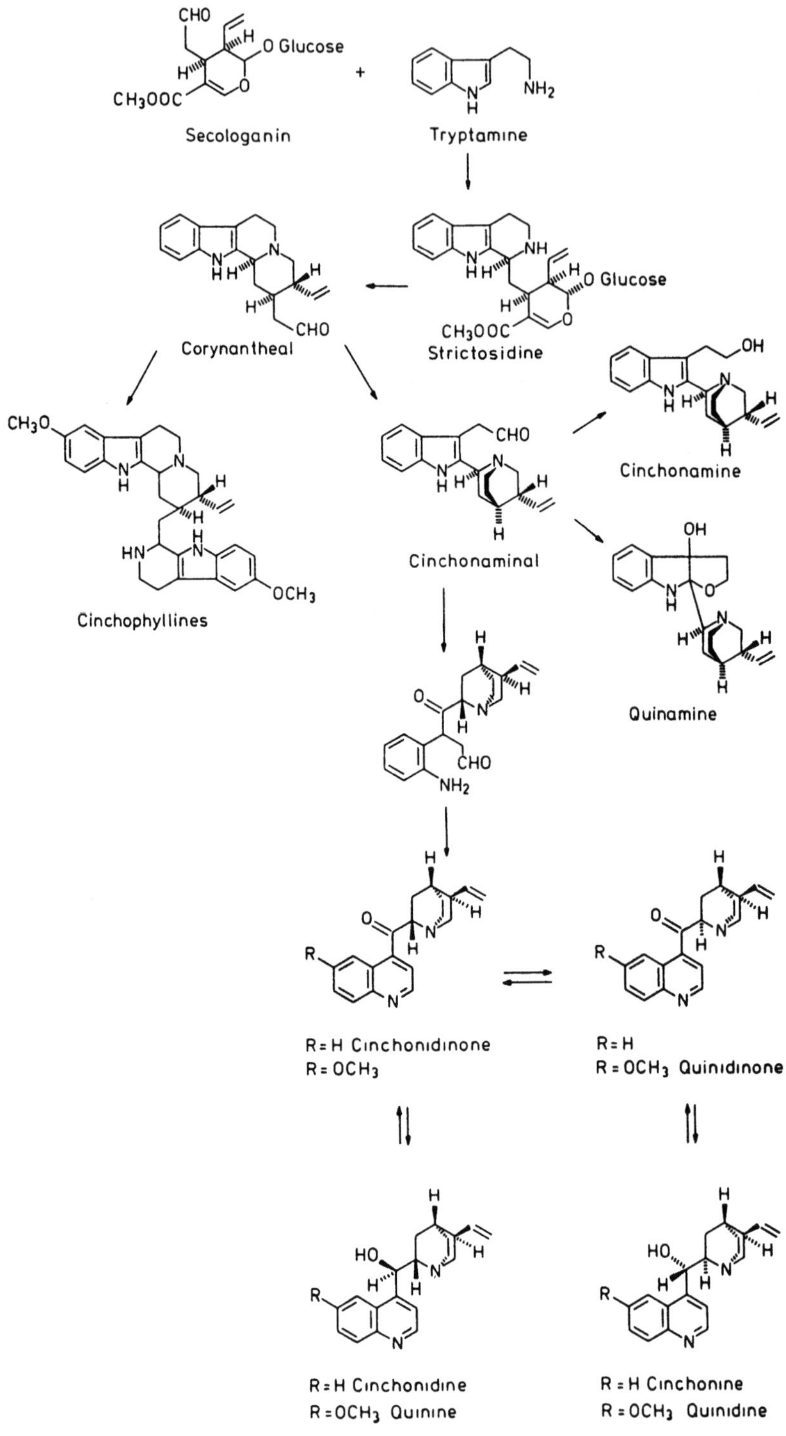

Scheme 1. Biosynthesis _Cinchona_ alkaloids.

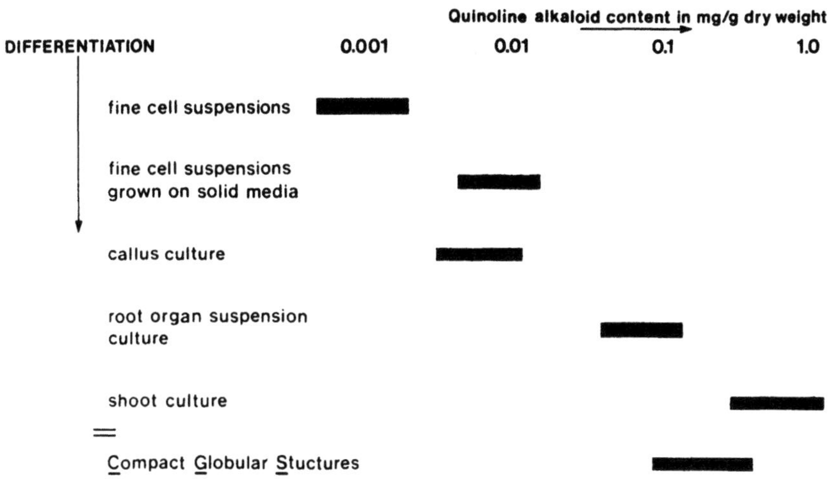

Fig. 2. Relationship between degree of differentiation and quinoline alkaloid production in *Cinchona* cell and tissue cultures.

TABLE 5. Cross-reactivity of *Cinchona* alkaloids in immunoassay methods. (Wijnsma et al., unpublished results)

Alkaloid	EMIT[a]	FIA[b]
Quinidine	100%	100%
Dihydroquinidine	46	91
Cinchonine	22	93
Cupreidine	20	74
Epiquinine	5	9
Cinchoninon	-	8
Quinidinon	-	4
Cinchonamine	-	3
Cinchonidine	-	-
Quinine	-	-
Epiquinidine	9	-
Hydroquinine	-	-
Detection limit Qd	70 ng ml^{-1}	40 ng ml^{-1}

a Enzyme immunoassay (SYVA)

b Fluorescent immunoassay (Ames TDA)

Optimization of media have been tried to improve alkaloid production in the cell suspension cultures (32,33). Although influence on the production could be noted, the levels remained too low for commercial interest. Therefore a number of precursors have been fed to the cultures, with the aim to improve the yield and to obtain information on the possible site of the block in the biosynthesis.

Feeding early precursors as secologanin, tryptophan and tryptamine did not improve the alkaloid production in our cell suspension cultures. Tryptophan even proved to be highly toxic to the cells. Tryptophan was converted in the presence of the cells, among others into norharman. This blue fluorescent product was also found to be formed from tryptophan dissolved in medium without cells upon sterilization (34,35).

Further down in the biosynthetic pathway corynantheal is a putative intermediate (Scheme 1) . Feeding of this compound to aggregates or cell suspension cultures did not result in the formation of quinoline alkaloids. The alkaloid was rapidly taken up by the cells ahd converted into corynatheol by reduction of the aldehyde group. Minor quantitities (less than 5%) of corynantheal and corynantheol are oxidized into the didehydro and tetradehydro derivatives (Scheme 2) (36). In one of the experiments also a cinchophylline like compound was isolated (3), the tentative structure is based upon its UV and MS. The enzyme responsible for the reduction of corynantheal does not reduce quinidinone. Whether corynantheal is not an intermediate or the corynantheal converting enzyme is lacking in our cultures remains an unanswered question.

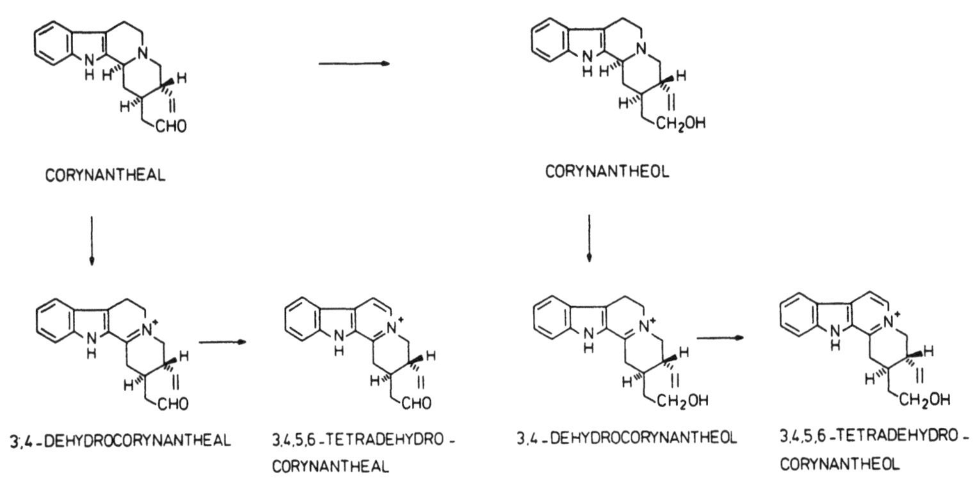

Scheme 2. Bioconversion of corynantheal in cell suspension cultures of Cinchona ledgeriana.

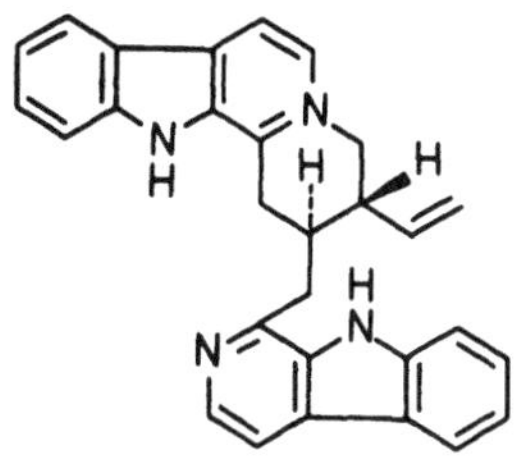

3

Lack of the enzyme can not be excluded as also tryptophan decarboxylase activity could not be determined in the same non-alkaloid producing cultures.
The alkaloid biosynthesis is thus not expressed in the cell suspension cultures. However, another secondary metabolite pathway is operative. The yellow-orange coloured products formed by the cell cultures were identified as anthraquinones, at that time not yet known as product from Cinchona trees. Further studies proved that these compounds probably acts as phytoalexins in Cinchona (37). They have antimicrobial activity, their production can be induced by microrganisms (e.g. fungi) and preparations thereof. Also in infected Cinchona trees we have been able to proof the presence of anthraquinones, whereas healthy ones are free of these compounds (38).

Finally we have been able to develop a cell culture system which does produce alkaloids. By changing the growth hormones in media cell aggregates were obtained: the compact globular structures (CGS). The CGS showed differentiation similar as found in the outer layers of the stem, i.e. xylem, cambium and phloem can be distinguished. This outer layer contained about 90% of all the alkaloids of the aggregate, whereas it only comprises about 10% of the biomass (39). The alkaloid levels in the CGS can be as high as 0.4%. No selection was performed, which means that probably higher levels can be reached upon selection.

Why the cells are capable of expressing the anthraquinone biosynthesis in in-vitro cell suspension cultures and not the alkaloid biosynthesis has still to be answered. Further studies on the genetic regulation of the secondary metabolism in Cinchona are needed to answer this question and to eventually open the way for a biotechnological production method.

3. TABERNAEMONTANA CELL CULTURES

The genus Tabernaemontana (Apocynaceae) is a wellknown source of a variety of indole alkaloids, many of which have pharmacological interest (40). Cell cultures of plants from this genus have been initiated (41) and were found to produce alkaloids. In order to be able to identify the alkaloids an isolation method has been elaborated, using droplet counter current chromatography (DCCC).

TABLE 6
Gradient ion-pair droplet counter current chromatography *Tabernaemontana* alkaloids (42).

McIlvaine buffer pH 4.2 containing 0-1 M $NaClO_4$	3
Methanol	5
Chloroform	5

Mobile phase: upper layer; stationary phase: lower layer.
Gradient see Fig. 3.

This system consisted of an ion-pair gradient solvent system for the DCCC (Table 6), enabling the efficient separation of a series of alkaloids. The type of separation was very different from that obtained with straight-phase silica gel LC systems. This made a combination with preparative TLC a very useful tool for the purification of the alkaloids (Fig. 3) (42). In this way a number of alkaloids have been isolated and their spectral data were recorded for identification. During this work we encountered several compounds which had ^{1}HNMR spectra similar to known alkaloids, however with a number of signals considerably shifted. The first thought, stereoisomers of the known alkaloids, was abandonned when it was found that these alkaloids had similar chromatograhic behaviour as the supposed stereoisomer. Finally it was found that small traces of acid present in the samples were responsible for large changes in the NMR spectra. Fig. 4 shows for example the spectra of appariclne and of a compound isolated, the changes only being due to the presence of acidic impurities. In the case of tubotaiwine (4) not only the ^{1}HNMR was changed, but also the MS showed a completely different fragmentation. The shifts under the influence of an increasing amount of trifluoroacetic acid are shown in Fig. 5. There is a linear relationship between the amount of acid and the shifts observed, until equimolar concentrations are reached (Fig. 6). The protonation of a second nitrogen, e.g. in quinine, might occur than inducing also shifts of protons close to the secondly protonated nitrogen.

From this it can be concluded that such shifts can be used as a tool in structure elucidation. Overlapping signals can be resolved and also information on protons close to the protonation site can be obtained, as both protons adjacent to, or close in space to the nitrogen are affected most. This makes trifluoroacetic acid into a useful NMR shift reagent (43,44) in structure elucidation of alkaloids.

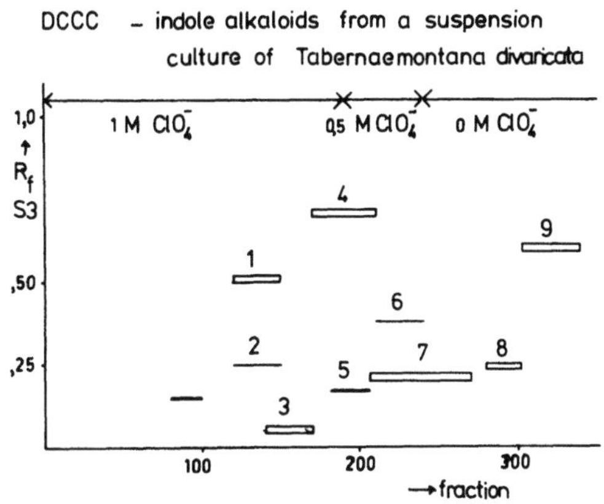

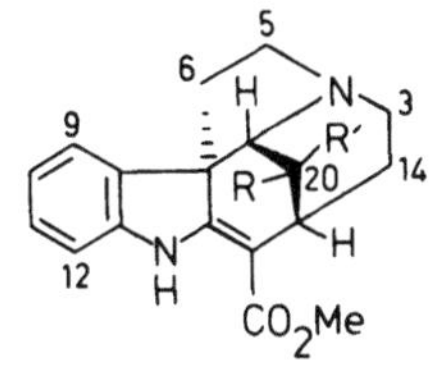

4 R=C_2H_5, R^1=H tubotaiwine

Fig. 3. **Separation of Tabernaemontana alkaloids by means of droplet counter current chromatography compared to a thin-layer chromatograhic separation.**
Y-axis: Rf value in TLC system $CHCl_3$-MeOH (9:1) on silica plates; X-axis: separation in DCCC-system (Table 6).
1 = vobasine; 2 = perivine; 3 = voaphylline hydroxy indolenine; 4 = voaphylline; 5 = vallesamine; 6 = pericyclivine; 7 = apparicine; 8 = tubotaiwine; 9 = coronaridine.

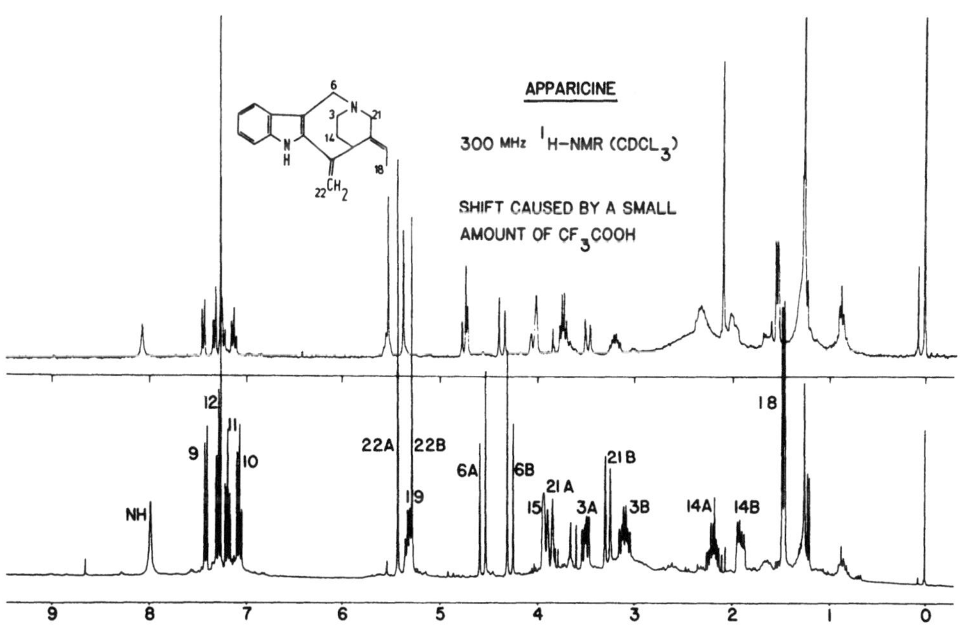

Fig. 4. ^{1}HNMR-spectrum of apparicine and apparicine contaminated with a trace of acid.

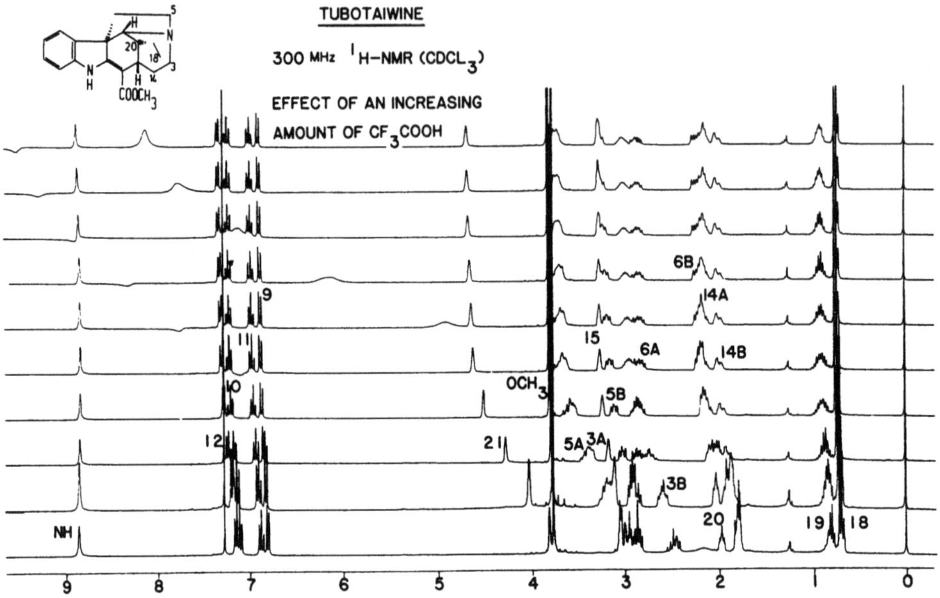

Fig. 5. Changes in the ^{1}HNMR-spectrum of tubotaiwine (4) upon addition of an increasing amount of trifluoro acetic acid.

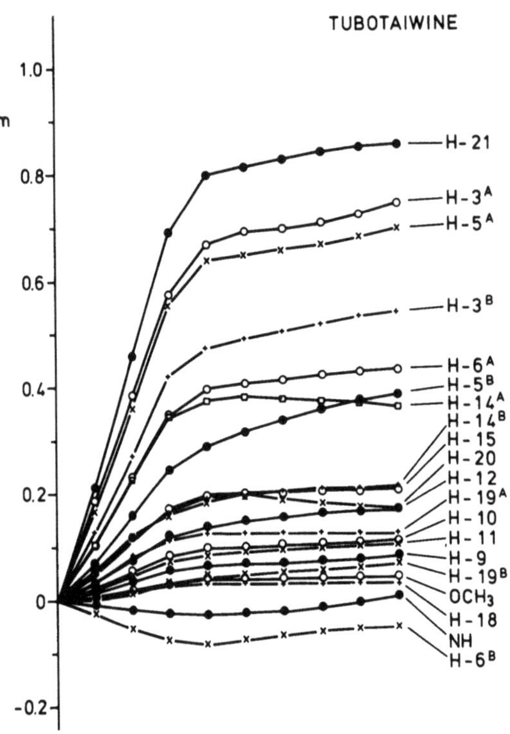

Fig. 6. Relationship of concentration of trifluoroacetic acid added and the changes in the ^{1}HNMR chemical shifts for the protons in tubotaiwine (4).

TABLE 7
Indole alkaloids from suspension cultures of Tabernaemontana species.

	T.africana	T.elegans	T.divaricata	T.orientalis
apparicine A*	+	+	+	+
tubotaiwine A		+	+	+
vallesamine A			+	
O-acetyl-vallesamine A		+		
16-OH-16,22-dihydro-apparicine A			+	
voaphylline P			+	+
voaphylline hydroxy-indolenine P			+	+
vobasine C		+	+	
vobasinol C		+		
perivine C		+	+	
tabernaemontanine C			+	
pericyclivine C			+	
isositsirikine C		+		+
coronaridine I			+	
19-S-heyneanine I			+	
catharanthine I			+	
12-OH-akuammicine S			+	

* A = aspidospermatan-type of alkaloid
P = plumeran-type of alkaloid
C = corynanthean-type of alkaloid
I = ibogan-type of alkaloid
S = strychnan-type of alkaloid

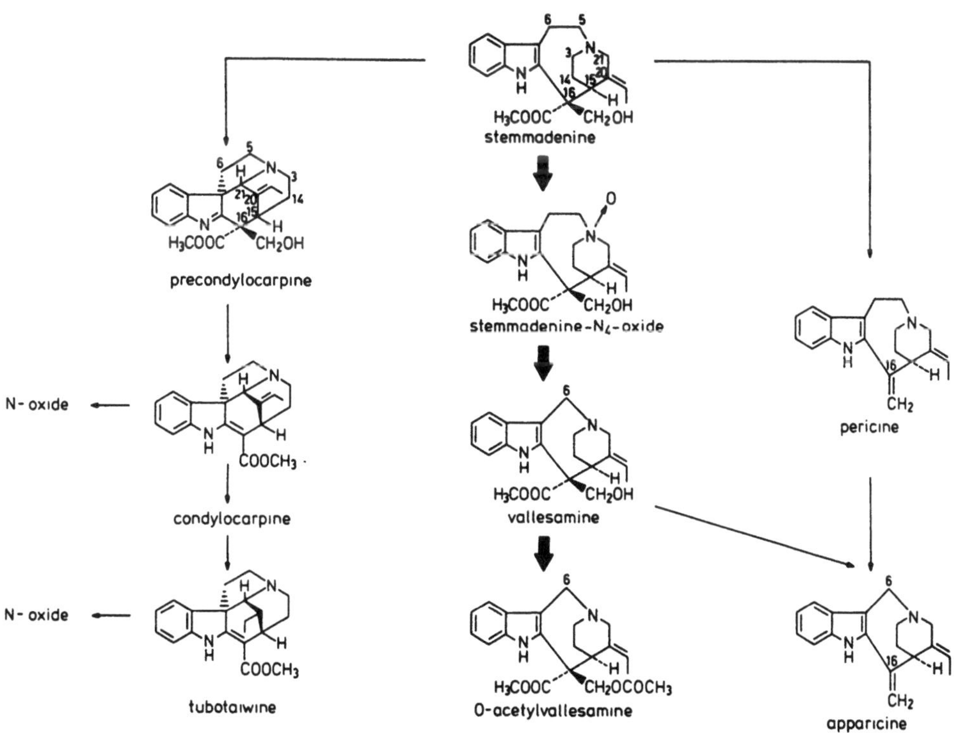

Scheme 3. Pathway from stemmadenine to various related alkaloids.

In Table 7 the alkaloids identified in the cell cultures of some Tabernaemontana species are summarized. Dimeric alkaloids, frequently found in plants, have so far only been found in callus cultures (45). In cell suspension cultures they occur in very low amounts, not allowing their identification with certainty. The low levels of ibogan alkaloids, part of the usual type of dimers, could be an explanation for this. The other part of the dimers, vobasinol and related alkaloids have been found in several cultures. Most common are alkaloids as apparicine, vallesamine and tubotaiwine, all derived from stemmadenine (Scheme 3). Intermediates as pericine has been identified at some occassions by means of UV and MS. Feeding of stemmadenine to cell free extracts of T. divaricata showed some conversion into vallesamine, however most of the precursor was spontaneously converted into condylocarpine (48). Precondylocarpine was tentatively identified by its UV, using HPLC equiped with a diodearray detector in analysing the incubation mixtures (49).

In fact stemmadenine might be a major constituent in the cells, which is rapidly converted upon destruction of the cellular organization, i.e. enzymes from other compartments in the cell may cause its conversion in among others condylocarpine.

4. CONCLUSIONS

Cell cultures of terpenoid-indole alkaloid producing plants from the genus Apocynaceae are capable of producing considerable amounts of alkaloids in batch cultures, e.g. Tabernaemontana cell cultures produce up to 0.2 g/l of alkaloids, without any cell line selection procedure. However cultures of the Rubiaceae genus Cinchona are very poor producers of terpenoid-indole and related quinoline alkaloids. Why this clear difference occurs is still an unanswered question. Studies into the molecular biology, i.e. the genetic regulation of this biosynthetic answer are needed to eventually find an answer. That means that further studies are needed of the biosynthesis. Most biosynthetic studies in plant secondary metabolism have been on the level of the chemistry: feeding various intermediates resulted in putative routes to a certain secondary product. Future research should go one step further, the isolation and characterization of the enzymes catalyzing the various steps. The enzymes being the key to genes responsible for secondary metabolism. The past years in several fields studies on the level of the enzymes of secondary metabolism have been started (50 and references cited therein, and the contribution of M. Rueffer in these proceedings). Further knowledge in this field will not only contribute to develop methods to improve production of secondary metabolites in cell cultures. Probably even more importantly, it will in future be of interest in plant breeding, by improving resistancy against e.g. microorganisms or insects, by genetic manipulation of secondary metabolism.Secondary metabolism in cell cultures of some terpenoid-indole alkaloid producing plants.

REFERENCES

1 M. Tabata and Y. Fuyita, Production of shikonin by plant cell cultures. Biotechnology in Plant science. Academic Press Inc., (1985) 207-218.

2 K. Ushiyama, H. Oda and Y. Miyamoto, Large scale tissue culture of Panax ginseng root, Abstracts VI International congress of plant tissue and cell culture, Minneapolis (1986)252.

3 M.A. de Jong and G.C. Stefess, Fabrieksvoorontwerp alkaloidproduktie door plantecellen, University of Technology, Delft, 1984.

4 W.E. Goldstein, L.L. Lasure and M.B. Ingle, Product cost analysis, in: E.J. Staba (Ed.), Plant tissue culture as a source of biochemicals. CRC Press, Boca Raton, (1980), pp. 191-234.

5 M.J.C. Rhodes and B.H. Kirsop, Plant cell cultures as source of valuable secondary products, Biologist 29 (1982) 134-140.

6 J.J. Meijer et al., unpublished results.

7 B. Ulbrich, W. Wiesner and H. Arens, Large scale production of rosmarinic acid from plant cell cultures of Coleus blume: Benth. In: K.H. Neumann, W. Barz and E. Reinhard (Eds), Primary and secondary metabolism of plant cell cultures, Springer Verlag, Berlin, 1985, pp 293-303.

8 W. De-Eknamkul and B.E. Ellis, Effect of auxins and cytokinins on growth and rosmarinic acid formation in cell suspension cultures of Anchusa officinalis, Plant Cell Rep., 4 (1985) 50-53.

9 Y. Fujita, S. Takahashi and Y. Yamada, Selection of cell lines with high productivity of shikonin derivatives by protoplast culture of Lithospermum erythrorhizon cells, Agric. Biol. Chem., 49 (1985) 1755-1759.

10 M.H. Zenk, H. El-Shagi and U. Schulte, Anthraquinone production by cell suspension cultures of Morinda citrofolia, Planta Med. Suppl., (1975) 79-101.

11 U. Schulte, H. El-Shagi and M.H. Zenk, Optimization of 19 Rubiaceae species in cell cultures for the production of anthraquinones, Plant Cell. Rep., 3 (1984) 51-54.

12 B.E. Ellis, Production of hydroxyphenylethanol glycosides in suspension cultures of Syringa vulgaris, Phytochemistry, 22 (1983) 1941-1943.

13 F. Sato and Y. Yamada, High berberine-producing cultures of Coptis japonica cells, Phytochemistry, 23 (1984) 281-285.

14 K. Nakagawa, A. Konagai, H. Fukni and M. Tabata, Release and crystallization of berberine in the liquid medium of Thalictrum minus cell suspension cultures, Plant Cell. Rep., 3 (1984) 254-257.

15 H. Hinz and M.H. Zenk, Production of protoberberine alkaloids by cell suspension cultures of Berberis species, Naturwiss., 67 (1981) 620-621.

16 H. Schübel and J. Stöckigt, RLCC-isolation of Raucaffricine from its most efficient source-cell suspension cultures of Rauwolfia serpentina Benth, Plant Cell. Rep., 3 (1984) 72-74.

17 B. Deus-Neumann and M.H. Zenk, Instability of indole alkaloid production in Catharanthus roseus cell suspension cultures, Planta Med., 50 (1984) 427-431.

18 Y. Kitamura, H. Miura and M. Sugii, Alkaloid composition and atropine esterase activity in callus and differentiated tissue of Duboisia myoporoides R.Br., Chem. Pharm. Bull., 33 (1985) 5445-5448.

19 Y. Mano, S. Nabeshima, C. Matsui and H. Okkawa, Production of tropane alkaloids by hairy root cultures of Scopolia japonica, Agric. Biol. Chem., 50 (1986) 3725-2722.

20 H.E. Flores and P. Filner, Metabolic relationships of putrescine. GABA and alkaloids in cell and root cultures of solanaceae, In: K.H. Neumann, W. Barz and E. Reinhard (Eds.), Primary and Secondary Metabolism of Plant Cell Cultures. Springer Verlag, Berlin, 1985, pp. 174-185.

21 T. Endo and Y. Yamata, Alkaloid production in cultured roots of three species of Duboisia, Phyto Chemistry, 24, (1985) 1233.

22 T. Hartmann, L. Witte, F. Oprach and G. Toppel, Reinvestigation of the Alkaloid composition of Atropa belladonna planrs, root cultures and cell suspension cultures, Planta Med., (1986) 390-395.

23 S.W. Zito and E.J. Staba, Thebaine from root cultures of Papaver bracteatum, Planta Med., 45 (1982) 53-54.

24 T. Yoshikawa and T. Furya, Morphinan alkaloid production by tissues differentiated from cultured cells of Papaver somniferum, Planta Med., 51 (1985) 110-113.
25 Y. Mi-ura, Preparation of vinblastine using Vinca rosea tissue culture in plant growth factor containing culture medium (Ja), Patent J5 9088-096: 09-11-82-JP-197412 (21.05.84) 09.11.82.
26 D. McHale, The Cinchona tree, Biologist, 33 (1986) 45-53.
27 E.H.D. Smit, Verleden, heden en perspectieven van Cinchona spp., Pharm. Weekbld., 119 (1984) 159-162.
28 R. Verpoorte, J. Schripsema and T. van der Leer, Cinchona alkaloids, The Alkaloids, Brossi (Editor). Academic Press. In press.
29 R. Wijnsma and R. Verpoorte, Quinoline alkaloids in cell and tissue cultures of Cinchona species. Cell cultures and somatic genetics of plants. Vol. 5. I.K. Vasil and F. Constabel (Eds.), Academic Press. in press.
30 R. Wijnsma, T.B. van Vliet, P.A.A. Harkes, H.J. van Groningen, R. van der Heijden, R. Verpoorte and A. Baerheim Svendsen, An improved method for the extraction of alkaloids from cell and tissue cultures of Cinchona species, Planta Medica, 53 (1987) 80-84.
31 S.S. Hoekstra, M. Meijerink, R. Visse, K.R. Libbenga, P.A.A. Harkes and R. Verpoorte, Selection of in vitro cultures of Cinchona ledgeriana and of Cinchona robusta for production of quinoline alkaloids, Proc. 4th European Congress on Biotechnology 1987, vol. 2. O.M. Neijssel, R.R. van der Meer and K.Ch.A.M. Luyben (Eds.), Elsevier, Amsterdam, (1987) 413-416.
32 P.A.A. Harkes, L. Krijbolder, K.R. Libbenga, R. Wijnsma and R. Verpoorte, Influence of various media constituents on the growth of Cinchona ledgeriana tissue cultures and the production of alkaloids and anthraquinones therein, Plant Cell, Tissue and Organ Culture, 4 (1985) 199.
33 R. Wijnsma, R. Verpoorte, P.A.A. Harkes, T.B. van Vliet, H.J.G. ten Hoopen and A. Baerheim Svendsen, The influence of initial sucrose and nitrate concentrations on the growth of Cinchona ledgeriana cell suspension cultures and the production of alkaloids and anthraquinones therein, Plant Cell, Tissue and Organ Culture, 7 (1986) 21-29.
34 P.A.A. Harkes, P.J. de Jong, R. Wijnsma, R. Verpoorte and T. van der Leer, Influence of production media on Cinchona cell cultures; spontaneous formation of B-carbolines from L-tryptophan, Plant Science, 47 (1986) 71-76.
35 T. van der Leer et al., unpublished results.
36 T. van der Leer et al., unpublished results.
37 R. Wijnsma, J.T.K.A. Go, I.N. van Weerden, P.A.A. Harkes, R. Verpoorte and A. Baerheim Svendsen, Anthraquinones as phytoalexines in cell and tissue culturesof Cinchona, Plant Cell Rep., 4 (1985) 241.
38 R. Wijnsma, I.N. van Weerden, R. Verpoorte, P.A.A. Harkes, Ch.B. Lugt, J.J.C. Scheffer and A. Baerheim Svendsen, Anthraquinones in Cinchona ledgeriana bark infected with Phytophthora cinnamonii, Planta Medica, (1986) 211-212.
39 S.S. Hoekstra et al., unpublished results.
40 T.A. van Beek, R. Verpoorte, A. Baerheim Svendsen, A.J.M. Leeuwenberg and N.G. Bisset, Tabernaemontana, a review of its alkaloids, botany, pharmacology and ethnopharmacy, J. Ethnopharmacol., 10 (1984) 1.
41 R. van der Heijden, R. Wijnsma, R. Verpoorte, P.A.A. Harkes and A. Baerheim Svendsen, The influence of nutrient media, auxins and cytokinins on the initiation of alkaloid producing callus on leaf explants of eight species of Tabernaemontana, Fitoterapia, 57 (1986) 415-421.
42 R. van der Heijden, A. Hermans-Lokkerbol, R. Verpoorte and A. Baerheim Svendsen, Pharmacognostical studies on Tabernaemontana XX. Ion-pair droplet counter-current chromatography of indole alkaloids from suspension cultures, J. Chromatogr. 396(1987)410-415.
43 R. Verpoorte. Methods in structure elucidation of alkaloids. J. Nat. Prod., 49 (1986) 1-25.
44 J. Schripsema, R. Verpoorte, and A. Baerheim Svendsen, Trifluoroacetic acid, a ^{1}HNMR shift reagent for alkaloids, Tetrahedron Letters, 27 (1986) 2523-2526.

45 R. van der Heijden, R.L. Brouwer, R. Verpoorte, R. Wijnsma, T.A. van Beek, P.A.A. Harkes and A. Baerheim Svendsen, Indole alkaloids from a callus culture of Tabernaemontana elegans, Phytochemistry, 25 (1986) 843.
46 R. van der Heijden, A. Hermans-Lokkerbol, L.P.J. de Kool, P.A.A. Harkes and R. Verpoorte, Indole alkaloids from suspension cultures of Tabernaemontana divaricata, I. Isolation and identification of the alkaloids, Planta Medica. Submitted.
47 R. van der Heijden et al. unpublished results.
48 R. van der Heijden, G.M. van der Graaf, E.J.M. Pennings and R. Verpoorte, Formation and degradation of some aspidospermatan indole alkaloids in suspension cultures of Tabernaemontana divaricata, Zeitschrift f. Naturforsch. C. Submitted.
49 R. van der Heijden, P.J. Lamping, P.P.Out, R. Wijnsma and R. Verpoorte, High-performance liquid chromatographic determination of indole alkaloids in a suspension culture of Tabernaemontana divaricata, J. Chromatogr. 396 (1987) 287-295.
50 R. Verpoorte, Enzymen in de terpenoid-indool alkaloid biosynthese. Een overzicht, Pharm. weekblad, 121 (1986) 277.

Synthetic, Multi-Deuteriated Cholesterol as a Quantitative Probe of the Formation of Cholesterol Oxidation Products

B. A. WASILCHUK, P. FEIBUSH, P. W. LE QUESNE AND PAUL VOUROS

Department of Chemistry and Barnett Institute of Chemical Analysis, Northeastern University, Boston, MA, 02115 USA

ABSTRACT

The synthesis of 24,24,25,26,26,26,27,27,27-$^2H_9$-cholesterol is described. A method for use of the compound as an internal monitor for determination of the origin of oxidation products of cholesterol and their quantitation in naturally occurring media is discussed. In the analytical procedure the labeled cholesterol is added to the sample immediately upon collection thus subjecting the labeled and unlabeled substrates to identical oxidation conditions during the analyte isolation process and corrects for procedural artifacts. Conventional capillary gas chromatography-flame ionization detection (GC-FID) and combined gas chromatography-mass spectrometry (GC-MS) have been used. Quantitation with an accuracy of 5% or better at the low part per million level for some of the common oxidation products is demonstrated.

1. INTRODUCTION

The determination of cholesterol oxidation products in prepared foods or biological samples is hindered by artifacts arising from the well documented (ref. 1) strong susceptibility of cholesterol itself to undergo oxidation. For example, oxidation products such as 7-ketocholesterol and 25-hydroxycholesterol can be produced quite readily upon exposure of cholesterol to air at room temperature (refs. 2,3). Interest in these oxides (structures summarized in Fig. 1) has increased since the original reports of their presence in aortic tissues from normal and atherosclerotic patients by Brooks (refs. 4,5) and Smith (refs. 6,7). This early work was qualitative in nature and, as acknowledged by the authors, the procedures used for sample isolation may have been responsible for the production of artifacts. More recent reports (refs. 8,9), raise the strong possibility that oxygenated sterols, and not necessarily cholesterol, may play the primary role in arterial wall injury and lesion development. In a series of carefully controlled experiments these investigators (ref. 8) showed that intravenous injection of oxygenated sterols -- among them, cholestan-3β,5α,6β-triol, 25-hydroxycholesterol and

Cholesterol

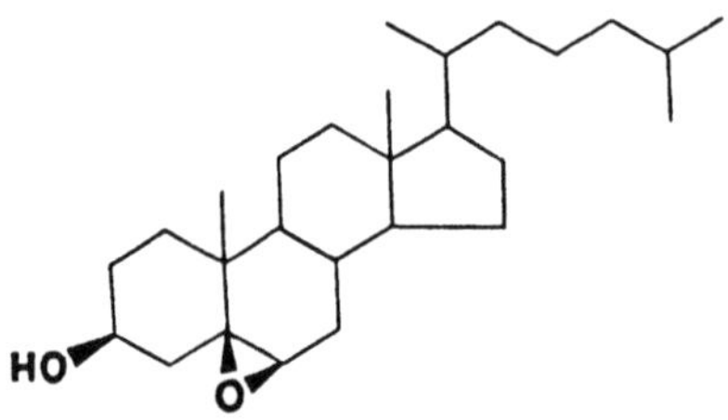

5β,6β-Epoxycholesterol

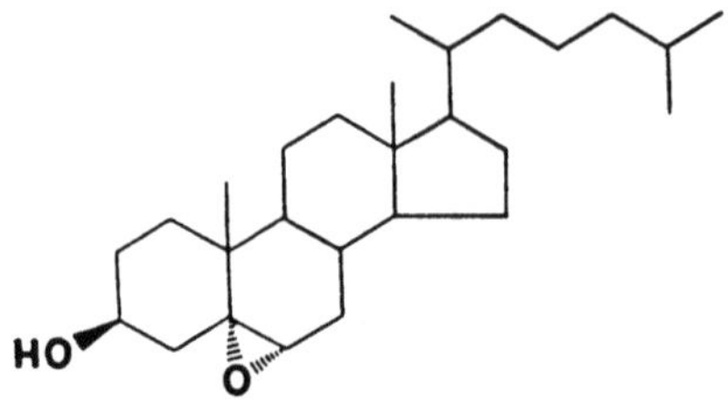

5α,6α-Epoxycholesterol

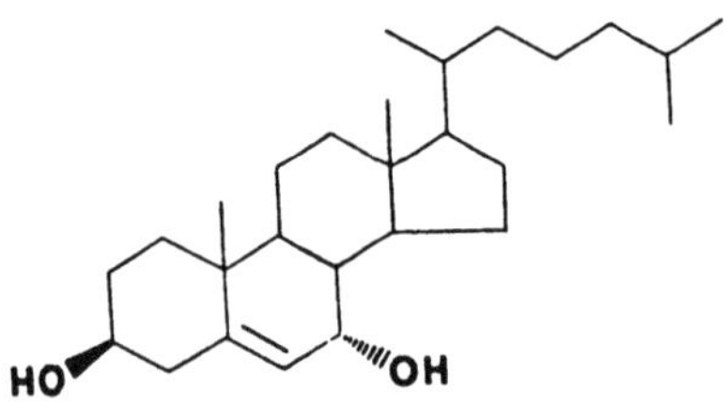

7α-Hydroxycholesterol

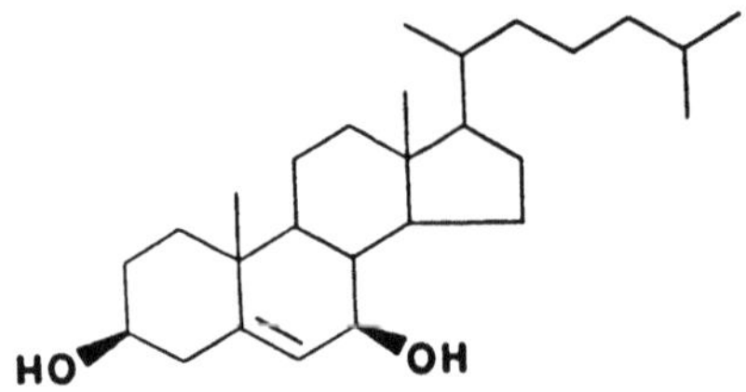

7β-Hydroxycholesterol

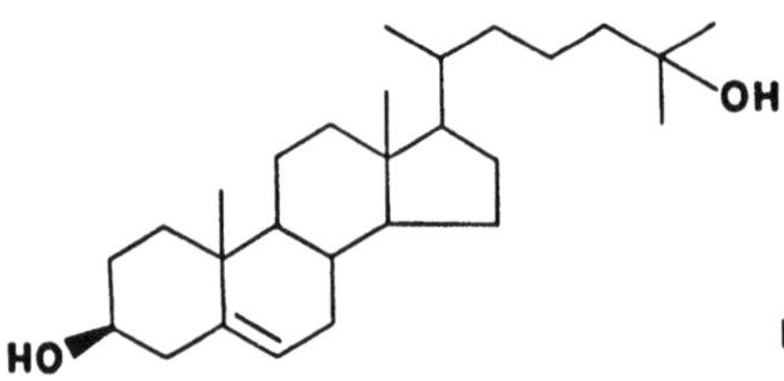

25-Hydroxycholesterol

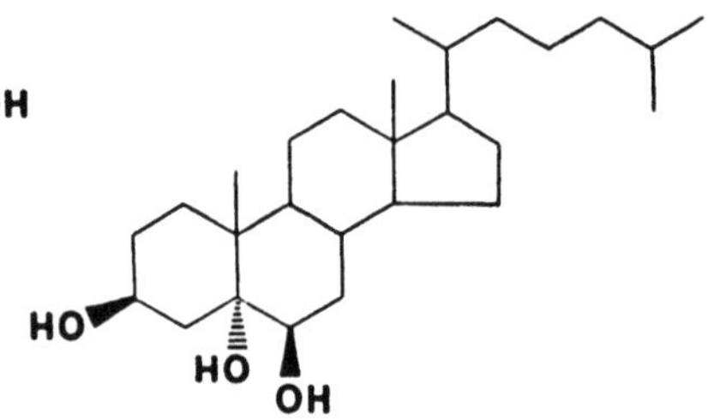

5α-Cholestane-3β,5,6β-triol

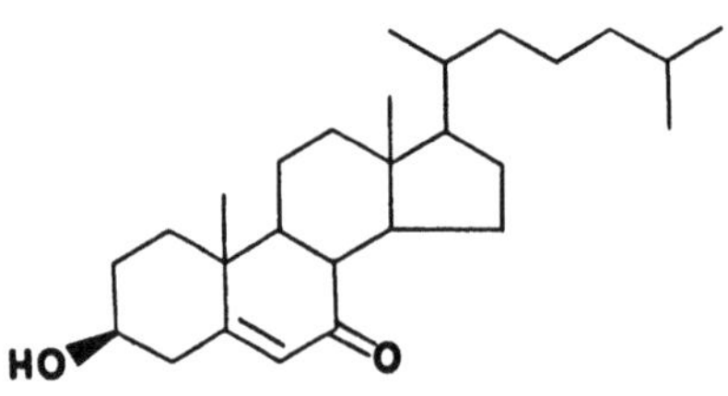

7-Ketocholesterol

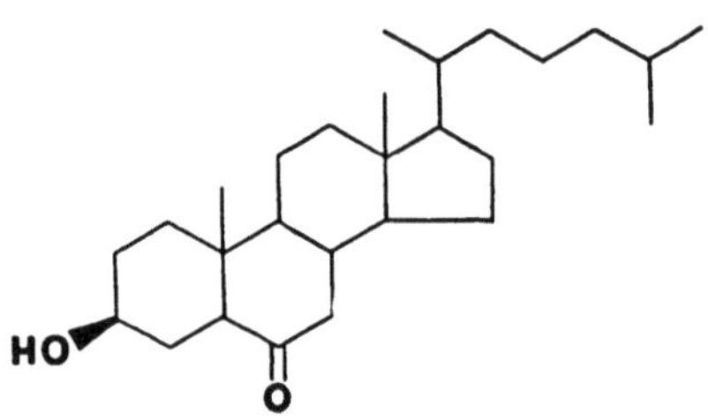

6-Ketocholestanol
(internal standard)

Figure 1. Molecular Structures of Cholesterol Oxidation Products of Interest

cholesterol-5α,6α-epoxide -- led to characteristic patterns of cell death, inflammation and repair in rabbits' aortas and pulmonary arteries as early as 24 hours after administration. No such drastic effects were observed with subjects administered highly purified cholesterol. These trends are consistent with the reported relative toxicities of cholesterol and some of its oxidized derivatives as summarized in Table 1. Further evidence of the importance of oxidized sterols is also lent by the report of Kandutsch et al. (ref. 10) that certain oxygenated derivatives of cholesterol are potent inhibitors of cholesterol biosynthesis in mammalian cells. Given the considerations outlined above, we undertook this study in order to develop analytical methodology which would address in an unequivocal manner the issue of the presence and level of occurrence of cholesterol oxidation products in a given biological or other related type of sample rich in cholesterol.

TABLE 1

Cytoxicity of Oxidized Derivatives of Cholesterol in Aortal Smooth Muscle Cell Cultures (ref. 11)

	Cytotoxicity (Dying or Dead Cells) %			
Sterol	10μg/mL	20μg/mL	50μg/mL	100μg/mL
25-Hydroxy	5-25	25-50	75-100	75-100
3β,5α,6β-triol	5-25	25-50	75-100	75-100
7α-Hydroxy			5-25	25-50
7β-Hydroxy			5-25	50-75
7-Keto			5-25	50-75
α-Epoxide	Non-toxic			
Cholesterol (pure)	Non-toxic			

2. EXPERIMENTAL

Steroid standards were purchased from the following sources: Cholestane-3β,5α,6β-triol and 7α-hydroxycholesterol from Research Plus Laboratories, Inc. (Denville, NJ); Cholesterol, 25-hydroxycholesterol, 7-ketocholesterol, 7β-hydroxycholesterol, 6-ketocholestanol and cholesterol-5α,6α-epoxide from Steraloids, Inc. (Wilton, NH); Desmosterol from Supelco, Inc. (Bellefonte, PA) and cholesterol-5β,6β-epoxide from the College of Pharmacy, Northeastern University (Boston, MA). A series of 1.0 μg/g stock solutions of the above steroid standards were prepared, by weight, in ethyl acetate for subsequent use in calibration curve construction and spiking.

Gas chromatography (GC) was performed on a Varian Associates Inc. (Palo Alto, CA) model 3700 gas chromatograph equipped with a flame ionization detector (FID) and modified to accommodate a laboratory-made capillary on-column injector (ref. 12). All gas chromatographic separations were carried out on a J&W Scientific (Folsom, CA) bonded phase fused silica capillary column (60 m × 0.32 mm ID, DB-1 stationary phase, 0.25 μm film thickness) unless otherwise noted. The GC oven temperature was programmed from 50°C to 300°C at 20°C/min.

Gas chromatography-mass spectrometry (GC-MS) was performed with a Finnigan MAT (San Jose, CA) model 4021B quadrupole mass spectrometer interfaced to a Hewlett Packard 5890 GC equipped with an on-column injector. Mass spectra during GC-MS were obtained by chemical ionization with ammonia reagent gas at a source pressure of 0.3 torr.

2.1 Procedure

Solutions of $^2H_9$-cholesterol and $^2H_0$-cholesterol were oxidized in the following manner unless otherwise stated: Appropriate amounts of labeled and unlabeled cholesterol (1-2 mg) were combined and then purified by reversed phase HPLC to eliminate initial oxide impurities in the mixture. An aliquot of this purified cholesterol solution was then taken in order to establish the initial $^2H_0/^2H_9$ ratio. An aqueous cholesterol suspension was then placed in a heating mantle, heated to 50°C, continuously purged with oxygen (<10 psi) and UV irradiated from a mercury arc lamp for 2.25 hours. These reaction conditions were experimentally found to produce the following cholesterol oxidation products in good yield: 7α-hydroxycholesterol, 7β-hydroxycholesterol, β-epoxide, α-epoxide and 7-ketocholesterol.

3. RESULTS AND DISCUSSION

3.1 Experimental Approach

Our fundamental approach to addressing the problem of potential artifacts arising from the procedural oxidation of cholesterol is pictorially summarized in Fig. 2. For illustration purposes, this hypothetical example assumes a 1:1 ratio of labeled to unlabeled cholesterol. Isotopically labeled $^2H_9$-cholesterol is added to the analytical sample immediately upon collection and prior to initiation of any extraction or other isolation procedures. As shown in Fig. 2, the measured ratio of $^2H_0/^2H_9$ cholesterol recovered from the sample serves as the reference index as to whether endogenous oxides were present in the sample or not. Three possibilities may be considered. First, if the detected oxides contain no deuterium label, trace (A), it is clear that they are likely to be constituents of the original sample and, certainly, not procedurally derived. Second, if the ratio of labeled to unlabeled oxides is identical to that shown for cholesterol, trace (B), it is obvious that the isolation procedure is responsible for their formation. In the event that there are contributions from both sources -- i.e., both procedural and endogenous oxides -- a third trace, (C), would be expected. The level of endogenous oxides can be calculated from the difference, as shown in Fig. 2. These principles were applied in this study.

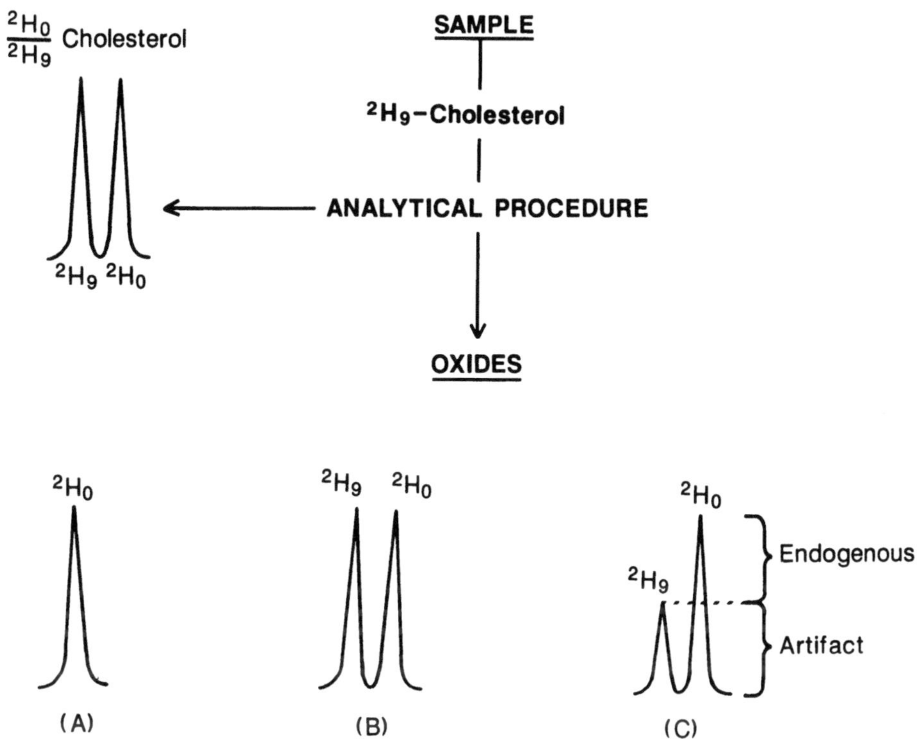

Figure 2. Use of Deuterium-Labeled Cholesterol to Monitor the Formation of Artifacts Arising from Procedural Oxidation

In developing the methodology described in the preceding paragraph, we focussed on the following most commonly encountered oxidized derivatives of cholesterol: 5β,6β-epoxycholesterol (β-epoxide); 5α,6α-epoxycholesterol (α-epoxide); 5α-cholestane-3β,5,6β-triol (triol); 7α-hydroxycholesterol (7α-hydroxy); 7β-hydroxycholesterol (7β-hydroxy); 25-hydroxycholesterol (25-hydroxy); and 7-ketocholesterol (7-keto). The related compound, 6-ketocholestanol (6-keto), which has not been reported to form in the spontaneous oxidation of cholesterol, has been used as an internal standard for quantitation purposes. The names in parentheses refer to the common abbreviation terms for these compounds. Their structures are given in the text.

Two of these compounds, the α-epoxide and 7-ketocholesterol, were used as test solutes in a model experiment to establish whether a typical procedure for their isolation from dried egg powder (ref. 13), can lead to their formation from pure cholesterol. A sample of cholesterol, 40 mg, was purified by

HPLC and the oxide fraction collected and analyzed by GC. The purified cholesterol was again subjected to extraction, repurified and the α-epoxide and 7-keto content determined. Care was exercised to carry out these experiments under inert conditions and with solvents degassed with pure nitrogen gas. Five such passes of the cholesterol failed to yield a sample void of the oxides as indicated from the results summarized in Table 2. Significantly, after a relatively high oxide content in the initial analysis, a leveling out is observed in the subsequent passes. The data confirm the strong propensity of the cholesterol molecule to undergo oxidation during routine sample cleanup and illustrate the need for rigorously controlled conditions -- i.e., use of an internal monitor -- when analyzing for its oxidized by-products.

TABLE 2

Quantity of Oxidized Product Found in 40 mg Cholesterol After Each Repeated Pass

Compound	Pass #1	Pass #2	Pass #3	Pass #4	Pass #5
α-Epoxide	5 μg	4 μg	4 μg	3 μg	4 μg
7-Keto	7 μg	4 μg	4 μg	3 μg	4 μg

3.2 **Synthesis of 24,24,25,26,26,26,27,27,27-$^{2}H_{9}$-Cholesterol**

Use of isotopically labeled analogues as internal standards for quantitative analysis is an established procedure in GC-MS (ref. 14). Less common is the use of these analogues as internal monitors for purposes such as the one described in this report. Selection of the appropriate labeled analogue is important in either case. For example, in quantitative GC-MS, retention of the label(s) with a major fragment ion is an important requirement. Our decision to prepare a counterpart of cholesterol labeled on the side chain instead of the more readily accessible steroid nucleus was motivated by the desire to eliminate the possibility of isotope effects, since the majority of the oxidation products of interest are formed by attack on the "A" and "B" rings. Indeed, this proved to be a wise decision as illustrated from the example of 25-hydroxycholesterol discussed below. Moreover, we felt that preparation of a multiply labeled cholesterol molecule might provide the additional option of carrying out the final analysis by conventional GC as well as GC-MS. This belief was reinforced by the report of Hindriks, et al. (ref. 15), who showed good (*ca.* 50% valley) separation between cholesterol and 2,2,4,4,6-$^{2}H_{6}$-cholesterol by capillary GC analysis.

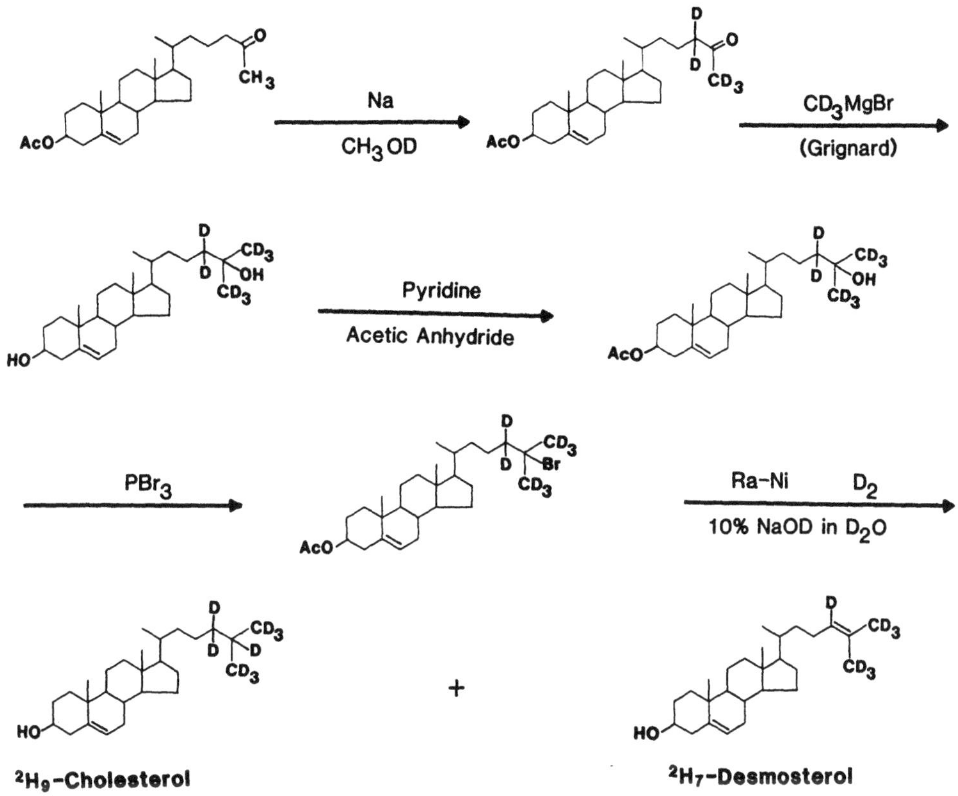

Figure 3. Synthesis of $^{2}H_{9}$-Cholesterol

The synthesis of $^{2}H_{9}$-cholesterol is outlined in Fig. 3. The starting material, 3β-acetoxy-26-norcholest-5-ene-25-one (ref. 16), was labeled by simple exchange of the hydrogens alpha to the carbonyl in O-$^{2}H_{1}$-methanol. The subsequent Grignard reaction produced $^{2}H_{8}$-25-hydroxycholesterol some of which was recovered for future use. The acetyl group was reintroduced to protect the 3β-hydroxy group and the 25-hydroxy group replaced with bromine by reaction with PBr_3. The final debromination step resulted in a nearly 1:1 mixture of $^{2}H_{9}$-cholesterol and $^{2}H_{7}$-desmosterol which were separated and collected by reversed phase HPLC.

The deuterium isotope incorporation of the labeled cholesterol was calculated by electron impact mass spectrometry of its trimethylsilyl ether. The molecular ion regions of the spectra of the labeled and unlabeled derivatives, obtained at an ionizing energy of 15 eV, are compared in Figs. 4(a) and 4(b),

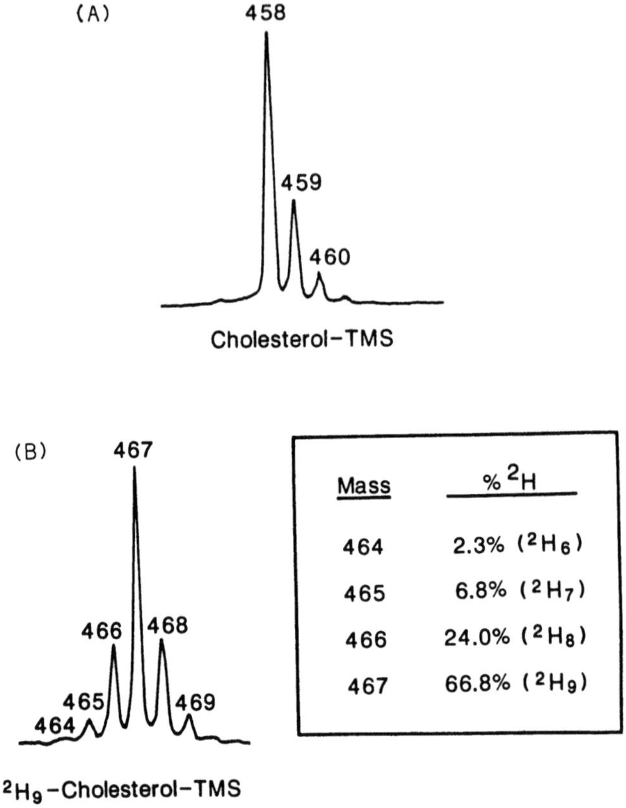

Mass	% ^{2}H
464	2.3% ($^{2}H_{6}$)
465	6.8% ($^{2}H_{7}$)
466	24.0% ($^{2}H_{8}$)
467	66.8% ($^{2}H_{9}$)

Figure 4. Molecular Ion Regions of Electron Impact (15 eV) Spectra of Cholesterol. (a) $^{2}H_{0}$-cholesterol-TMS; (b) $^{2}H_{9}$-cholesterol-TMS

respectively. On the basis of the peak ratios, nearly 70% purity in $^{2}H_{9}$- and 24% in $^{2}H_{8}$-incorporation is calculated. Significantly, only minor amounts of products with lower deuterium content and no unlabeled derivative are present in the final product.These are important requirements for the projected use of the compound.

The value of multiple isotope content in the internal monitor is realized both in terms of the mass spectrometry and the chromatographic analysis. In terms of mass spectrometry, incorporation of several deuterium atoms eliminates the possibility of spectral overlap of the signals from the labeled and unlabeled compounds in case of coeluting or overlapping chromatographic peaks. From the point of view of the chromatography, the mass increment introduced by the multiple number of labels offers the opportunity for better resolution of the deuteriated and undeuteriated cholesterols and their respective oxides. The capillary GC chromatogram (Fig. 5) of a mixture of $^{2}H_{9}$-cholesterol and $^{2}H_{8}$-25-hydroxycholesterol isolated from the synthetic scheme (Fig. 3) and their unlabeled counterparts illustrates the validity of this hypothesis. Excellent resolution of the peaks into well defined doublets is indicated, thus supporting the possibility for development of a capillary GC-based analytical procedure. The accomplishment of these goals is outlined below.

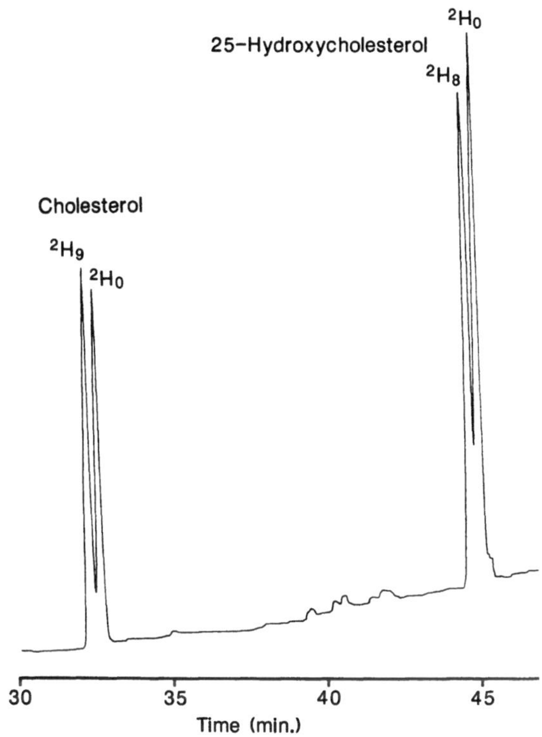

Figure 5. Capillary GC Separation of $^2H_9$-cholesterol-TMS and $^2H_8$-25-hydroxycholesterol-TMS from Their Unlabeled Analogs

3.3 <u>Results</u>

In order to confirm the validity of the proposed approach, i.e., use of deuterium labeled cholesterol as an internal monitor for the correction of procedural artifacts, we first had to establish that the relative rates of oxidation of labeled and unlabeled cholesterol were identical. In order to assess this relationship an approximately 1:1 mixture of $^2H_9$- and $^2H_0$-cholesterol, void of any oxides, was subjected to oxidation as described in the experimental section above. A gas chromatogram of the mixture is shown in Fig. 6. Notable is the excellent separation of the constituents and the close correlation of the ratios of labeled and unlabeled oxides in the mixture. GC-MS analysis of the same solution yielded a similar pattern. The results are summarized in Table 3. The same experiment was conducted with a *ca.* 3:1 mixture of unlabeled to labeled cholesterol. The GC-MS chromatogram of the mixture is shown in Fig. 7 and the corresponding ratios in each doublet indicated on the figure. Again the ratios of the oxides are extremely comparable to that of cholesterol confirming the validity of the use of a labeled internal monitor. An additional noteworthy feature of this chromatogram is the nearly baseline chromatographic separation of the labeled and unlabeled analogs.

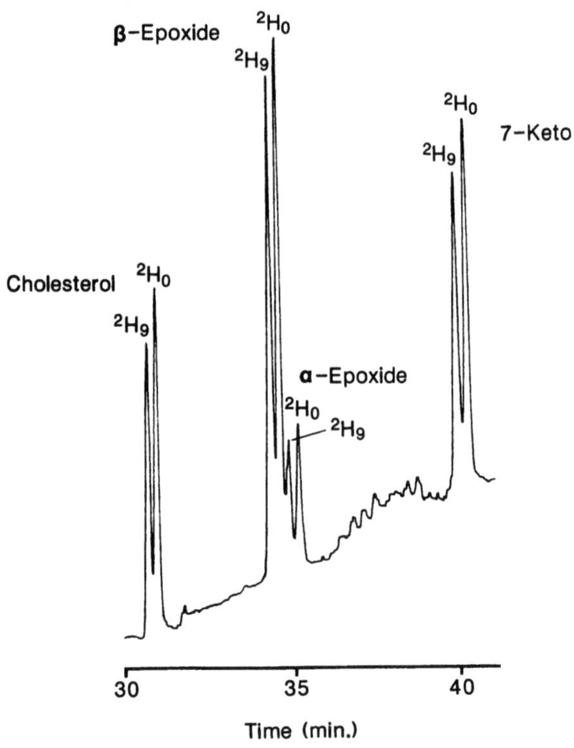

Figure 6. Capillary GC Separation of Labeled and Unlabeled Cholesterol Oxides Obtained from the Oxidation of a *ca.* 1:1 $^2H_0/^2H_9$ Cholesterol Solution

TABLE 3

Comparison of Ratios of Unlabeled to Labeled Cholesterol and Corresponding Oxidation Products

Compound	GC	GC-MS
Cholesterol	1.07 ± 0.07	1.26 ± 0.04
β-Epoxide	1.16 ± 0.07	1.29 ± 0.04
α-Epoxide	1.15 ± 0.07	1.35 ± 0.02
7-Keto	1.13 ± 0.06	1.26 ± 0.02

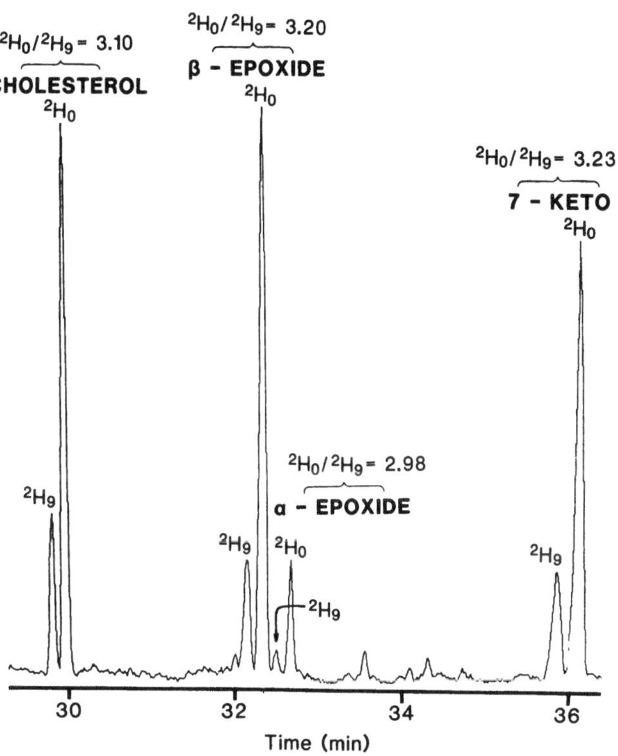

Figure 7. Capillary GC-MS Separation of Labeled and Unlabeled Cholesterol Oxides Obtained from the Oxidation of a *ca.* 3:1 $^2H_0/^2H_9$ Cholesterol Solution

In both of the aforementioned oxidation experiments, significant was the lack of formation of any 25-hydroxycholesterol. Since this compound is one of the more significant cholesterol oxidation products, we examined conditions more favorable for its formation. A mixture of cholesterol and $^2H_9$-

cholesterol (*ca.* 2:3), was purified by HPLC. The cholesterol fraction was evaporated to dryness in a 40 mL beaker so that solid cholesterol coated the bottom of the beaker. The beaker was then placed on a hot plate and heated to 90-100°C while irradiating the cholesterol sample under UV (254 nm) for *ca.* 5 hours. The oxidized cholesterol was purified by HPLC and the oxide fraction along with some of the cholesterol collected, derivatized with BSTFA and analyzed by GC. The chromatogram, Fig. 8, shows only a single peak for 25-hydroxycholesterol, corresponding to the unlabeled analog as verified by GC-MS. It is reasoned that the deuterium isotope effect inhibits oxidation at the 25-position. Given the severity of the isotope effect, it is clear that an alternative internal monitor is needed to correct for any procedural artifacts leading to the formation of 25-hydroxycholesterol. Referring to the synthetic scheme (Fig. 3), such a compound can be prepared by carrying out the debromination step in a hydrogen-containing medium or, alternatively, via the hydrogenation of $^{2}H_{7}$-desmosterol. With the exception of the 25-hydroxycholesterol case, the oxidation experiments established the feasibility of the proposed use of labeled internal monitors for determination of the origin of cholesterol oxidation products.

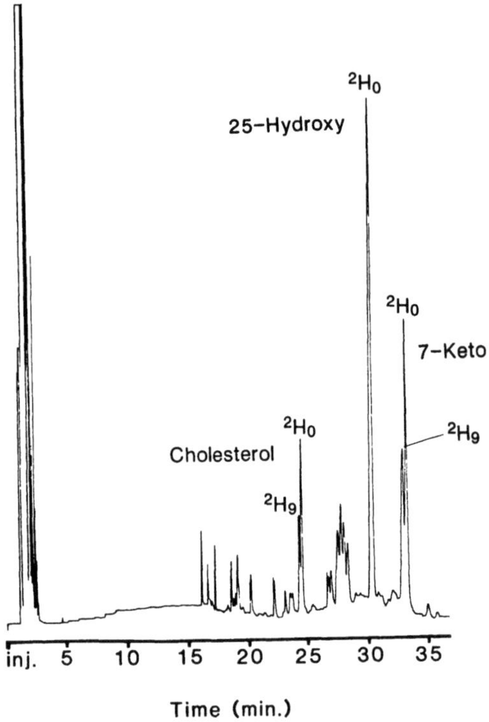

Figure 8. Capillary GC Trace of Oxides Obtained from the Oxidation of a Mixture (*ca.* 2:3) of solid $^{2}H_{0}$-cholesterol and $^{2}H_{9}$-cholesterol

The final step in the development of the internal monitor technique was a confirmation that the methodology could be used to accurately quantify the amount of endogenous oxides in a sample, in addition to simply monitoring for their formation from procedural oxidation. A quantification test was thus used for this purpose based on the following series of steps:

(i) A mixture of $^{2}H_9$- and $^{2}H_0$-cholesterol was purified by HPLC.

(ii) The cholesterol fraction was collected and the oxides discarded.

(iii) The ratio of $^{2}H_9$- to $^{2}H_0$-cholesterol was measured by GC and GC-MS.

(iv) The cholesterol fraction was subjected to oxidation.

(v) The oxidized mixture was spiked with measured amounts of 6-keto (internal standard) and the test solutes 7-keto and α-epoxide.

(vi) The mixture was processed and analyzed by GC and GC-MS.

The "endogenous" oxides in this experiment are represented by the spiked test solutes, α-epoxide and 7-keto. The internal standard is used to calculate the total quantity of each oxidized product found in the final analysis. A calibration curve for each analyte in question is constructed for this purpose. Using the experimentally measured ratio of labeled to unlabeled cholesterol as a reference, it is possible to calculate the endogenous oxide content of a given sample by subtracting the procedurally formed oxides from the total amount. A representative gas chromatogram of a test mixture prepared as outlined in (i)-(vi) above is shown in Fig. 9. Note the close correspondence of the ratios of doublets corresponding to the 7α-hydroxy, β-epoxide, and 7β-hydroxy compounds to the doublet generated from the labeled and unlabeled cholesterol. On the other hand, the ratio is distorted in the case of the "endogenous" oxides (α-epoxide and 7-keto) which were added to the sample in the unlabeled form. The amounts of the two test solutes found by this procedure match closely the spiked quantities. The results of this experiment, obtained both by GC and GC-MS analysis, are summarized in Table 4.

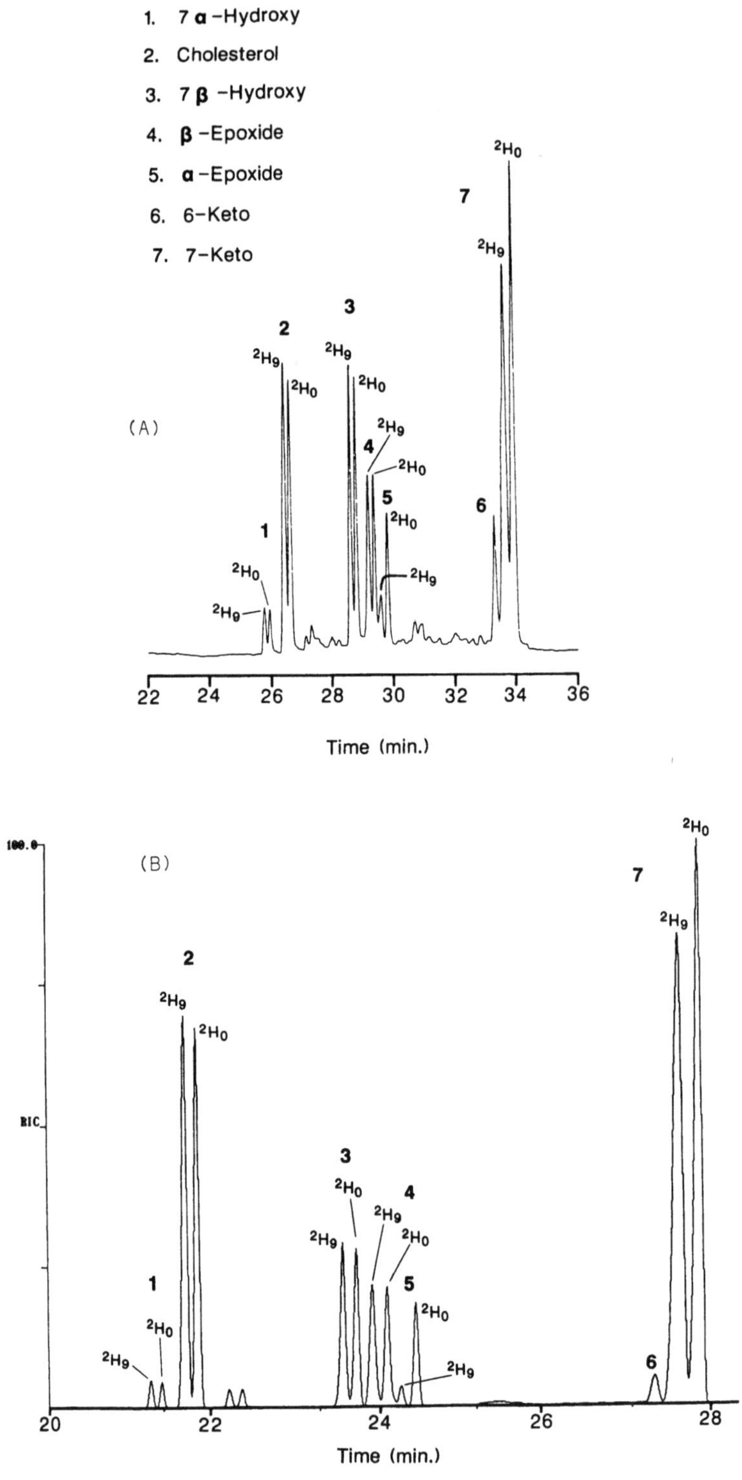

Figure 9. Test for Quantitation of "Endogenous" Oxides (7-keto, α-epoxide) in the Presence of Oxidation Artifacts. (A) GC; (B) GC-MS.

TABLE 4

Results from Quantitation Test. Ratio of Unlabeled to Labeled Analogs (refer also to Fig. 9)

Compound	GC-MS	GC
Cholesterol	0.80 ± 0.01	0.92 ± 0.05
7β-Hydroxy	0.83 ± 0.01	0.93 ± 0.04
β-Epoxide	0.81 ± 0.02	0.94 ± 0.04
7α-Hydroxy	0.82 ± 0.03	0.87 ± 0.06
α-Epoxide[a]	5.34 ± 0.10	2.81 ± 0.05
7-Keto[b]	1.19 ± 0.02	1.08 ± 0.06

[a]Amount of a-Epoxide spiked: 10.01 μg
Determined: 10.40 ± 0.27 μg (GC); 9.22 ± 0.30 μg (GC-MS)

[b]Amount of 7-keto spiked: 4.99 μg
Determined: 4.89 ± 0.21 μg (GC); 5.04 ± 0.24 μg (GC-MS)

4. CONCLUSION

The data presented here demonstrate a reliable methodology for monitoring the formation and quantifying the presence of oxidized products of cholesterol. This approach can now be applied towards the analysis of these materials in any type of medium. The synthesis of nona-labeled cholesterol and its use as an internal monitor permit reliable analysis by capillary GC as well as GC-MS. For the analysis of oxides isolated from complex matrices, GC-MS is the desirable analytical method because of its high selectivity. However, the demonstrated applicability of conventional capillary GC provides an inexpensive and more readily available alternative which can be used as an initial screening method. These results provide a basis for new analytical procedures which can be applied to the analysis of sterols of universally recognized medical significance.

ACKNOWLEDGMENT. This work was supported, in part, by a grant from the Council for Tobacco Research and funds from a Biomedical Research Support Grant (RR07143). Publication No. 333 from the Barnett Institute.

DEDICATION. We dedicate this publication to the memory of our late colleague and friend Nicholas T. Werthessen whose inspiration and commitment were instrumental in our pursuit of this work.

REFERENCES

1. L.L. Smith, Cholesterol Autoxidation, Plenum Press, 1981, pp.13-41.
2. L.L. Smith, J.I. Teng, M.J. Kulig and F.L. Hill, J. Org. Chem., 38, 1763 (1973).
3. N.D. Weiner, P. Noomnont and A. Felmeister, J. Lipid Res., 13, 252 (1972).
4. C.J.W. Brooks, W.A. Harland and G. Steel, Biochem. Biophys. Acta, 125, 620 (1966).
5. C.J.W. Brooks, G. Steel, J.D. Gilbert and W.A. Harland, Atherosclerosis, 13, 223 (1970).
6. L.L. Smith and J.E. VanLier, Atherosclerosis, 12, 1 (1970).
7. L.L. Smith, D.R. Day, J.A. Moody, J.D. Wells and J.E. VanLier, J. Neurochem., 19, 899 (1974).
8. H. Imai, N.T. Werthessen, V.S. Subramanyam, P.W. Le Quesne, H.A. Soloway and M. Kanisawa, Science, 207, 651 (1980).
9. Anonymous, The Lancet, 1980, p.964.
10. A.A. Kandutsch, H.W. Chen and H.J. Heniger, Science, 201, 498 (1978).
11. S.-K. Peng, P. Tham, C.B. Taylor and B. Mikkelson, Am. J. Clin. Nutrition, 32, 1033 (1979).
12. B.A. Wasilchuk and P. Vouros, J. Chromatogr., 407, 305 (1987).
13. S.R. Missler, B.A. Wasilchuk and C. Merritt, Jr., J. Food Sci., 50, 595 (1985).
14. B.M. Milard, Quantitative Mass Spectrometry, Heyden & Son, Ltd., London, 1978.
15. B.G. Wolthers, F.R. Hindriks, F.A.J. Muskict and A. Croen, Clin. Chim. Acta, 103, 305 (1980).
16. T.A. Nerwik, K.E. Cooney and M. Uskokovich, Helv. Chim. Acta, 53, 771 (1974).

Discovery and Chemistry of Naturally-Occurring Anticancer Agents

JOHN M. CASSADY, CHING-JER CHANG, AND R. GRAHAM COOKS (Affiliations see page 304)

1. ABSTRACT

Recent advances in the discovery of novel anticancer agents from higher plants are reported. These compounds are uncovered based on bioassay-directed fractionation and tandem MS analysis to detect analogs of significant lead compounds. The evolution and current status of the screening program of the U.S. National Cancer Institute is reviewed.

Structural work and chemistry is reported for active compounds from *Annona densicoma*, *Cudrania cochinchinensis*, *Desfontainia spinosa*, *Pararistolochia flos-avis*, *Podocarpus* species, *Psorospermum febrifugum*, and *Spathelia sorbifolia*. A number of these novel compounds exhibit sufficient activity to initiate studies of their mechanism of action and analog synthesis.

2. INTRODUCTION

Further advances in cancer chemotherapy require the continued development and testing of new and improved antineoplastic agents. This drug development effort has a relatively short history tracing back about forty years to the observations that nitrogen mustard and folic acid antagonists could inhibit tumor growth. Since that time a rapidly increasing number of clinically useful compounds have been developed through screening of a variety of synthetic and natural products. Much of this effort has been stimulated and coordinated by the National Cancer Institute through its comprehensive drug development program (ref. 1-15). This program includes the screening of compounds obtained from natural products and has led to the testing of over 200,000 microbial cultures and 120,000 plant extracts for cytotoxicity and activity against various mouse leukemias (ref. 10,12). The total number of plant species tested to date in the NCI program is estimated to be about 10 percent of the 230,000 to 500,000 available species (ref. 9,10).

The need for novel and more selective anticancer drugs continues, in spite of continuing efforts to understand how cancer develops and spreads. If efforts to prevent the disease by identifying causal agents or by using chemical modifiers were immediately instituted and successful it is likely that cancer would continue to occur at the present rate for several decades. Rational design of novel and effective anticancer drugs is limited by our still scant knowledge of cancer cell biology and biochemistry. Thus, it appears likely that development of novel agents from a variety of sources based on screening in animal tumor models will remain as an important research area in the foreseeable future.

3. ANTICANCER DRUGS FROM HIGHER PLANTS

Natural products have a history of providing novel, clinically useful anticancer drugs and a number of these have come from higher plants (ref. 1-21). In addition, many of these compounds have served as prototypes for the development of novel analogs of clinical importance. Some specific examples are colchicine (ref. 22) and its derivatives, the podophyllotoxins (ref. 23) including the clinically-effective epipodophyllotoxin derivatives etoposide and teniposide (ref. 24), and the *Vinca* alkaloids, vincristine, vinblastine, and vindesine (ref. 25,26). More recently indicine N-oxide, homoharringtonine and taxol have entered clinical trials (ref. 13,21). The discovery of natural products with potential as anticancer drugs has recently been reemphasized by the NCI.

Thus, plant natural products continue to provide 1) novel, clinically useful antineoplastic drugs, 2) novel structural prototypes for the development of analogs and 3) biochemical tools which often elucidate novel mechanisms of growth control. Chemical variation of known structural classes does and should continue since it provides a means for obtaining improved drugs, increased understanding of the biochemistry of the cancer cell, and novel structure-growth inhibitory activity relationships. At the same time, however, it is clear that continued screening of natural products leading to the discovery of novel prototypes is essential to continuing progress in cancer chemotherapy and therein lies the significance of this project.

The key to the success of the plant anticancer drug discovery program resides in the bioactivity-directed isolation procedure. This approach has been demonstrated to be effective. It has several advantages over the phytochemical approach which generally leads to the isolation of the most abundant, easily separated and easily crystallized components in a crude plant extract. The activity-directed approach often yields highly potent, minor, novel components which would be missed in a clasical approach. In order to detect potentially useful compounds which are present in minute amounts a number of *in vitro* prescreens are used in early confirmation and fractionation stages. These systems have enhanced sensitivity and are less costly than *in vivo* screens. The addition of *in vitro* prescreens to detect agents with specific mechanisms, such as mitotic inhibition using the 9ASK system, or with specific action against stable lines from human lung or colon adds further significance to this approach.

4. SCREENING SYSTEM

The Division of Cancer Treatment, National Cancer Institute (NCI) has operated a program for the discovery of potential anticancer drugs since 1955. The screen used has undergone almost continuous evaluation and modification since then. Initially the choice was to use an *in vivo* disease model, the L1210 mouse leukemia. In 1971 this was replaced by a more sensitive line, the P388 mouse leukemia. This model was used to prescreen 10,000 compounds per year. A secondary panel of costly solid tumor models, including some human xenografts was utilized from 1975-85. This compound oriented approach led to the discovery of many useful clinical agents for a narrow spectrum of human tumors, primarily leukemias and lymphomas. The failure to find effective chemotherapeutic agents for adult solid tumors has led to a recent decision to switch to a disease oriented screen (ref. 14,15). The primary screen consists of an *in vitro* system which comprises a panel of up to 100 human cancer cell lines, including resistant disease types such as lung, colon, breast and melanoma. The goal is to identify agents with selective toxicity toward cell types of interest. Compounds of interest are evaluated in animal models derived from the specific target cell. Our strategy is to duplicate a sub-set of cell lines for our in-house testing. This is essential in order to minimize the delay in obtaining results from the bioassays which will guide fractionation of confirmed active extracts (see Figures 1 and 2). In our initial phase we have decided to include one cell line from five major human tumor types. The lines in place include a colon cancer, HT29 and a non-small cell lung cancer, A549. The MCF7 breast line will be introduced in 1988 followed by a small cell lung and melanoma line.

Figure 2. Purdue Cancer Center *In Vitro* Screen

Disease	**Cell (1987)**	**(1988)**
Leukemia	P388	
Non-Small Cell Lung Cancer	A549	
Small Cell Lung Cancer		TBD
Colon Cancer	HT29	
Breast Cancer		MCF7
Melanoma		TBD

5. MS/MS AS AN ISOLATION PROBE

Mass spectrometry has long been an important tool in natural products research (ref. 27), in part because of its sensitivity and also because mass spectral fragmentations often serve to class a compound with related structural groups. The technique also provides molecular weights (and in the high resolution mode, molecular formulas) together with fragmentation data

Figure 1. Disease Oriented *In Vitro* Screen, NCI

Psorospermin Diol (NSC 600964)

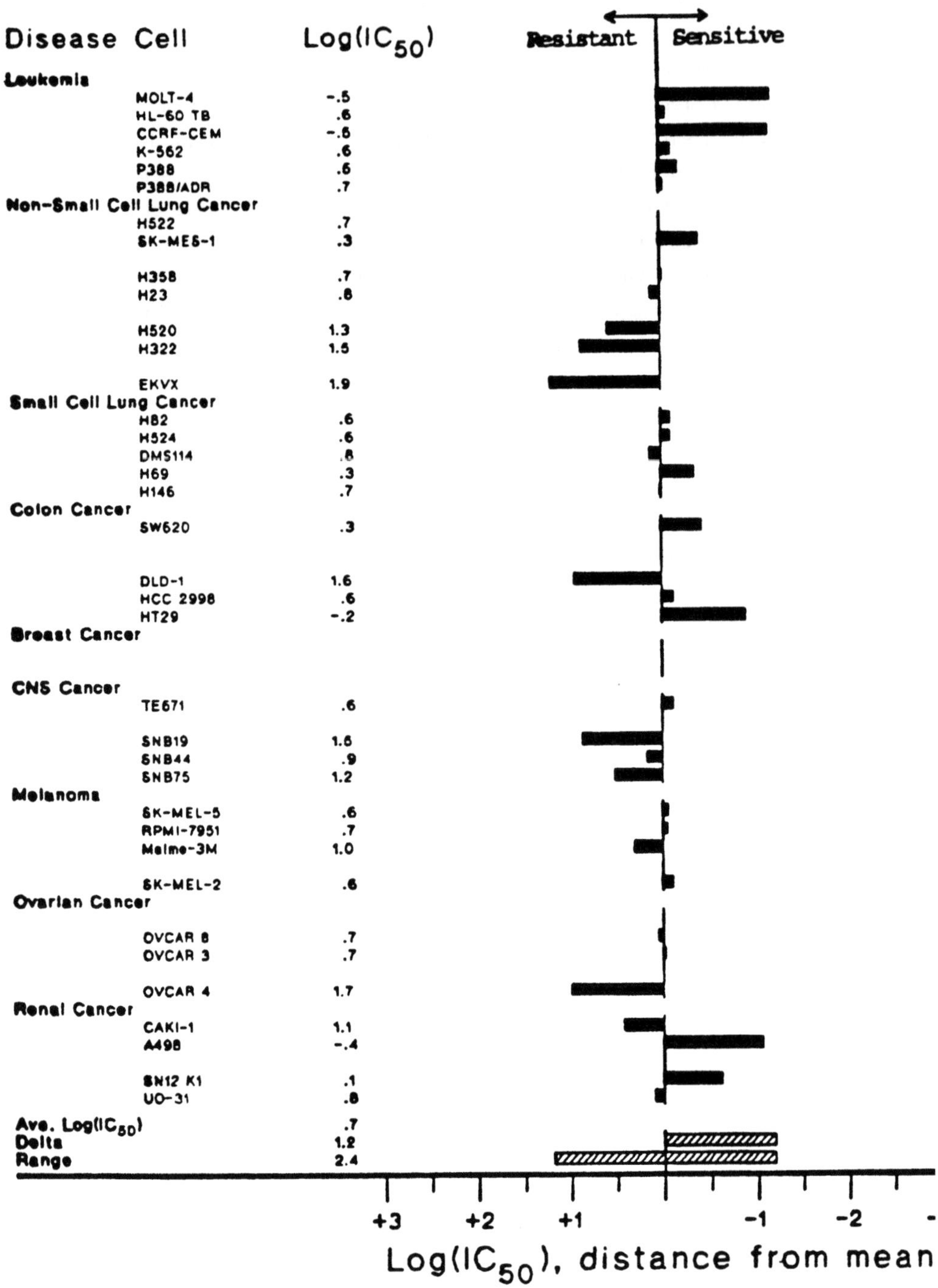

which are a unique fingerprint of the individual molecule. Two recent developments add greatly to these capabilities: one, the heralded new ionization methods which extend the grasp of MS to include fragile, non-volatile and ionic compounds (ref. 28-33). Second, and probably as important for natural products research, is the tandem mass spectrometry family of methods (ref. 34-37).

Tandem mass spectrometry offers:

(i) qualitative and quantitative mixture analysis with reduced sample work-up (ref. 38-39).

(ii) improved detection limits over those achieved by single stage mass spectrometry for many samples (ref. 40,41).

(iii) structurally diagnostic fragmentations, especially useful when mixtures are analyzed or with ionization methods such as fast atom bombardment in which fragmentation may be obscured or limited (ref. 42).

These procedures are valuable in natural products research because they can dramatically decrease the work required to characterize natural compounds; at the same time their sensitivity and structural specificity allow new compounds to be discovered in trace concentrations not amenable to conventional approaches. They allow rapid screening of crude extracts for particular compounds or sets of related compounds with all that this implies for the natural products chemist. Extension of these techniques will allow the use of ion/molecular reactions in structural characterization of natural products (ref. 43-45).

6. *ANNONA DENSICOMA*

Members of the family Annonaceae, the custard apple family, are widely distributed and the subject of numerous phytochemical investigations. Earlier studies of various Annonaceae have resulted in the isolation and identification of a number of aporphine alkaloids. Recently a series of novel, cytotoxic polyketides containing tetrahydrofuran and γ-lactone groups have been discovered (ref. 46-49).

In preliminary screening, extracts of *Annona densicoma* Mart. (Annonaceae) collected in Peru (South America) were active in the 9ASK and 9PS cell culture systems. Activity-directed fractionation of an ethanolic stem-bark extract resulted in the isolation of a highly cytotoxic compound named annonacin (1, ref. 50).

Annonacin had several structural features (see Figure 3) which were novel among compounds of this type such as hydroxyl groups at C-4 and C-10 and a single tetrahydrofuran group flanked by hydroxyls. It was also the first member of this group reported to contain 35 carbons. In addition, it was unique in that it affected reversal of ASK cells. Preliminary results indicated that this was not a result of direct binding to tubulin (private communication from Dr. Ernest Hamel, NCI, Bethesda, MD, 1987).

Further fractionation has now resulted in the isolation and structural elucidation of three novel analogs of **1**. These compounds, named annonacinone (**2**), isoannonacin (**3**), and isoannonacinone (**4**) were elucidated by chemical interrelationships (see Figure 4) and by detailed analysis of NMR and HRMS data (ref. 51).

Figure 3. Cytotoxic Polyketides of the Annonaceae

Extracts of *Annona densicoma* ⟶ Annonacin (1) $C_{35}H_{64}O_7$

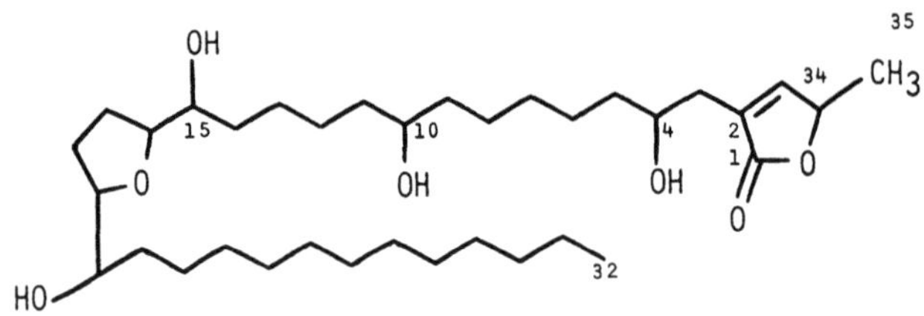

9KB	1×10^{-3} μg/ml
9PS	1×10^{-5} μg/ml
9ASK	51% Reversal

Figure 4. Cytotoxic Polyketides from *Annona*

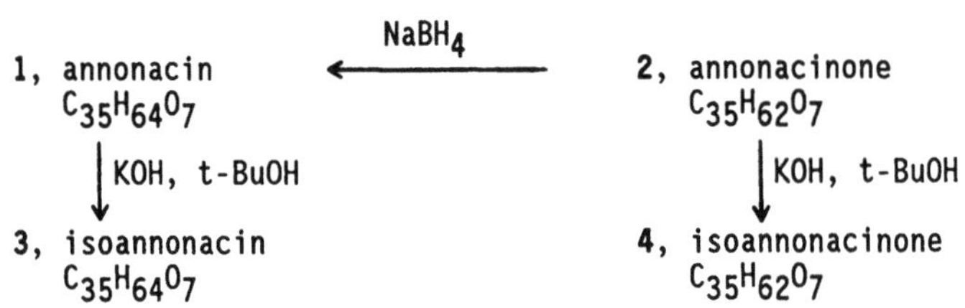

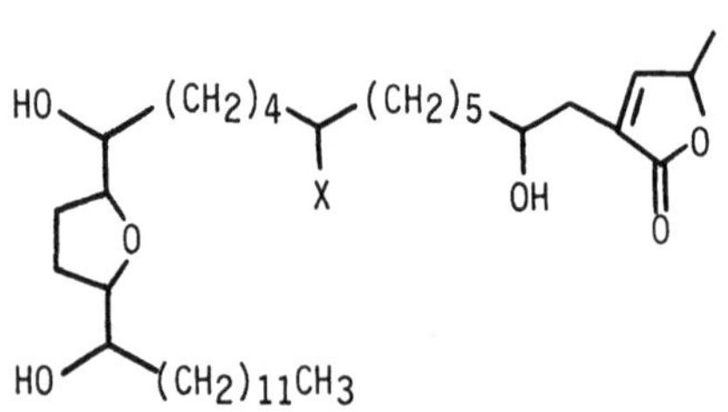

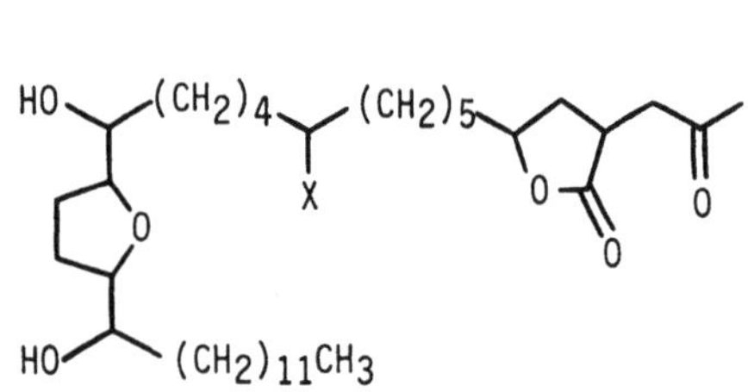

1, Annonacin (X = OH)

2, Annonacinone (X = $\overset{O}{\parallel}$)

3, Isoannonacin (X = OH)

4, Isoannonacinone (X = $\overset{O}{\parallel}$)

Further research is underway to establish the structure-activity relationships in this series and to determine the stereochemistry of **1-4** (see Figure 4). The preliminary cytotoxicity data (Table 1) is very interesting since the iso series appears to affect colon cells more specifically than leukemia cells. Based on this preliminary data **1** and analogs are undergoing further evaluation by NCI in the disease oriented tumor panel.

Table 1. Cytotoxicity of Annonacin (1) and Analogs

Compound ED_{50} in μg/ml in	**9PS**	**A549**	**HT29**
Annonacin (1)	$1x10^{-5}$	$1x10^{-3}$	NT
Annonacinone (2)	$1x10^{-6}$	NT	NT
Isoannonacin (3)	<$1x10^{-1}$	$2x10^{-2}$	$2x10^{-3}$
Isoannonacinone (4)	$5x10^{-1}$	$7x10^{-2}$	$9x10^{-3}$

NT - not tested

7. *CUDRANIA COCHINCHINENSIS*

The plant *Cudrania cochinchinensis* has been used in Chinese traditional medicine for the treatment of hepatitis (ref. 52). Prior studies of this plant uncovered the agent cudranone (ref. 53). Fractionation of the ethanol extract of the twigs and leaves, which was found to be cytotoxic and active in the P388 mouse leukemia system (ref. 54) led to isolation of a new isoflavone, named cudraisoflavone-A (**5**) which was cytotoxic to PS cells in culture at a dose of 1 x 10° μg/ml (ref. 55). The structure of **5** was elucidated on the basis of NMR, MS, UV and IR spectral data. In addition to **5** the known compounds 3-O-methylorobol, dehydrocostus lactone, methyl linoleate and β-sitosterol were isolated and identified. Dehydrocostus lactone and methyl linoleate showed cytotoxicity at 0.9 x 10° and 1.0 x 10° μg/ml against 9PS cells in culture. Compound **4** is undergoing further tests versus the human tumor *in vitro* panel at NCI.

Figure 5. Cudraisoflavone A (**5**), the Cytotoxic Constituent of *Cudrania*

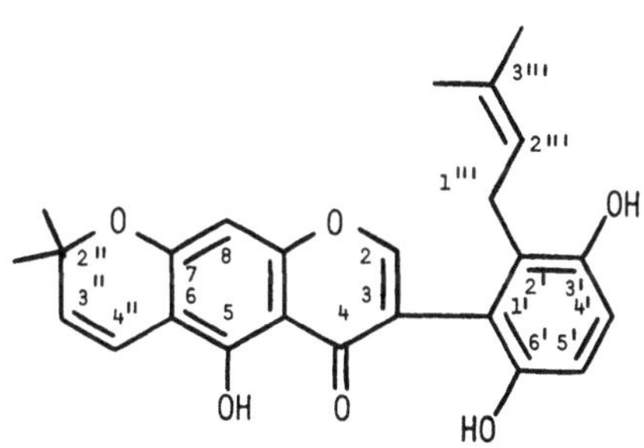

5

8. *DESFONTAINIA SPINOSA*

A novel cytotoxic triterpene (9KB ED_{50}, 1.0 x 10^{-2} μg/ml) was isolated from the 95% ethanol extract of *Desfontainia spinosa* by a sequence of purification steps including column and preparative-layer chromatography on normal and reversed phase silica gels, and crystallization. The structure was determined to be 11-deoxocucurbitacin I (**6**) by extensive analysis of the high field ^{1}H NMR, ^{13}C NMR and field desorption MS. This was the first example of a cucurbitacin lacking an 11-keto functionality (ref. 56). The corresponding 23,24-dihydro analog **7** was also isolated and characterized (ref. 57).

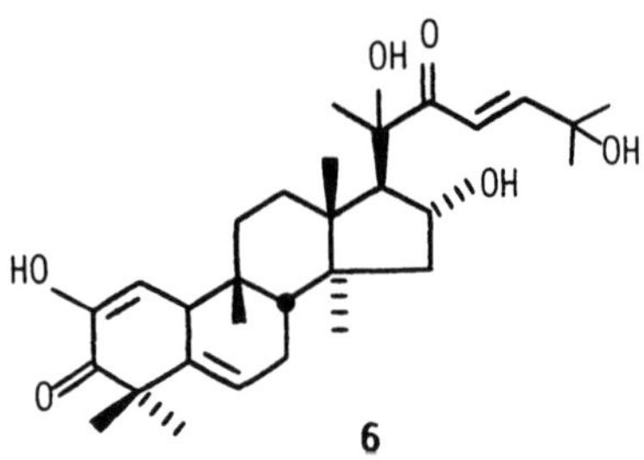

6

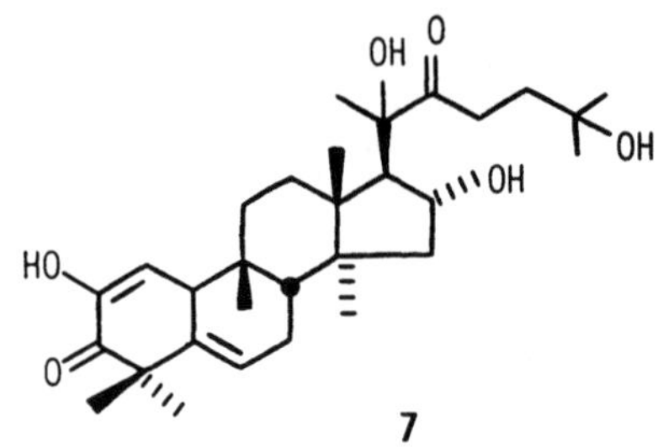

7

Further fractionation of the active ethanol extract of *Desfontainia spinosa* based on cytotoxicity has led to the discovery of two novel, cytotoxic glycosides of the parent, 11-deoxocucurbitacin I, named spinoside A (**8**) and spinoside B (**9**). The structures of **8** and **9** were established on the basis of extensive analysis of their high field ^{1}H NMR, ^{13}C NMR, high resolution mass spectra (FAB, FD, CI, EI) and chemical interconversions (ref. 57). The novel arabinopyranoside derivatives of **6** were less cytotoxic than the parent compound in the 9PS system. Cucurbitacins although highly cytotoxic have never demonstrated significant anticancer effects *in vivo*.

Figure 6. Cytotoxic Glycosides from *Desfontainia spinosa*

D. spinosa, Lf-St Costa Rica ⟶ **6**, 11-Deoxocucurbitacin I; **8**, Spinoside A; **9**, Spinoside B

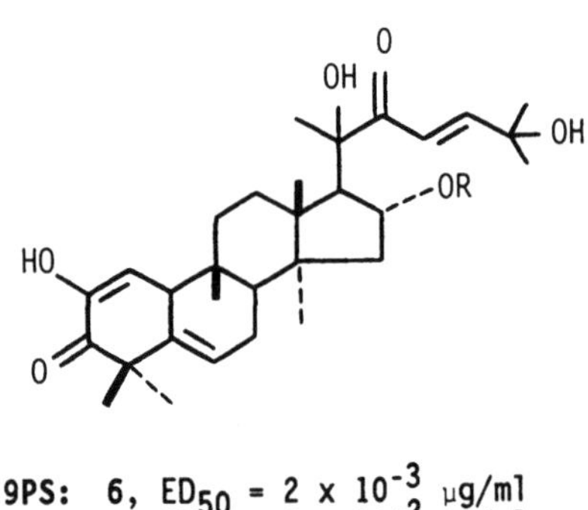

7, R = H

8, R =

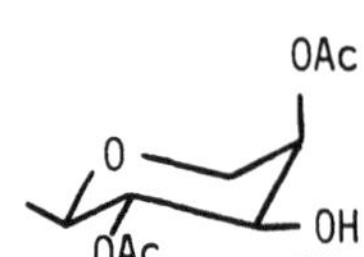

9, R =

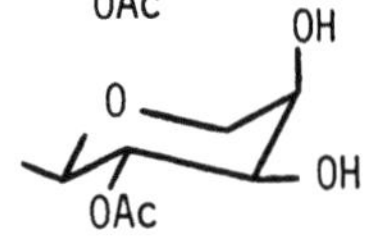

9PS: **6**, ED_{50} = 2 x 10^{-3} μg/ml
8, ED_{50} = 4 x 10^{-2} μg/ml
9, ED_{50} = 2 x 10^{-1} μg/ml

9. *PARARISTOLOCHIA FLOS-AVIS*

Fractions derived from the ethanol extract of this plant have shown activity in 3PS and 9PS. In addition to a series of cytotoxic aristolactams, we have isolated and elucidated the structure of a novel α-tetralone named flossonol (**10**). This compound showed borderline cytotoxicity versus 9PS cells at $ED_{50} = 3 \times 10^{\circ}$ µg/ml (ref. 58).

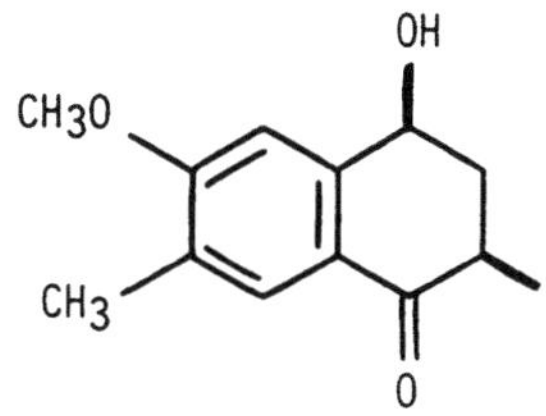

flossonol (**10**)

10. PODOCARPUS SPECIES

Activity-directed fractionation of *Podocarpus milanjianus* Rendle yielded a series of norditerpenoid dilactones. Four of the seven actives, nagilactones F and G, and milanjilactones A and B, were cytotoxic but not active *in vivo*. The P388 *in vivo* actives were isolated and identified as nagilactone C, sellowin A, and podolactone C. Podolactone C (**11**) previously proposed to be a 1,2-α-epoxide in ring A, was unambiguously revised to a 2,3-β-epoxide by single crystal X-ray analysis (ref. 59,60).

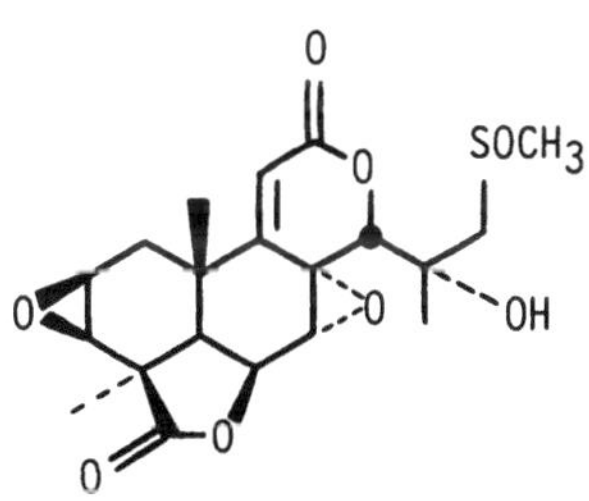

podolactone C (**11**)

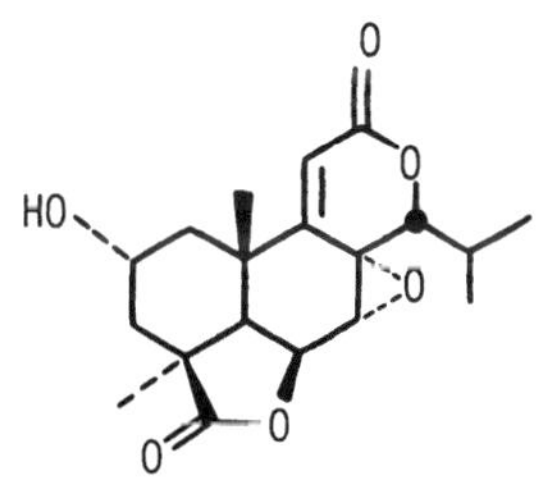

2-α-hydroxynagilactone G (**12**)

Using tandem mass spectrometry as a probe for the isolation of analogs further studies were carried out on *Podocarpus wallichianus*, *P. gracilior* and *P. purdieanus*. *P. wallichianus* yielded the novel compound **12** (ref. 61). Compounds **11** and **12** both show activity in 3PS (151% at 20 and 10 mg/kg respectively) and are undergoing further evaluation at NCI.

Recently we have extended the MS/MS studies of the *Podocarpus* lactones to include analysis employing reactive collisions with ethyl vinyl ether. Minor differences in structure led to different reactivities with the reagent (ref.

45). This method promises the possibility of identifying specific functionality and stereochemistry in complex natural products by analyzing the selective ion-molecule reactions which occur in the tandem mass spectrometer.

11. *PSOROSPERMUM FEBRIFUGUM*

Psorospermum febrifugum is an African tropical woody plant which belongs to the tribe Vismiae in the family Guttiferae. Previous chemical investigation of different species of this genus has led to the isolation of cytotoxic vismiones (ref. 62-66), anthraquinones (ref. 65,67), anthrones (ref. 66, 68-70), and xanthonolignoids (ref. 71). The cytotoxic and antitumor activity associated with the ethanolic extract of *Psorospermum* has led to extensive work to isolate the active compounds. The first account of the isolation of bioactive compounds from *Psorospermum febrifugum* was reported by Kupchan (ref. 72). The furanoxanthone psorospermin (**13**) was found to possess significant cytotoxic and antitumor properties. In connection with this study of the structure, stereochemistry, synthesis and interconversion among members of this interesting class of compounds have been thoroughly investigated in our laboratory (ref. 73-76). Further activity-directed fractionation of the plant extract yielded a number of active fractions which did not contain **11** and from which an additional series of cytotoxic and antitumor xanthones has been isolated (ref. 77).

The cytotoxicity and antitumor activity of **13** and related analogs is summarized in Table 2. Compounds **13**, **14**, **21** and **22** are under further evaluation at NCI.

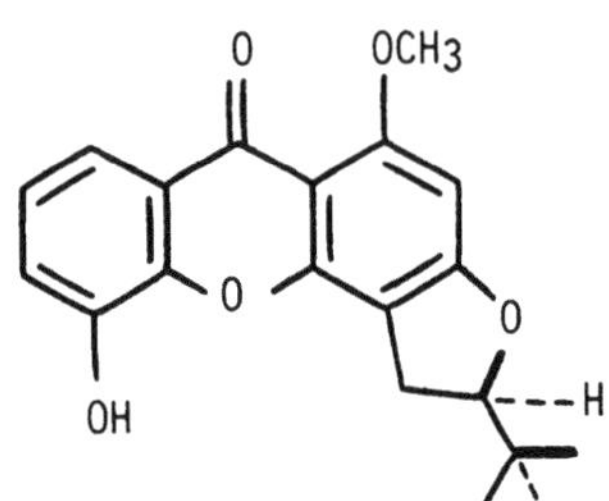

13, psorospermin

Table 2. Antitumor Activity of Psorospermin (**13**) and Related Xanthones

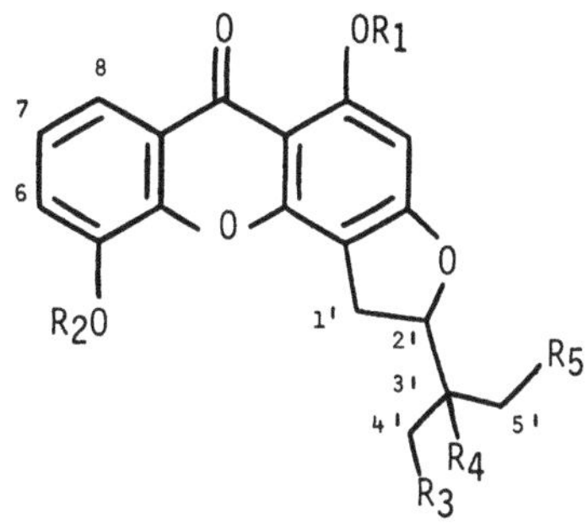

	R_1	R_2	R_3	R_4	R_5	9PS ED_{50}	A549 ED_{50}	HT29 ED_{50}	3PS T/C % at mg/kg	B1	C6
13	CH_3	H	--O--		H	$1x10^{-4}$	Under Test		160 at 4	138 at 2	135 at 4
14	CH_3	H	OH	OH	H	$4x10^{1}$	NT	$1x10^{-1}$	inactive	NT	NT
15	H	H	OH	OH	H	$1x10^{-2}$	NT	NT	inactive	NT	NT
16	H	H	OH	OH	H	$1x10^{-2}$	NT	NT	NT	NT	NT
17	CH_3	CH_3	OH	OH	H	$9x10^{1}$	NT	NT	NT	NT	NT
18	CH_3	H	OCH_3	OH	H	$2x10^{-4}$	$1x10^{1}$	$4x10^{-1}$	NT	NT	NT
19	CH_3	H	OC_2H_5	OH	H	$3x10^{-3}$	NT	NT	NT	NT	NT
20	H	CH_3	OH	OH	OH	$3x10^{\circ}$	NT	NT	NT	NT	NT
21	CH_3	H	--O--		OH	$7x10^{-5}$	$1.5x10^{\circ}$	$6x10^{-4}$	NT	NT	NT
22	CH_3	CH_3	Cl	OH	OH	$1x10^{-5}$	NT	NT	132 at 8	NT	NT

NT - not tested

12. *SPATHELIA SORBIFOLIA*

Activity-directed fractionation of *Spathelia sorbifolia* has led to the isolation and structural elucidation of three new chromones, spatheliol (**23**), sorbifoliol (**24**) and 6-(3′-methyl-3′-hydroxybut-1-enyl)-5-O-methylisospatheliachromen (**25**, ref. 78).

Compounds **23** and **25** were cytotoxic to PS cells in culture (ED_{50} 1 x 10^{-1}) and are under further evaluation in the NCI *in vitro* panel.

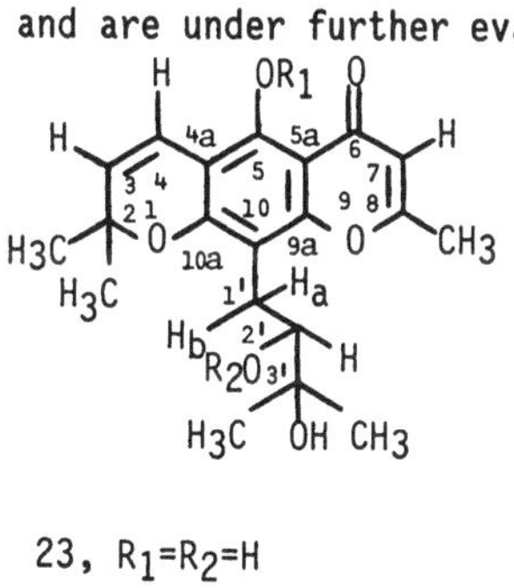

23, R_1=R_2=H

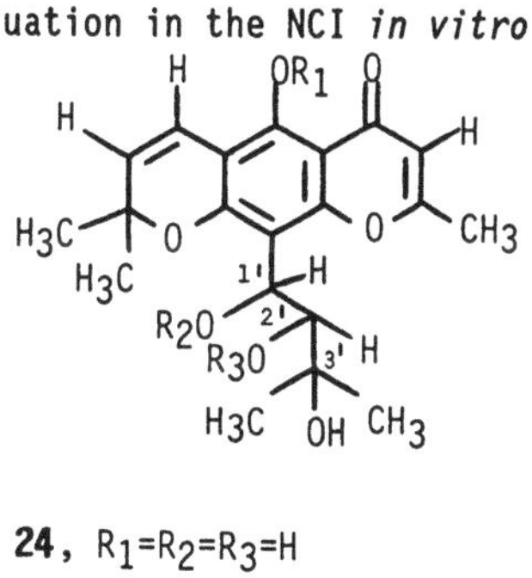

24, R_1=R_2=R_3=H

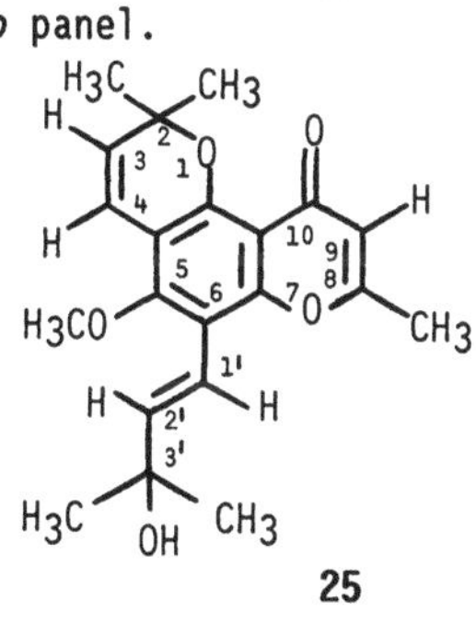

25

13. ACKNOWLEDGEMENTS

This research was supported by grant CA33326 from the Division of Cancer Treatment, National Cancer Institute, National Institute of Health, Bethesda. *In vitro* testing was performed by Drs. Linda Jacobsen and Peggy Criswell in the Cell Culture Laboratory, Purdue Cancer Center. High resolution (470 MHz) proton spectra was recorded at the Purdue University Biochemical Magnetic Laboratory (NIH grant RR1077). The animal testing data are the results of screening performed under the auspices of the Developmental Therapeutic Program, Division of Cancer Treatment, National Cancer Institute, Bethesda, Maryland.

14. REFERENCES

1. J.D. Douros and M. Suffness, *Cancer Chemotherapy and Pharmacology* 1 (1978), 91.
2. G.R. Pettit, Biosynthetic Products for Cancer Chemotherapy, Vol. 1, Plenum, N.Y. and London (1977).
3. G.R. Pettit and G.M. Cragg, Biosynthetic Products for Cancer Chemotherapy, Vol. 2, Plenum, N.Y. and London (1978).
4. G.R. Pettit and R.H. Ode, Biosynthetic Products for Cancer Chemotherapy, Vol. 3, Plenum, N.Y. and London (1979).
5. The 1978 Report of the Division of Cancer Treatment, NCI, Vol. 1 Summary Reports Sept., 1978.
6. A. Aszalos, Antitumor Compounds of Natural Origin, Vol. 1, CRC Press, Boca Raton, Florida (1981).
7. A. Aszalos, Antitumor Compounds of Natural Origin, Vol. 2, CRC Press, Boca Raton, Florida (1981).
8. J.M. Cassady and J.D. Douros, Anticancer Agents Based on Natural Product Models, Academic Press, Medicinal Chemistry, Vol. 16 (1980).
9. M. Suffness and J. Douros, Drugs of Plant Origin, Methods in Cancer Research, Vol. XVI, Chapter 3, Academic Press, N.Y. (1979).
10. J. Douros and M. Suffness, *Cancer Treatment Reviews* **8** (1981), 63-87.
11. J.M. Cassady, C.-j. Chang, and J.L. McLaughlin, Recent Advances in the Isolation and Structural Elucidation of Antineoplastic Agents of Higher Plants, In: Natural Products as Medicinal Agents, ed. J.L. Beal and E. Reinhart, Hippokrates, Verlag, Stuttgart, 1981, pp. 93-124.
12. M. Suffness and J. Douros, *J. Nat. Prod.* **45** (1982), 1-14.
13. J.S. Driscoll, *Cancer Treatment Rep.* **68** (1984), 63-75.
14. M. Boyd, R.H. Shoemaker, G.M. Cragg, and M. Suffness, Future Perspectives on Anticancer Drug Discovery Program of NCI, Sea Pharm Conference, Oct. 24-25, 1985, Ft. Pierce, Florida.
15. M. Suffness, New Approaches to Discovery of Antitumor Agents, Symposium on Biologically Active Natural Products, Phytochemical Society of Europe, Lausanne, Switzerland, Sept., 1986.
16. J.D. Douros, Lower Plants as a Source of Anticancer Drugs, *Cancer Treatment Repts.* **60** (1976), 1069.
17. J.L. Hartwell, Types of Anticancer Agents Isolated from Plants, *Cancer Treatment Repts.* **60** (1976), 1031.
18. S.M. Kupchan, *Cancer Treatment Repts.* **60** (1976), 1115.
19. G.A. Cordell and N.R. Farnsworth, *J. Nat. Prod. (Lloydia)* **40** (1977), 1.
20. M.E. Wall and M.C. Wani, *Ann. Rev. Pharmacol. Tox.* **17** (1977), 117.
21. M. Suffness, in: Advances in Medicinal Plant Research, Proceedings of 32nd Congress of the Arzneipflanzenforschung, Antwerp, Belgium, July (1984), Edited by A.J. Vlietinck and R.A. Dommisse, Wissenschaftliche Verlagsgesellschaft, Stuttgart (1985).

22. E.M. Lessman and J.E. Sonal, *J. Amer. Med. Assn.* **175** (1961), 741.
23. A. Gonzelez, J. Bermejo, J.L. Breton, G.M. Massenet, B. Dominquez, and J. Amaro, *J.C.S. Perkin Trans. I*, (1976), 1663.
24. J.A. Sinkale, *Pharmacotherapy* **4** (1984), 61-73.
25. N. Neuss, M. Gorman, and I.S. Johnson, in: H. Busch (ed.) Methods in Cancer Research, Volume III, Chapter 10, N.Y., Academic Press (1967).
26. C.J. Barnett, G.J. Callinam, K. Gerzon, R.C. Hoying, W.E. Jones, W.M. Newton, G.A. Poore, R.L. Robison, and M.J. Sweeney, *J. Med. Chem.* **21** (1978), 88.
27. R.A. Roush and R.G. Cooks, *J. Nat. Prod.* **47** (1984), 197-214, and references therein.
28. W. Alberth, K.M. Straub, and A.L. Burlingame, *Anal. Chem.* **54** (1982), 2029-2034.
29. M. Barber, R.S. Bordoli, G.J. Elliot, R.D. Sedgwick, and A.N. Tyler, *Anal. Chem.* **54** (1982), 645A-657A.
30. R.J. Day, S.E. Unger, and R.G. Cooks, *Anal. Chem.* **52** (1980), 557A-572A.
31. D.F. Hunt, *Int. J. Mass Spectrom Ion Phys.* **45** (1982), 111-123.
32. G.R. Pettit, C.W. Holzapfel, G.M. Cragg, C.L. Herald, and P. Williams, *J. Nat. Prod.* **46** (1983), 917-922.
33. M.L. Vestae, *Mass Spectrom. Rev.* **2** (1983), 447-480.
34. M.P. Barbalas, F.W. McLafferty, and J.L. Occolowitz, *Biomed. Mass Spectrom.* **10** (1983), 258-261.
35. R.G. Cooks, *Am. Lab.* (1983), 111-123.
36. R.G. Cooks, K.C. Busch, and G.L. Glish, *Science* **222** (1983), 273-291.
37. F.W. McLafferty, *Science* **214** (1981), 280-287.
38. F.W. McLafferty, *et al.*, *Int. J. Mass Spectrom. Ion Phys.* **45** (1982), 323-329.
39. R.A. Yost, H.O. Brotherton, and R.J. Perchalski, *Int. J. Mass Spectrom. Ion Phys.* **48** (1983), 77-80.
40. M.T. Cheng, M.P. Barbalas, R.F. Pegue, and F.W. McLafferty, *J. Am. Chem. Soc.* **105** (1983), 1510-1513.
41. F.W. McLafferty, *Biomed. Mass Spectrom.* **8** (1981), 446-448.
42. M.L. Gross, *et al.*, *Tetrahedron Lett.* **23** (1982), 5381-5384.
43. J.N. Louris, L.G. Wright, R.G. Cooks, and A.E. Schoen, *Anal. Chem.* **57** (1985), 2918.
44. J.R. O'Lear, L.G. Wright, N.J. Louris, and R.G. Cooks, *Org. Mass Spectrom.* **22** (1987), 343-358.
45. R.R. Pachuta, H.I. Kenttamaa, R.G. Cooks, T.M. Zennie, C. Ping, C.-j. Chang and J.M. Cassady, *Org. Mass Spectrom.*, in press (1988).
46. S.D. Jolad, J.J. Hoffmann, K.H. Schram, J.R. Cole, M.S. Tempesta, G.R. Kriek, and R.B. Bates, *J. Org. Chem.* **47** (1982), 3151.
47. T.T. Dabrah and A.T. Sneden, *Phytochemistry* **23** (1984), 2013.
48. S.D. Jolad, J.J. Hoffmann, and J.R. Cole, *J. Nat. Prod.* **48** (1985), 644.
49. J.K. Rupprecht, C.-j. Chang, J.M. Cassady, J.L. McLaughlin, K.L. Mikolajczak, and D. Weisleder, *Heterocycles* **24** (1986), 1197.
50. T.G. McCloud, D.L. Smith, C.-j. Chang, and J.M. Cassady, *Experientia* **43** (1987), 947.
51. J.-G. Yu, L. Xu, T. McCloud, C.-j. Chang, and J.M. Cassady, in preparation.
52. Dictionary of Chinese Traditional Drugs, II, Jiang-su New Medical College (1977), p. 1731.
53. B.L. Chang, F.S. El-Feraly, and N.J. Doorenbos, *J. Pharmaceutical Sciences* **66** (1977), 908.
54. R.I. Geran, N.H. Greenberg, M.M. MacDonald, A.M. Shumacher, and B.J. Abbott, *Cancer Chemother.* **3** (1972), 1.
55. N.-j. Sun, C.-j. Chang, and J.M. Cassady, *Phytochemistry* **27** (1988), 0000.
56. A.A. Amonkar, T.G. McCloud, C.-j. Chang, J.A. Saenz-Renauld, and J.M. Cassady, *Phytochemistry* **24** (1985), 1803-1805.
57. K.S. Reddy, A.M. Amonkar, T.G. McCloud, C.-j. Chang, and J.M. Cassady, *J. Chem. Soc. Perkin I*, submitted for publication (1988).
58. N.-j. Sun, C.-j. Chang, and J.M. Cassady, *Phytochemistry* **26** (1987), 3051.

59. J.A. Hembree, C.-j. Chang, J.L.McLaughlin, J.M. Cassady, D.J. Watts, E. Wenkert, *Phytochemistry* **18** (1979), 1691.
60. J.M. Cassady, T.K. Lightner, T.G. McCloud, J.A. Hembree, S.R. Byrn, and C.-j. Chang, *J. Org. Chem.* **49** (1984), 942.
61. C. Ping, C.-j. Chang, and J.M. Cassady, manuscript in preparation (1988).
62. A.M. Habib, C.-j. Chang, and J.M. Cassady, *Phytochemistry*, in preparation (1988).
63. G. Cassinelli and C. Geroni, *J. Nat. Prod.* **49** (1986), 929.
64. B. Botta, F. Delle Monache, G. Delle Monache, and K. Kabangu, *Phytochemistry* **25** (1986), 766.
65. B. Botta, F. Delle Monache, G. Delle Monache, G.B. Marini Bettolo, and J.U. Oguakwa, *Phytochemistry* **22** (1983), 539.
66. B. Botta, F. Delle Monache, G. Delle Monache, G.B. Marini Bettolo, and J.D. Msonthi, *Phytochemistry* **24** (1985), 827.
67. A. Marston, J.-C. Chapuis, B. Sordat, J.D. Msonthi, and K. Hostettman, *Planta Medica* (1986), 208.
68. A. Amonkar, C.-j. Chang, and J.M. Cassady, *Experientia* **37** (1981), 1138.
69. F. Delle Monache, B. Botta, G. Delle Monache, and G.B. Marini Bettolo, *Phytochemistry* **24** (1985), 1855.
70. B. Botta, F. Delle Monache, and G. Delle Monache, *Tet. Lett.* **28** (1987), 567.
71. M. Abou-shoer, A.M. Habib, C.-j. Chang, J.M. Cassady, in preparation.
72. S.M. Kupchan, D.R. Streelman, and A.T. Sneden, *J. Nat. Prod.* (Lloydia) **43** (1980), 296.
73. A.M. Habib, D.K. Ho, S. Masuda, T. McCloud, K.S. Reddy, M. Abou-shoer, A. McKenzie, S.R. Byrn, C.-j. Chang, and J.M. Cassady, *J. Org. Chem.* **52** (1987), 412, and references therein.
74. R.R. Pachuta, R.G. Cooks, J.M. Cassady, P. Cong, T. McCloud, and C.-j. Chang, *J. Nat. Prod.* **49** (1986), 412.
75. D.K. Ho, A.T. McKenzie, S.R. Byrn, and J.M. Cassady, *J. Org. Chem.* **52** (1987), 342.
76. K.S. Reddy, O.-H. Ko, D. Ho, P.E. Persons, and J.M. Cassady, *Tet. Lett.* **28** No. 27 (1987), 3075.
77. M. Abou-shoer, C.-j. Chang, and J.M. Cassady, manuscript in preparation.
78. K. Suwanborirux, C.-j. Chang, and J.M. Cassady, *J. Nat. Prod.* **50** (1987), 102-107.

Affiliations

Dr. John M. Cassady
Department of Medicinal Chemistry
and Pharmacognosy
School of Pharmacy and Pharmacal
Sciences
Purdue University
West Lafayette, IN 47907

Dr. Ching-jer Chang
Department of Medicinal Chemistry
and Pharmacognosy
School of Pharmacy and Pharmacal
Sciences
Purdue University
West Lafayette, IN 47907

Dr. R. Graham Cooks
Department of Chemistry
Purdue University
West Lafayette, IN 47907

Stereochemical Studies on Pre-Anthraquinones and Dimeric Anthraquinone Pigments

GÜNTER BILLEN, ULRICH KARL, THOMAS SCHOLL, KLAUS DIETER STROECH
and WOLFGANG STEGLICH*
* Affiliation see page 315

INTRODUCTION

Labelling studies have shown that the biosynthesis of emodin (**6**) and related anthraquinone pigments in fungi and higher plants occurs via the acetate/malonate pathway (ref. 1). Details of the early steps in this biosynthesis, however, remain unknown. It can be assumed that after assembling of the octaketide chain condensation to a carboxylic acid **1** takes place, which can either be decarboxylated to atrochrysone (**4**) or dehydrated to endocrocin anthrone (**2**) (Scheme 1). Atrochrysone then may undergo water elimination to emodin anthrone (**5**) followed by oxidation to emodin (**6**), which is the precursor of many neutral anthraquinones. The possibility that **5** is formed from **1** by a concerted loss of water and CO_2 has also to be considered. On the other hand, **2** can be converted into endocrocin (**3**) which serves as precursor to other anthraquinone carboxylic acids.

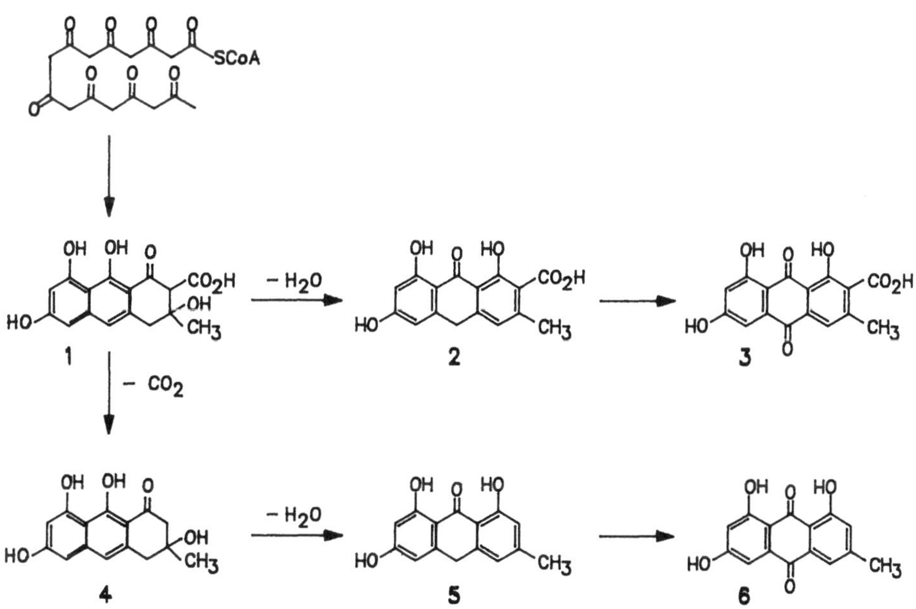

Scheme 1. Anthraquinone biosynthesis

By feeding radioactively labelled **3** to fruit bodies of Dermocybe species it could be demonstrated that endocrocin is not decarboxylated to emodin (**6**) (ref. 2) in these toadstools. Therefore, this step has to appear at an earlier stage in emodin biosynthesis.

In recent years atrochrysone (**4**) and several other pre-anthraquinones have been isolated from fungi and higher plants (Scheme 2). The parent compound **4** occurs in the toadstools Cortinarius atrovirens and C. odoratus (ref. 3). The monomethyl ethers torosachrysone (**7**) (ref. 4) and asperflavin (**8**) (ref. 5) are produced by Cassia torosa and Aspergillus flavus, respectively. Torosachrysone-8-O-methyl ether (**9**) has been found in fruitbodies of several toadstools along with its trans-4-hydroxy derivative **10** (ref. 3). Further compounds of this type, e.g. vismione A (**11**) and B (**12**), have been isolated from Vismia species (ref. 6). With the only exception of **7**, the absolute configuration of these pre-anthraquinones remains unknown.

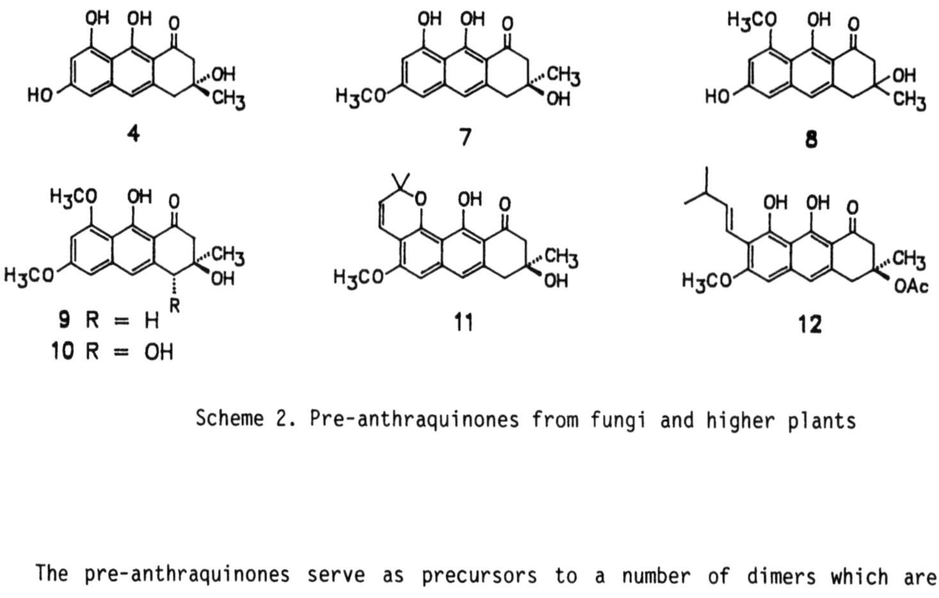

Scheme 2. Pre-anthraquinones from fungi and higher plants

The pre-anthraquinones serve as precursors to a number of dimers which are depicted in Scheme 3. During studies on the pigments of Cortinarius we have detected several structural types which are apparently formed from the monomers by oxidative coupling at different positions. According to the coupling mode, pigments of the atrovirin (**13**), flavomannin (**14**), pseudophlegmacin (**15**) and phlegmacin type (**17**) can be defined (ref. 3). Interestingly, a dimer of the biarylether type has been found in Cortinarius elegantior (refs. 3,7). This compound, which we have named cortinarin (**16**), exhibits doubling of its ^{1}H n.m.r. signals due to hindered rotation around the biarylether linkage.

Some pre-anthraquinone dimers have also been isolated from higher plants (Scheme 4). Thus, stereoisomers of the fungal phlegmacins are found in seedlings of *Cassia torosa* (ref. 8) and a new type of dimerisation is observed in torosanin (**18**) (ref. 9) from unripe seeds of *Cassia torosa*. *Karwinskia humboldtiana*, a Mexican shrub of the Rhamnaceae family, produces a number of interesting dimers including compound **19** which causes 'Buckthorn neuropathy', a nervous disorder which has led to many fatalities (ref. 10). Similiar toxic effects have been demonstrated for flavomannin and phlegmacin derivatives (ref. 11).

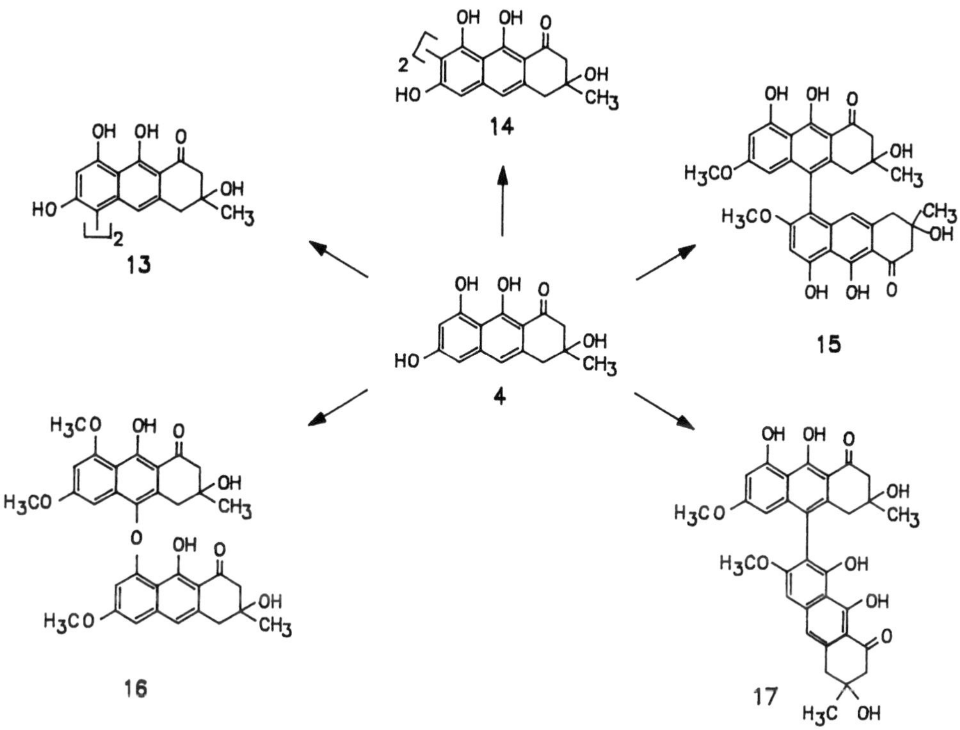

Scheme 3. Pre-anthraquinone dimers from fungi

Due to the presence of two centres of chirality in the cyclohexenone rings and a chiral axis at the biaryl linkage, the dimeric pre-anthraquinones present considerable stereochemical problems. The chiroptical effects of the axial chirality dominate in the CD spectrum and cause the appearance of two Cotton effects centred around 275 nm. The form of this couplet has been used for a preliminary stereochemical correlation of these dimers. Those compounds which exhibit a negative Cotton effect to longer wave length and a positive Cotton effect to shorter wave length in the vicinity of 275 nm in the CD spectrum are designated type A, while in the CD spectrum of type B the signs of the Cotton effects are inverted (ref. 3).

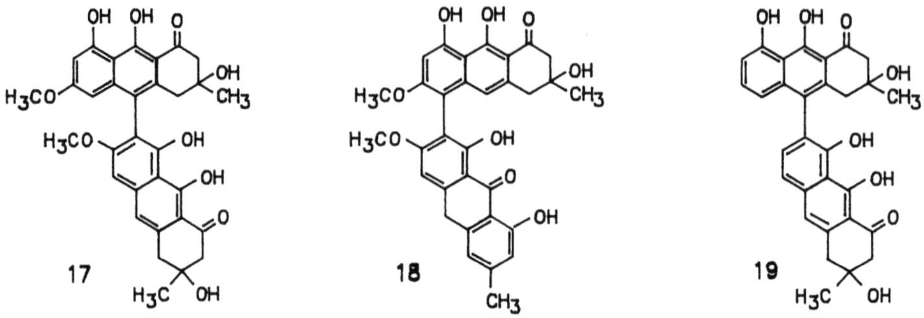

Scheme 4. Pre-anthraquinone dimers from higher plants

DETERMINATION OF THE ABSOLUTE CONFIGURATION OF MONOMERIC PRE-ANTHRAQUINONES

The (3*S*)-configuration of torosachrysone (**7**) has been determined by application of the exciton chirality method (ref. 12) to its 3-*O*-benzoate (ref. 13). In order to obtain rigorous proof of the absolute configuration of the pre-anthraquinones, we have carried out a synthetic correlation with (-)-quinic acid (**20**) (Scheme 5).

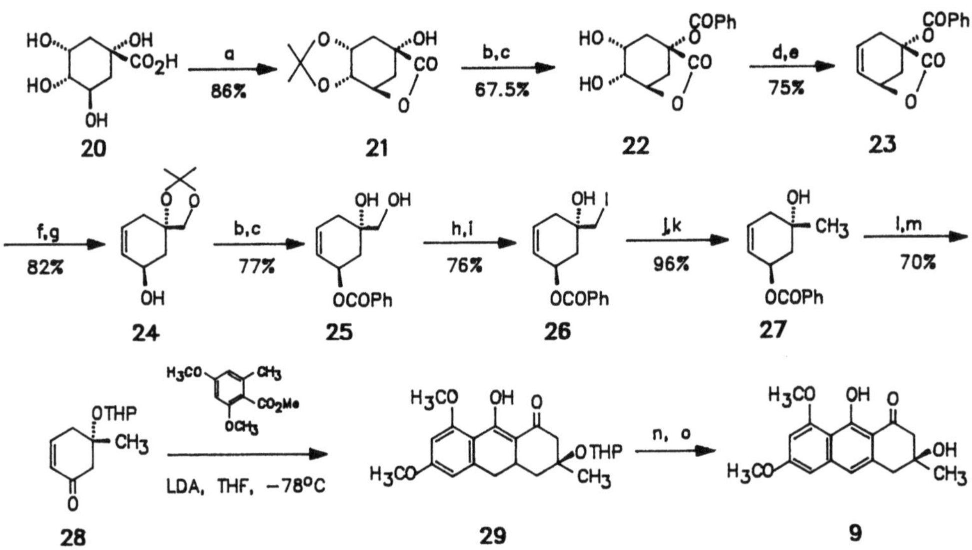

a. Acetone, H^+; b. PhCOCl, Py, DMAP; c. AcOH, pyridinium tosylate (PPTS); d. $HC(OEt)_3$, H^+, e. I, H^+; f. $LiAlH_4$; g. 2,2-dimethoxypropane, H^+; h. TsCl, P i. NaI, acetone; j. Bu_3SnH; k. dihydopyran, PPTS; l. KOH, MeOH; m. pyridinium chlorochromate, NaOAc, CH_2Cl_2; n. DDQ, benzene; o. PPTS, EtOH.

Scheme 5. Synthesis of torosachrysone-8-*O*-methyl ether (**9**) from (-)-quinic acid (2

The synthesis starts with the known lactone **21** (ref. 14) which is easily obtained from (-)-quinic acid (**20**) by treatment with acetone in the presence of a trace of acid. After benzoylation and removal of the acetal protecting groups, **21** yielded the diol **22** which was converted into the cyclohexene derivative **23** by application of Eastwood's method (ref. 15). Reduction of **23** with $LiAlH_4$ followed by acid catalysed acetalisation with 2,2-dimethoxypropane afforded the cyclohexenol derivative **24**.

In order to remove the primary hydroxy group, the protected triol **24** was acetylated and treated with pyridinium tosylate to yield the monoester **25**. Selective tosylation of the primary hydroxy group followed by Finkelstein exchange for iodine yielded the iodo derivative **26** which was transformed into the protected cyclohexene diol **27** by reductive removal of the iodine with tributyltin hydride followed by protection of the tertiary hydroxy group with dihydopyran. After removal of the benzoyl group by alkaline saponification the resulting allylic alcohol was oxidised to the desired enone **28** with pyridinium chlorochromate in buffered solution (ref. 16).

Condensation of enone **28** with methyl 4,6-di-O-methylorsellinate yielded the tricyclic compound **29**, which after dehydrogenation with DDQ and cleavage of the THP group under mild conditions afforded (-)-torosachrysone-8-O-methyl ether (**9**). This was in every respect, including its CD spectrum, identical with the natural compound which therefore has the (3S)-configuration. Comparison of the CD-spectrum of **9** (Figure 1) with those of authentic samples of atrochrysone (**4**), torosachrysone (**7**), trans-4-hydroxy-5-O-methyltorosachrysone (**10**), vismione A (**11**) and vismione B (**12**) clarified the absolute configuration of these pre-anthraquinones as indicated in the formula of Scheme 2.

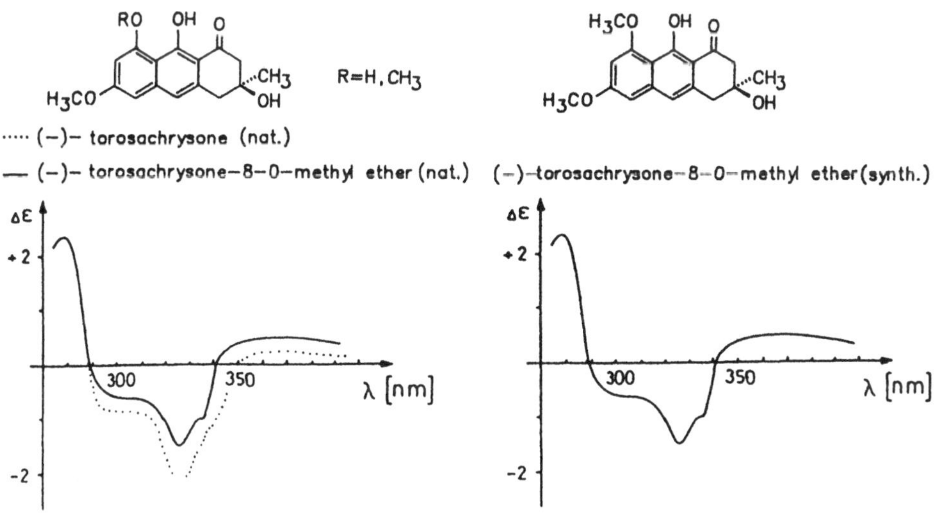

Figure 1. CD-Spectra of natural (-)-torosachrysone (**7**) and synthetic (-)-torosachrysone-8-O-methyl ether (**9**) in MeOH

DETERMINATION OF THE ABSOLUTE CONFIGURATION OF DIMERIC PRE-ANTHRAQUINONES AND ANTHRAQUINONES

The absolute configuration at the biaryl linkage of pre-anthraquinone dimers has only been clarified in a single case by use of the exciton chirality method (ref. 13). Because the conformation of most dimeric pigments is unknown, the use of CD arguments in these cases is ambiguous. We therefore tried to use independent methods and selected skyrin (**30**) as an example. Skyrin occurs in several fungi and lichens, often as its (+)-enantiomer in differing optical purity (ref. 17).

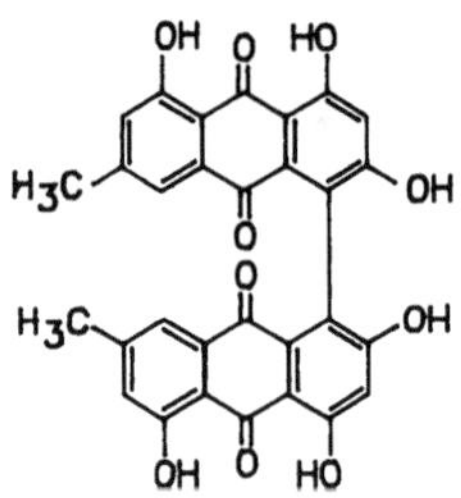

30

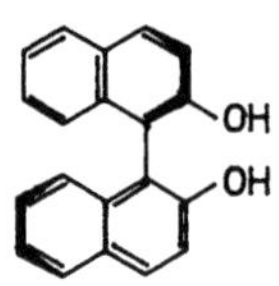

(+)-(R)-**31**

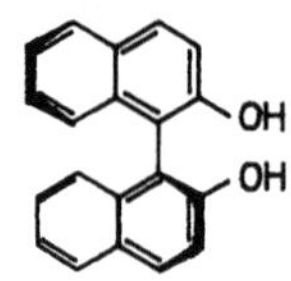

(−)-(S)-**31**

At first we tried to address the problem with Horeau's method of kinetic resolution which has been applied with great success to compounds of central chirality (ref. 18). To our knowledge this technique has not yet been used for compounds with axial chirality. To be on safe ground, we had to calibrate the effects first with compounds of similar structure and known absolute configuration. We selected 2,2'-dihydroxy-1,1'-binaphthyl (**31**) as an example and reacted it in the usual way with an excess of racemic 2-phenylbutanoic anhydride in the presence of pyridine. The optical excess in the remaining anhydride was determined by means of the gas chromatographic variant of Brooks and Gilbert (ref. 19) (Scheme 6).

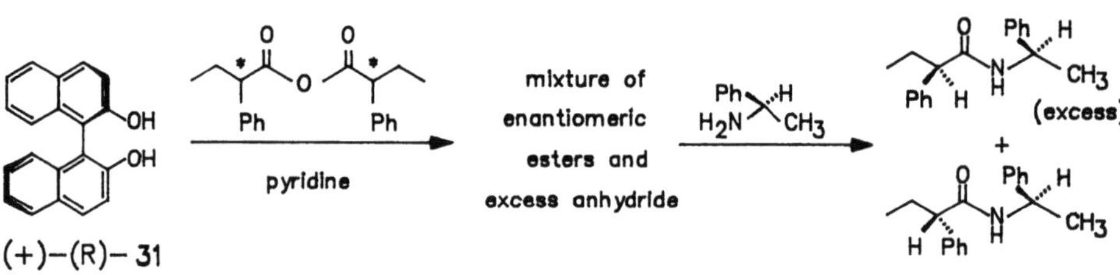

Scheme 6. Application of Horeau's method to (+)-(R)-binaphthol **31**

In this technique 2-phenylethylamine is added to the reaction mixture and the ratio of the two diastereomeric amides formed from the excess of anhydride is determined gaschromatographically. We obtained in this way a negative peak increment with (+)-(R)-binaphthol, whereas the (-)-(S)-compound gave a positive value (Table 1).

TABLE 1. Application of Horeau's method to compounds with axial chirality

compound	average peak increment*	compound	average peak increment*
(±)-binaphthol **31**	0	(±)-skyrin (**30**)	0
(+)-(R)-binaphthol **31**	+2.72	(+)-skyrin (**30**)	-0.70
(-)-(S)-binaphthol **31**	-2.75	(-)-atrovirin B (**13**)	-0.69
		(-)-flavomannin A (**14**)	-0.65

* Positive peak increment (ref. 19): (S)-2-phenylbutanoic acid reacts faster with the biaryl derivative; less amide of (S)-acid is formed after addition of (R)-2-phenylethylamine to the reaction mixture.

Because binaphthol **31** and skyrin (**30**) have the same gross structure around the biaryl linkage, the method can be used to determine the absolute configuration of the latter compound. The negative peak increment indicates the (S)-configuration for (+)-skyrin. The absolute value is slightly lower in comparison with the binaphthols which points to a 'dilution' through reaction of the 2-phenylbutanoic anhydride with the excess hydroxy groups of skyrin. (-)-Atrovirin B (**13**) also showed a negative peak increment which proves its (S)-configuration at the biaryl linkage. Because **13** co-occurs with (+)-skyrin in Cortinarius odoratus, this result is in accord with the close biogenetic relationship of these pigments (ref. 3).

The kinetic resolution should be considerably enhanced if a chiral dicarboxylic acid capable of ring formation to both ß-hydroxy groups of skyrin would be applied. We selected for this purpose 6,6'-dinitrodiphenic acid dichloride (**32**) which can easily be obtained in both enantiomeric forms. Reaction of (+)-(R)-binaphthol **31** with racemic **32** afforded the (+)-lactone **33**, $[\alpha]_D$ +213°, whereas the unreacted acid **34** showed a negative rotation (Scheme 7). Further investigations with the optically pure reagents indicated that (+)-(R)-**31** forms only a lactone **33** with (+)-(R)-**32** whereas (-)-(S)-**31** reacts only with (-)-(S)-**32**. The same results were obtained with tetrachlorodiphenic acid dichloride. Similar observations have recently been published by Miyano and co-workers (ref. 20) for 1,1'-binaphthyl-2,2'-dicarboxylic dichloride.

With these results in hand, we applied the method to (+)-skyrin (**30**) and (-)-flavomannin A (**14**). After the usual work-up and chromatographic purification,

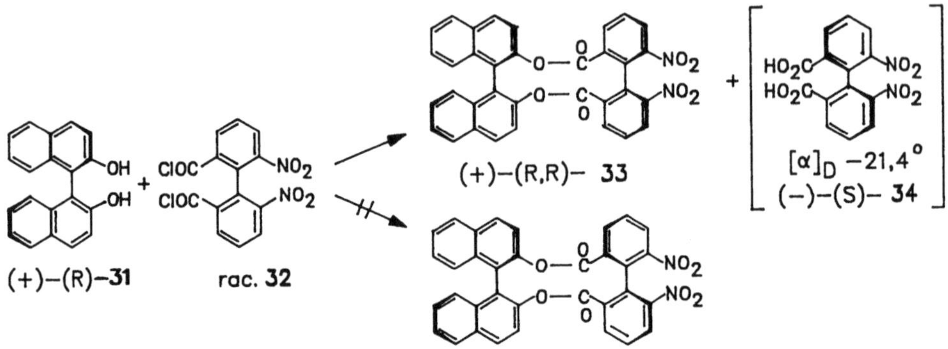

Scheme 7. Reaction of rac. 6,6'-dinitrodiphenic acid dichloride (**32**) with (+)-(R)-binaphthol **31**

the remaining dinitrodiphenic acid **34** exhibited specific rotations of $[\alpha]_D$ +26° and -8.2°, respectively. This shows that (+)-skyrin has reacted with (-)-(S)-**32** and (-)-flavomannin A with (+)-(R)-**32** indicating the S-configuration for (+)-skyrin and the (R)-configuration at the biaryl linkage for flavomannin A (Scheme 8). A similar approach to determine the stereochemistry of axially chiral compounds has been recently described by Yamamoto et al. (ref. 21).

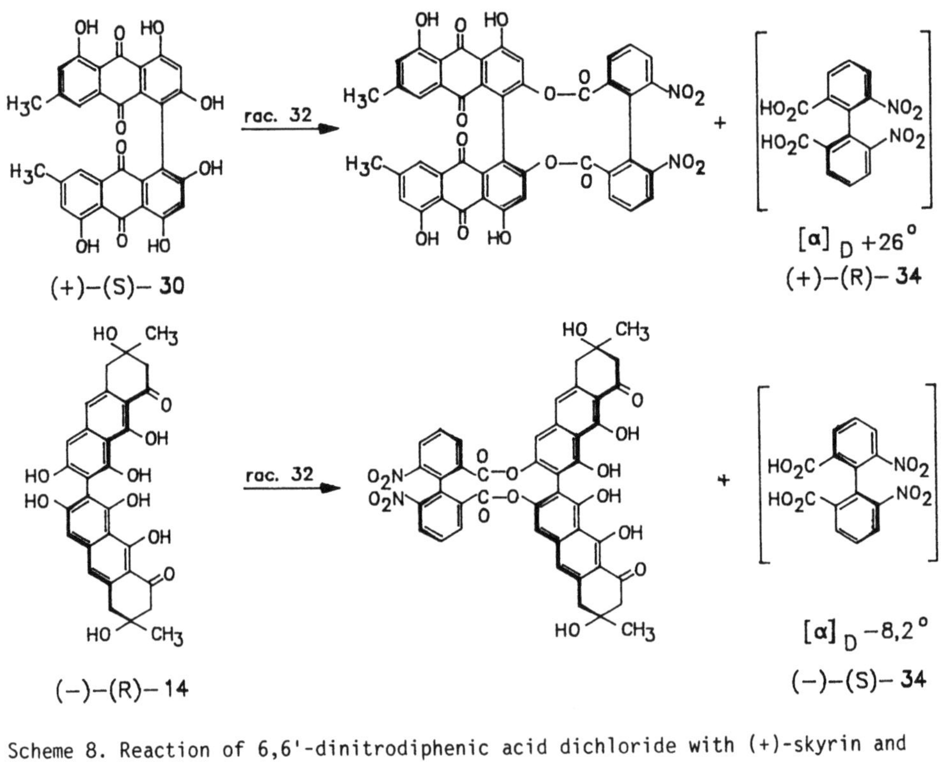

Scheme 8. Reaction of 6,6'-dinitrodiphenic acid dichloride with (+)-skyrin and (-)-flavomannin A

The (S)-configuration of (+)-skyrin is further supported by a ^1H n.m.r. comparison of the esters obtained from the (R)- and (S)-binaphthols **31** and skyrin (**30**) with Mosher's acid (ref. 22). Because the chemical shift differences observed with the esters of (-)-binaphthol and (+)-skyrin were in good agreement, both compounds possess the same stereochemistry.

To obtain additional proof for the absolute configuration of the flavomannin group, we tried to carry out a synthesis starting with a diphenic acid derivative of known absolute configuration (Scheme 9).

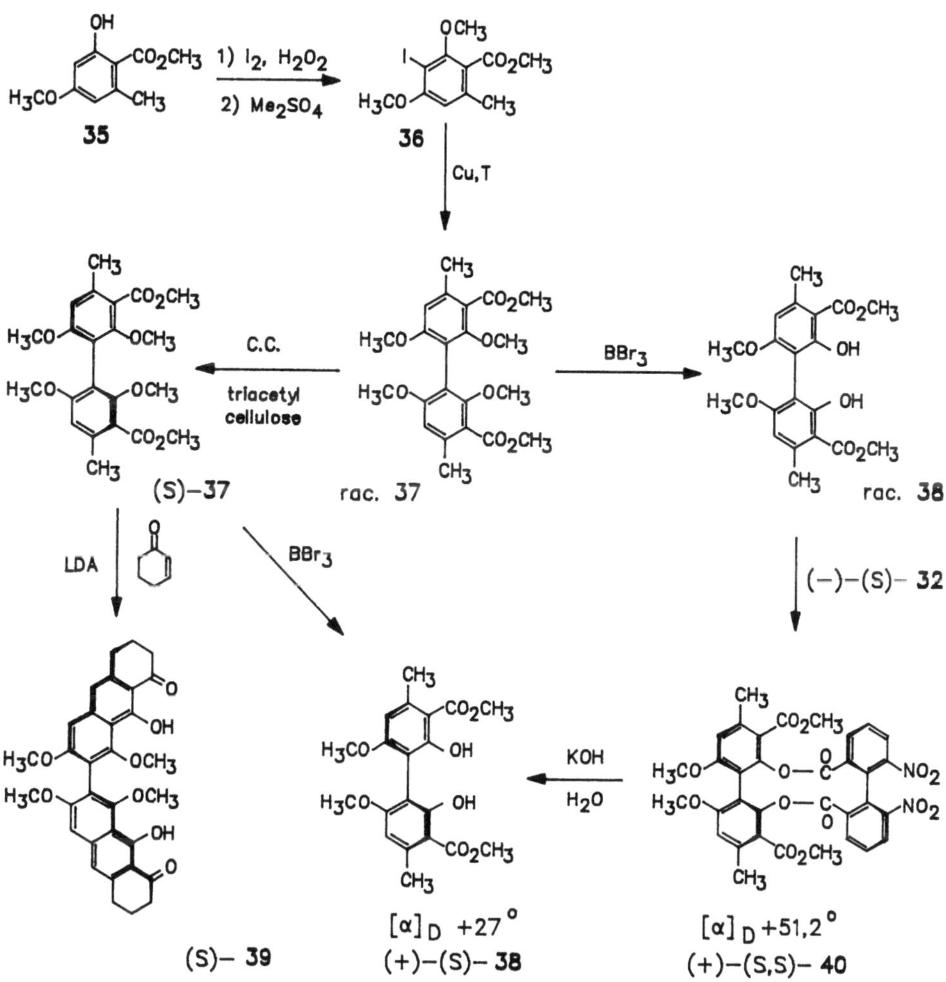

Scheme 9. Synthesis of an optically active model compound of the flavomannin typ

The desired compound was prepared from 4-O-methyl orsellinate (**35**) which on iodination followed by O-methylation yielded exclusively the 3-iodo derivative **36**. Ullmann coupling of **36** led to the dimer **37** in 65% yield. Selective cleavage of the 2-O-methyl groups was achieved by treatment with BBr_3 at -5°C. Reaction of the resulting racemic diphenol **38** with (-)-(S)-**33** yielded the optically pure (+)-(S,S)-lactone **40** and (-)-(R)-**38**. By this esterification and related transformations shown in Scheme 9 the absolute configuration of the dimeric orsellinic acid derivatives was secured. We then carried out a partial optical resolution of dimer **37** by column chromatography on triacetyl cellulose. The partially resolved (-)-(S)-ester was then reacted with two moles of cyclohexen-3-one which led to the formation of a flavomannin type compound **39**. Its CD spectrum exhibited a positive Cotton effect at 270 nm which agreed with the couplet observed in the spectrum of flavomannin B derivatives. These compounds therefore possess the (S)-configuration at the chiral axis, whereas those of the A series have the (R)-configuration. This is in agreement with the result of the kinetic resolution with racemic **32** described above. Experiments are under way to carry out the synthesis with optically pure dimer **37** and chiral enone **28** to obtain flavomannin derivatives which correspond to the natural products.

It can be summarised that the different methods of kinetic resolution and the synthetic correlations have led to the determination of stereochemistry at the biaryl linkage of atrivirin and flavomannin derivatives. Because the position and shape of the CD couplet around 275 nm is not much influenced by substituents present in the two halves of the dimers (refs. 3, 23), several other pigments can now be correlated. It will however need the development of further methods to elucidate the stereochemistry of the phlegmacin group and to determine the configuration of the chiral centers located in the alicyclic rings of the dimers.

Acknowledgement - We thank the Deutsche Forschungsgemeinschaft for continuous financial support. We are grateful to Prof. F. Delle Monache and Dr. M. Takido who very kindly provided us with authentic samples of vismiones and torosachrysone.

REFERENCES

This publication constitutes number 58 in the series 'Pigments of Fungi'; for 57 see: M. Brenner, H. Rexhausen, B. Steffan, W. Steglich, Tetrahedron, in print.

1 S. Gatenbeck, Acta Chem. Scand. 14 (1960) 296; ibid. 16 (1962) 1053.
2 W. Steglich, R. Arnold, W. Lösel and W. Reininger, J. Chem. Soc., Chem. Commun. 1972, 102.
3 W. Steglich and B. Oertel, Sydowia 37 (1984) 284; M. Gill and W. Steglich, Progr. Chem. Org. Nat. Prod. 51 (1987) 1-317.

4 M. Takido, S. Takahashi, K. Masuda and K. Yasukawa, Lloydia 40 (1977) 191.
5 J. F. Grove, J. Chem. Soc., Perkin I 1972, 2406.
6 F. Delle Monache, F. Ferrari, G. B. Marini-Bettolo, P. Maxfield, S. Cerrini, W. Fedeli, E. Gavuzzo and A. Vaciago, Gazz. Chim. Ital. 109 (1979) 301.
7 E. Jägers, Dissertation, Universität Bonn 1980.
8 S. Takahashi, S. Kitanaka, M. Takido, U. Sankawa and S. Shibata, Phytochemistry 16 (1977) 999.
9 S. Kitanaka and M. Takido, Phytochemistry 21 (1982) 2103.
10 D. L. Dreyer, I. Arai, C. D. Bachman, W. R. Anderson, Jr., R. G. Smith, and G. Doyle Daves, Jr., J. Am. Chem. Soc. 97 (1975) 4985; X. A. Dominguez, S. Templador and M. E. Cedillo, Rev. Latinoam. Quim. 7 (1976) 46.
11 N. Waksman de Torres, unpublished results.
12 N. Harada and K. Nakanishi, Acc. Chem. Res. 5 (1972) 257; N. Harada and K. Nakanishi, Circular Dichroic Spectroscopy, University Press, Oxford, 1983.
13 M. Endo and H. Naoki, Tetrahedron 36, 2449 (1980).
14 K.R. Hanson, Biochemistry 4 (1965) 2719.
15 F.W. Eastwood, K.J. Harrington, J.S. Josan and J.L. Pura, Tetrahedron Lett. 1970, 5223; T. Hiyama and H. Nozaki, Bull. Chem. Soc. Jpn. 46 (1973) 2248.
16 E.J. Corey and J.W. Suggs, Tetrahedron Lett. **1975**, 2647.
17 R.H. Thomson, Naturally Occurring Quinones, 2nd edn. Academic Press, London, 1971.
18 A. Horeau, Tetrahedron Lett. 1961, 506; A. Horeau, in: H.B. Kagan (Ed.), Stereochemistry, Vol. 3, Thieme Verlag, Stuttgart, 1977.
19 C.J.W. Brooks and J.D. Gilbert, J. Chem. Soc., Chem. Commun. 1973, 194.
20 S. Miyano, S. Handa, M. Tobita and H. Hashimoto, Bull. Chem. Soc. Jpn. 59 (1986) 235 and references cited therein.
21 K. Yamamoto, H. Yumioka, Y. Okamoto and H. Chikamatsu, J. Chem. Soc., Chem. Commun. 1987, 168.
22 J.A. Dale, D.L. Dull and H.S. Mosher, J. Org. Chem. 34 (1969) 2543; J.A. Dale and H.S. Mosher, J. Am. Chem. Soc. 95 (1973) 512; K. Kabuto, F. Yasuhara, S. Yamaguchi, Tetrahedron Lett. 21 (1980) 307; ibid. 22 (1981) 659.
23 U. Karl, Dissertation, Universität Bonn 1988.

Professor W. Steglich
Institut für Organische Chemie
Universität Bonn
Gerhard-Domagk-Straße 1
5300 Bonn 1
FRG

Peptide Siderophores from *Pseudomonas*

H. BUDZIKIEWICZ (Affiliation see page 333)

1. INTRODUCTION

A sufficient supply of Fe ions is essential for all living organisms. But although iron is amongst the most abundant elements making up about 3.5% of the earth crust it is hard to come by for microorganisms. In the original (reducing) atmosphere of the world iron existed in its divalent form and was thus sufficiently soluble, but when the atmosphere became oxidative due to the photosynthetic activity of certain procarionts iron was precipitated as polymeric ferric oxide hydrate. The consequence is that at the physiological pH of 7.4 $[Fe^{3+}]_{max}$ is 8.9 . 10^{-7} g/l, by far too low for survival (the optimum for microorganisms is about 5.10^{-6} g/l). Therefore, microroganisms did develop intricate chemical systems to make iron available for them. This is achieved by creating a reductive microbiotope or - more commonly - by giving off complexing agents (ref. 1).

The bacteria investigated by us belong to the genus *Pseudomonas*. They are characterized as straight or slightly bent gram negative rods with one or more polar flagella, not forming spors, usually living under aerobic conditions, and being able to utilize almost any carbon source. Many of them are soil bacteria, some are associated with higher plants (benign as well as pathogenic representatives are known), and *Pseudomonas aeruginosa* is amongst the most dangerous exitants of hospitalism. Since pseudomonads have to defend their habitat against many competitors they produce a variety of antibiotically active compounds (alkyl resorcinols (ref. 2), tropolones (ref. 3) and thiotropolones (refs. 4,5), alkyl quinolones - the so-called pyo-compounds (refs. 6-8), phenazine derivatives (ref. 9), pyridine (refs. 10, 11) and quinoline thiocarboxylic acids (ref. 12), unsaturated fatty acid derivatives (refs. 13, 14)) as well as iron complexing agents which are commonly referred to as siderophores.

Fe^{3+} due to its high charge density, small ion radius and low polarisability is a hard Lewis acid and forms stable bonds with hard Lewis bases as oxygen. Its octahedral complexes possess six coordination sites. They can accomodate three two-dentate ligands which for entropic reasons are frequently connected by more or less mobile chains keeping them in the right position. There are three main types of ligands, *viz.* a) catecholates (which form very stable complexes at $pH \geq 7$, but they are acid labile), b) hydroxamic acids (ref. 15) and c) carboxylic acid derivatives with a suitably located O- or N-functionality as the second coordination site (refs. 13, 14, 16, 17).

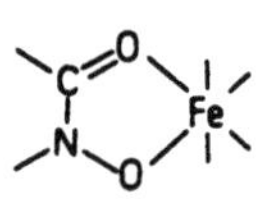

Hydroxamate

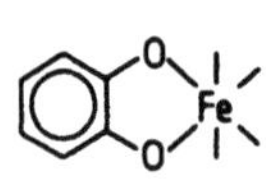

Catecholate

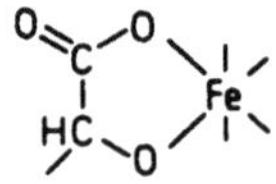

α-Hydroxy-carboxylate

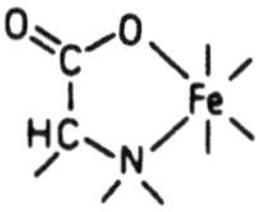

α-Amino-carboxylate

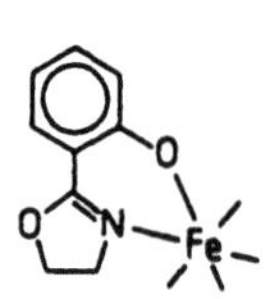

2-(o-Hydroxyphenyl)-oxazoline

The chemically and biologically most fascinating group of *Pseudomonas* siderophores are the pyoverdines and pseudobactins. They contain an amino *o*-dihydroxy quinoline system responsible for their fluorescence. It is amidically bound to a small dibasic acid and connected to a peptide chain containing about ten amino acids (both L and D; partially uncommon respresentatives as, e.g., N^5-hydroxy ornithine, ß-hydroxy aspartic acid, *allo*-threonine). Non-peptidic amide as well as ester bonds and cyclic structures have been encountered. One of the Fe^{3+} complexing ligands is the catecholic quinoline, the other two are located in the peptide chain (ß-hydroxy aspartic acid which is charateristic for pseudobactins, c-N^5-hydroxy ornithine or N^5-acetyl (or formyl)-N^5-hydroxy-ornithine).

2. STRATEGY OF STRUCTURE DETERMINATION

2.1 Physical Data

Pyoverdines and pseudobactins show a UV spectrum characterized by the following absorption bands (ref. 18):

~230 nm amide groups

~400 nm quinoline system (pH dependant for the free ligand)

~460 nm } charge transfer bands, missing in the iron-free
~550 } compounds

The molecular mass of the free ligands can be determined readily by FAB mass spectrometry (formation of $[M + H]^+$ ions). It may be verified by an analogous measurement of the Fe^{3+}

complexes. Occasionally, however, the mass of the latter will be found one mass unit too high due to a reduction of Fe^{3+} to Fe^{2+}. From the examples available by now it is not obvious when and why this reduction may occur.

Important information can be obtained from the electrophoretic mobility at different pH values regarding the acidic and basic groups present in the molecule. Thus, no change is observed in the mobility for acid amides between pH 6.5 and 2.0, weak acids as succinic acid will be protonated completely at pH 3.0 (loss of a negative charge as compared with pH 6.5) while α-keto glutaric acid will be protonated completely only at pH 2.0 (cf. Section 2.3) (refs. 18, 19).

The complexing constant K = [FePyo]/[Fe][Pyo] can be determined via the competing equilibrium with EDTA. It is in the vicinity of 10^{25} at pH 7.0 and of 10^{17} at pH 5 (ref. 20).

2.2 The Chromophore

The structure of the chromophore (1) has been elucidated first for pseudobactin by X-ray analysis (ref. 21). It may be obtained in substance by mild hydrolysis (ref. 20) while more drastic conditions lead to a replacement of the amino function by a hydroxyl group (2) (ref. 18). A complete set of mass and NMR spectral data is available for the chromophore itself as well as for several derivatives (refs. 18-20) which allow its ready identification. Regarding its stereochemistry see Section 2.5.

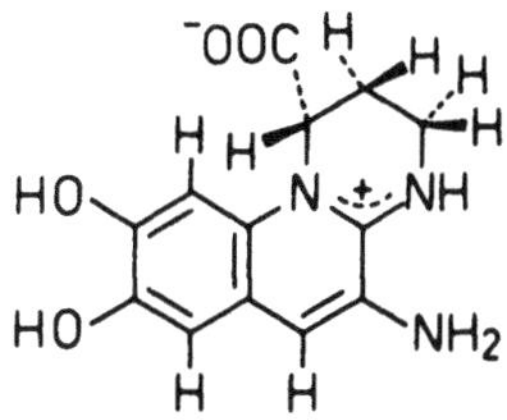

1

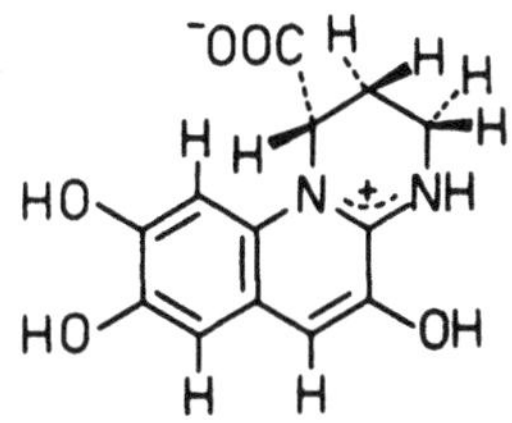

2

2.3 Dicarboxylic Acids Bound to the Chromophore

The amino group of the chromophore (1) is substituted amidically by small dibasic acids which all stem from the glutamic acid metabolism. In addition to the latter (ref. 20) succinic acid and its amide (refs. 18-21), malic acid (ref. 22) and α-keto glutaric acid (refs. 18-20) have been encountered (succinic acid seems to be an artefact stemming either from its amide or from α-keto glutaric acid). The mass of the acid present in the molecule can be deduced from a characteristic retro-Diels-Alder fragment of the quinoline nucleus in the FAB spectra of the free ligands (see *m/z* 985 in Fig. 1).

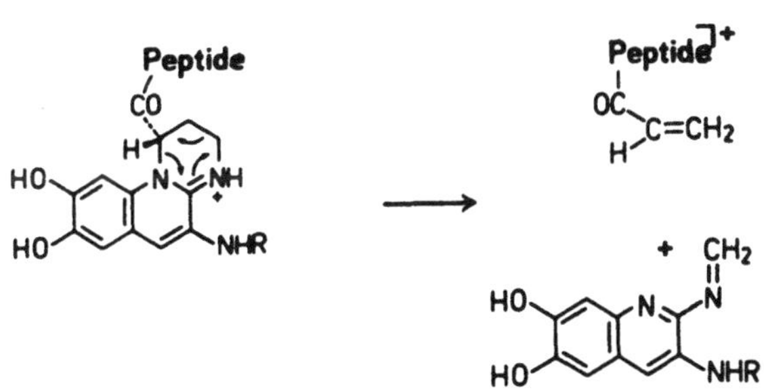

With the exception of α-keto glutaric acid the acids can be identified after total hydrolysis together with the amino acids. A distinction between succinic acid and its amide is possible by reduction with $NaBH_4$ after esterification: Only the ester will be reduced to γ-hydroxy butyric acid which can be identified as γ-butyrolactone (ref. 18).

α-Keto glutaric acid is instable after hydrolysis especially in the presence of amino acids. It may be identified after reduction to α-hydroxy glutaric acid or by oxidative decarboxylation with $Pb(OCOCH_3)_4$ or alcaline H_2O_2 of the complete siderophore to give the succinic acid analogue which also proves that the keto group is α to the free carboxylic acid (refs. 18, 19).

α-keto glutaric acid amides form equilibria with various cyclic and acyclic derivatives (see Scheme 1) which lie on the side of the latter. The relative contributions of the various derivatives depend on pH value of the solution (e.g.,the [13]C-

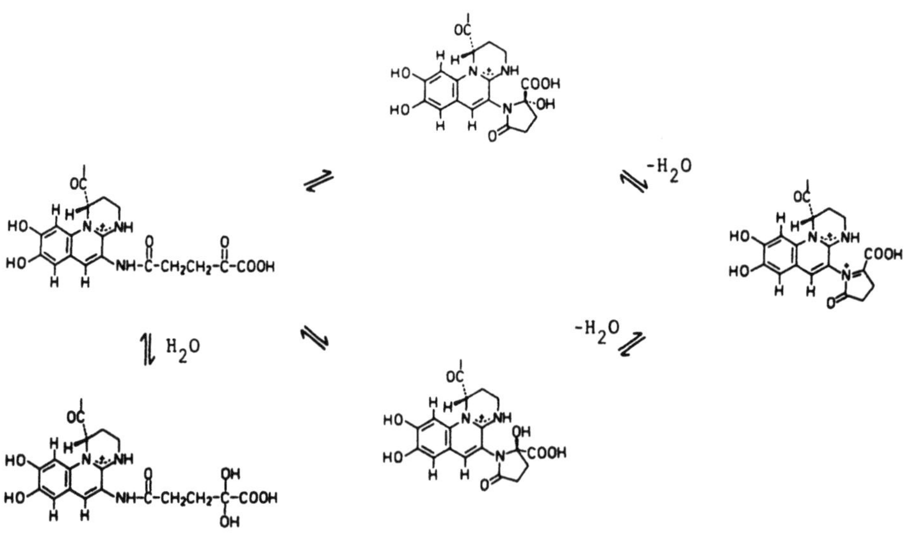

NMR signal for the α-carbonyl group at about 200 ppm can only be observed at about neutral pH values). These equilibria are reflected in the NMR-spectra: The CH_2-region (^1H- and ^{13}C-NMR) is rather complex, in the ^{13}C-NMR, i.a., signals at ~ 95 ppm (oxygenated quaternary C-atoms) can be observed, and some of the quinoline signals are broadened (refs. 18-20).

2.4 The Peptide Chain

The amino acids present can be identified and determined quantitatively by GC/MS analysis of the trifluoroacetyl n-butyl ("TAB") derivatives after total hydrolysis. It should be noted, however, that N-hydroxy amino acids will be reduced if the hydrolysis is effected by HI (but not if HCl is used) and that, e.g., N-acyl groups will be lost. The α-configuration (D or L) can be determined by chromatography of the heptafluoro-butyryl isopropyl derivatives on a chiral column (ref. 18).

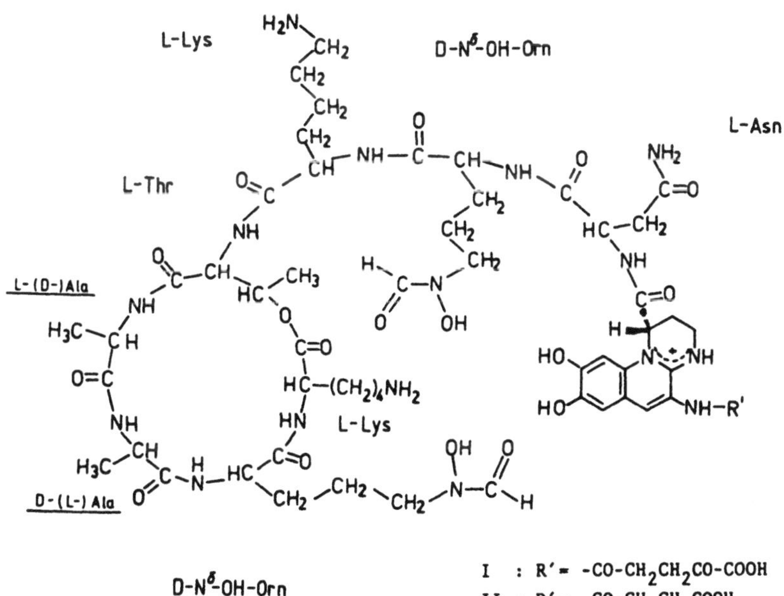

I : R′ = $-CO-CH_2CH_2CO-COOH$
II : R′ = $-CO-CH_2CH_2COOH$
III: R′ = $-CO-CH_2CH_2CONH_2$

3, Pyoverdine I, II and III

Since in pyoverdines and in pseudobactins both ester bonds and amidic linkages other than with the α-amino groups of the amino acids have been encountered it is essential to account for the functional groups of polyfunctional amino acids. Several examples relating to pyoverdine I-III (3) (ref. 18) will be given here:

(i) The presence of the hydroxamic acid moieties as N^5-formyl-N^5-hydroxy ornithine can be demonstrated by the split ^1H-NMR signals of the formyl groups in the vicinity of 8 ppm with an intensity ratio of about 3:1 (the high field signal related to the *cis* conformation being the more abundant one) and the facile hydrolysis (τ 1/2 ~1.5 h at pH 1 and 20°C) which can be followed by NMR (disappearance of the signals at 8 ppm and appearance of the formic acid signal at 8.22 ppm). In contrast, N^6-Formyl lysine, e.g., shows an intensity ratio of 1:8 (with a prevalent *trans* form signal at lower field) and a hydrolysis rate of several days under the conditions mentioned above.

(ii) The presence of the free N^6-amino groups of lysine can be demonstrated by dansylation and isolation of N^6-dansyl lysine after hydrolysis or by treatment of the pyoverdine with maleic anhydride at pH 9 to give the corresponding maleimides, the number of which can be determined by the mass increase in the FAB mass spectrum.

(iii) The presence of ester linkages can be shown by saponification with K_2CO_3 in CH_3OH (which does not attack amides or hydroxamic acids) or by reduction with $NaBH_4$ and isolation of the reduction product after complete hydrolysis (in the case of 3 2,6-diaminohexanol from lysine).

TABLE 1

Sequential Analysis of the Peptide Chain
by FAB Mass Spectrometry

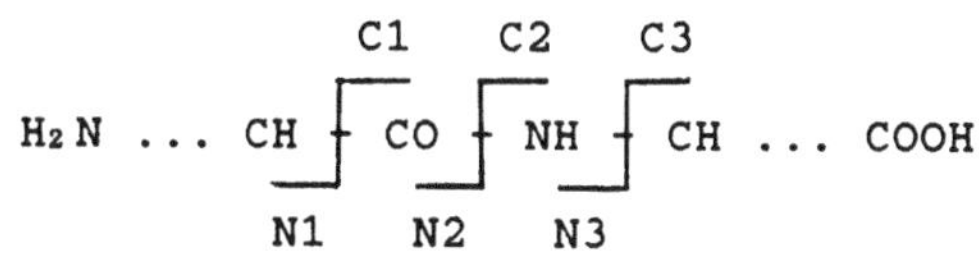

m/z C1	*m/z* C2	amino acids	difference	*m/z* N2	amino acids
$^+H_3N-$	$^+OCNH-$			$-CO^+$	
931	959	8	chromophore	473	1
816	842	7	Asn	630	2
685	711	6	N-Form-N-OH-Orn	758	3
530	556	5	Lys	858	4
447	473	4	$Thr-H_2O$		
375	401	3	Ala		
305	331	2	Ala		

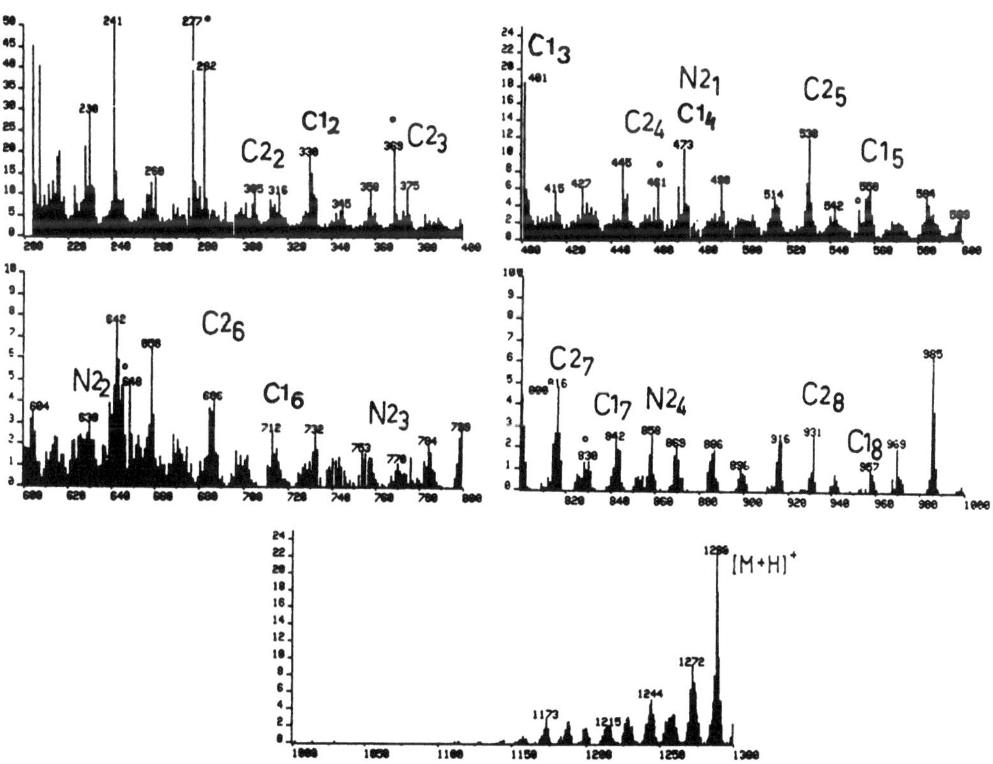

Fig. 1 FAB spectrum of pyoverdine II

Some first information regarding the sequence of the amino acids may be obtained from the FAB mass spectra. Peptides show three series of fragments by loss of parts of the N-terminus and three series by loss of parts of the C-terminus, though not all of the fragments may be recognizable with certainty (see Table 1 and Fig. 1). A certain amount of caution is necessary since ions of comparable intensity not belonging to any of the six series are present in the mass spectrum. It should be noted here that any sequencing involving Edman degradation or related techniques is not possible since in most cases the N-terminus is blocked by the quinoline moiety, and that enzymatic cleavages are impeded by the high percentage of D-amino acids. Partial hydrolyses, isolation of the hydrolysis products, their mass spectral investigation in combination with dansylation or Edman reactions are necessary to confirm the conclusions drawn from FAB spectra - especially when the peptide moiety comprises cyclic substructures.

2.5 Stereochemistry

Regarding the amino acids see Section 2.4. The stereochemistry of the chiral center in the quinoline portion has been a problem. For pseudobactin it has been determined as S by X-ray spectroscopy. At pH 7.2 pseudobactin (ref. 21) shows a *positive* Cotton effect at 410 nm ($\pi \rightarrow \pi^*$ transition of the quinoline), ferric pseudobactin a *positive* one at 400 nm. Ferric pyoverdines I and II (ref.18) - in contrast - show a *negative* Cotton effect at 400 nm from which an R configuration was deduced for the latter. Subsequently, it has been shown that a CD correlation is possible only for the iron free compounds: Pyoverdine(refs. 23, 24) Pa (= E, ref. 19) and pyoverdine I/II (refs. 18, 20) which upon hydrolysis give the identical quinoline chromophore (i.a., identical CD spectra) exhibit both *positive* Cotton maxima at 380 nm, and hence they are both S configurated.

The configuration of the iron complexes - with due caution - may be deduced from the sign of the Cotton effect at the charge-transfer-bands (ref. 1). From hydroxamic acid model compounds it has been deduced that a *negative* Cotton effect in the vicinity of 450 nm corresponds to a Δ configuration. The ferri pyoverdines I-III and Pa (= E) show a *negative* Cotton effect at 488 nm and 460 nm, resp., (ref. 20) and should, therefore, both be Δ configurated. Why for the latter one a Λ configuration has been postulated (ref. 24) is not obvious.

2.6 Structural Variations

In Table 2 pseudobactins and pyoverdines have been assembled for which complete or partial structures have been proposed. The most striking observation is the great variability of the peptide chain. Cross-feeding experiments have shown that the amino acid sequence is responsible for the recognition of the specific "own" siderophore of a *Pseudomonas* species: Fig.2a shows the increase in cell growth if the proper ferric siderophore is added to an iron deficient

TABLE 2

List of Pseudobactins and Pyoverdines

Pseudobactins			
Ps.	S-Chr-Lys-ß-OH-Asp-Ala-Thr-Ala-c-N^5-OH-Orn	X-ray	(ref. 21)
7SR1	M-Chr-Ser-Gly-Ala-N^5-Ac-N^5-OH-Orn-Ser-Thr-ß-OH-Asp-Ser	degradation, NMR	(ref. 22)
A 214	S-Chr-Ser-Ala-Gly-Ser-Ala-ß-OH-Asp-Thr-N^5-Ac-N^5-OH-Orn	degradation, NMR	(ref. 34)
17400	?-Chr-Ala-Lys-Gly-ß-OH-Asp-Gln-Ser-Ala-c-N^5-OH-Orn	only FAB	(ref. 24)
12633	?-Chr-Asp-Lys-ß-OH-Asp-Ser-Thr-Ala-Glu-Ser-c-N^5-OH-Orn	only FAB	(ref. 24)
GM-I	S-Chr-(5 Ala, ß-OH-Asp, Glu, 2 Gly, Lys, c-N^5-OH-Orn, Ser)	only hydrolysis	(ref. 20)
T2	S-Chr-(ß-OH-Asp, Dbu, Gln, Orn, X-N^5-OH-Orn, 3 Ser, Thr, *allo*-Thr)	only hydrolysis	(ref. 20)
Pyoverdines			
C, D, E (=Pa)	(G,S)-Chr-Ser-Arg-Ser-N^5-Form-N^5-OH-Orn-Thr-Thr-N^6-Form-Lys-c-N^5-OH-Orn	degradation and FAB	(refs.19,20,23)
I, II, III	(G,S)-Chr-Asn-N^5-Form-N^5-OH-Orn-Lys-Thr-Ala-Ala-N^5-Form-N^5-OH-Orn-Lys	degradation and FAB	(ref. 18)
3-I	S-Chr-(2 Glu, Gly, 2 N^5-OH-Orn)	only hydrolysis	(ref. 20)
7-I	(G,S)-Chr-(Glu, 2 Gly, 2 Lys, 2 N^5-OH-Orn, 3 Ser)	only hydrolysis	(ref. 20)
12633	?-Chr-Ser-Lys-Gly-N^5-X-N^5-OH-Orn-Ser-Lys-c-N^5-OH-Orn	only FAB	(ref. 24)
2192	?-Chr-Ser-Lys-Ser-Ser-Thr-Ser-N^5-X-N^5-OH-Orn-Thr-Ser-c-N^5-OH-Orn	only FAB	(ref. 24)
15692	?-Chr-Ser-Arg-Ser-N^5-X-N^5-OH-Orn-Thr-Thr-Lys-c-N^5-OH-Orn	only FAB	(ref. 24)
PSS	?-Chr-(2 ß-OH-Asp, 2 Ser, 2 Thr, Lys)	only hydrolysis	(ref. 30)
-	?-Chr-(Lys, X-N^5-OH-Orn, 3 Ser, 3 Thr)	only hydrolysis	(ref. 31)
-	?-Chr-(Gly, 2 Lys, 2 X-N^5-OH-Orn, 2 Ser)	only hydrolysis	(ref. 32)
-	?-Chr-(x Glu, x Lys, x Ser, x Thr)	only hydrolysis	(ref. 33)

S ... succinic acid/amide; M ... malic acid; G ... α-ketoglutaric acid

culture while no such effect is observed with a "strange" siderophore (Fig.2b; the slight increase can be explained by an iron exchange between the added ferric siderophore and the one produced by the bacteria)(ref. 20).

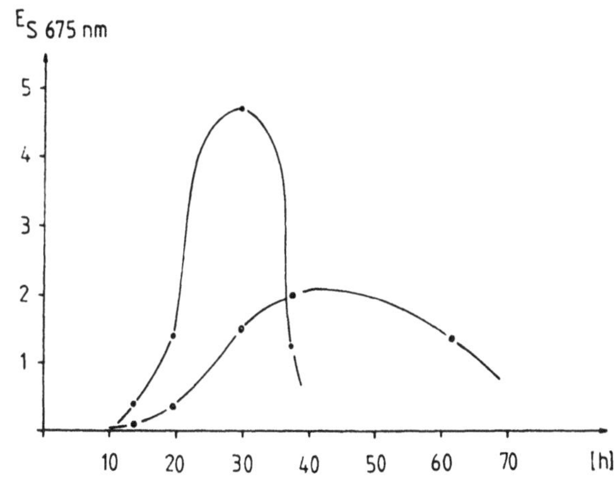

Fig. 2a Growth curves of *Pseudomonas aeruginosa* with (•) and without (°) ferric *Pseudomonas aeruginosa* siderophore

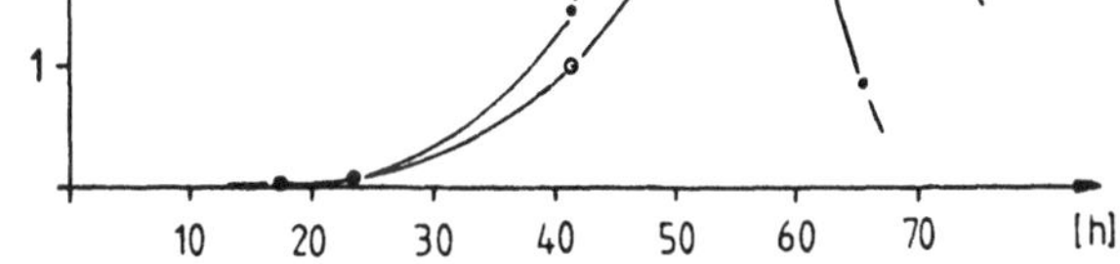

Fig. 2b Growth curves of *Pseudomonas aeruginosa* with(•) and without (°) *Pseudomonas fluorescens* siderophore

Concluding Remarks

The structure elucidation of these compounds is not only a challenge for the natural products chemist, it is also of practical importance. *Pseudomonas aeruginosa* is one of the most dangerous germs causing hospitalism resistant against all common antibiotics (ref. 25). The availability of iron seems to be an important factor for its virulence. As with increasing temperature the production of siderophores is reduced fever seems to be one answer of the human body against infection. One may speculate whether it is possible to bind antibiotically active compounds to the specific siderophore and thus cause their transport into the bacterial cells.

Siderophore producing bacteria have been shown to exhibit a growth promoting effect for certain higher plants (ref. 26). Two mechanisms seem to be responsible here: The plant may use th iron chelates for its own purposes, and the iron deficiency in its vicinity caused by the excretion of siderophores creates an environment unfavorable for other (e.g., plant pathogenic) bacteria (refs. 27, 28). Thus far, only bacterial extracts have been used. The availability of larger amounts of specific siderophores and the knowledge of their structures will facilitate experiments in this direction.

Still another effect of siderophores may be hinted here: Al^{3+} ions set free in the soil at low pH values caused by the so-called "sour rain" have been named as one of the causes for the recent damage of forest trees (ref. 29). As siderophores are able to bind also other trivalent cations (and specifically Al^{3+}) (ref. 15) they may act as the transport vehicle.

Acknowledgements

I wish to thank the following individuals engaged in the siderophore research in Cologne: Prof. Dr. G. Pulverer and Dr. H. Korth, Hygiene-Institut, Drs. K. Taraz, G. Briskot, K. Poppe, St. Winkler, Dipl.-Chem. K. Geisen, S. Gipp, I. Gwose, G. Mohn, Institut für Organische Chemie. Financial assistance by Deutsche Forschungsgemeinschaft and Fonds der Chemischen Industrie is greatfully acknowledged.

REFERENCES

1 K.N. Raymond, G. Müller, and B.F. Matzanke, Complexation of Iron by Siderophores. A Review of Their Solution and Structural Chemistry and Biological Function, Top. Curr. Chem. 123 (1984) 49-102.
2 H. Budzikiewicz, H. Scholl, W. Neuenhaus, G. Pulverer, and H. Korth, Dialkylresorcine aus *Pseudomonas aureofaciens*, Z. Naturforsch. 35b (1980) 909-910.
3 H. Korth, G. Pulverer, A. Römer, and H. Budzikiewicz, 7-Hydroxytropolon aus *Pseudomonas sp.*, Z. Naturforsch. 36c, (1981) 728-729.
4 K. Görler, W. Molls, H.U. Siehl, J. Strähle and Ch. Westphal, Struktur einer neuen schwefelhaltigen Tropolonverbindung aus *Pseudomonas cepacia*, Liebigs Ann. Chem. (1982) 1006-1011.
5 K. Kintaka, H. Ono, S. Tsubotani, S. Harada, and H. Okazaki, Thiotropocin, a New Sulfur-Containing 7-Membered-Ring Antibiotic Produced by a *Pseudomonas sp.*, J. Antibiot. 37 (1984) 1294-1300.
6 W. Neuenhaus, H. Budzikiewicz, H. Korth, and G. Pulverer, 3-Alkyltetrahydrochinolinderivate aus *Pseudomonas*, Z. Naturforsch. 34b (1979) 313-315.
7 H. Budzikiewicz, U.Schaller, H. Korth, and G. Pulverer, Alkylchinoline und deren N-Oxide aus *Pseudomonas aeruginosa*, Monatsh. Chem. 110 (1979) 947-953.

8 S. Kitumara, K. Hashizume, T. Iida, E. Miyashita, K. Shirahata, and H. Kase, KF8940 (2-n-Heptyl-4-hydroxy-quinoline-N-oxide), a Potent and Selective Inhibitor of 5-Lipoxygenase, Produced by *Psedomonas methanica*, J. Antibiot. 37 (1986) 1160-1166.
9 A. Römer, H. Scholl, H. Budzikiewicz, H. Korth, and G. Pulverer, Phenazine aus Pseudomonaden, Z. Naturforsch. 36b (1981) 1037-1046.
10 U. Hildebrand, J. Lex, K.Taraz, S. Winkler, W. Ockels, and H. Budzikiewicz, Untersuchungen zum Redox-System Bis-(pyridin-2,6-dicarbothioato)-ferrat(II)/-ferrat(III), Z. Naturforsch 39b (1984) 1607-1613.
11 U. Hildebrand, K.Taraz, H. Budzikiewicz, H. Korth, and G. Pulverer, Dicyano-bis(pyridin-2,6-dicarbothioato)-ferratII/ferrat(III), ein weiteres eisenhaltiges Redox-system aus der Kulturlösung eines *Pseudomonas*-Stammes, Z. Naturforsch. 40c (1985) 201-207.
12 W. Neuenhaus, H. Budzikiewicz, H. Korth, and G. Pulverer, 8-Hydroxy-4-methoxy-monothiochinaldinsäure, eine weitere Thiosäure aus *Pseudomonas*, Z. Naturforsch. 35b (1980) 1569-1571.
13 W.L. Parker, M.L. Rathnum, V. Seiner, W.H. Trejo, P.A. Principe, and R.B. Sykes, Cepacin A and Cepacin B, Two New Antibiotics Produced by *Pseudomonas cepacia*, J. Antibiot. 37 (1984) 431-440.
14 T. Kusumi, I. Ohtani, K. Nishiyama, and H. Kakisawa, Caryonencins, Potent Antibiotics from a Plant Pathogen *Pseudomonas caroyphilii*, Tetrahedron Lett. 28 (1987) 3981-3984.
15 S. Winkler, W. Ockels, H. Budzikiewicz, H. Korth, and G. Pulverer, 2-Hydroxy-4-methoxy-5-methylpyridin-N-oxid, ein Al^{3+} bindender Metabolit von *Pseudomonas cepacia*, Z. Naturforsch. 41c (1986) 807-808.
16 G.J. Bukovits, N. Mohr, H. Budzikiewicz, H. Korth, and G. Pulverer, 2-Phenylthiazol-Derivate aus *Pseudomonas cepacia*, Z. Naturforsch. 37b (1982) 877-880.
17 A. Zunnundzhanov, I.A. Bessonova, N.D. Abdullayev, and D.K. Ogai, Stroenie aerugina is *Pseudomonas aeruginosa*, Khim. prirod. soed. (1987) 553-558.
18 K. Poppe, K. Taraz, and H. Budzikiewicz, Pyoverdine Type Siderohpres from *Pseudomonas fluorescens*, Tetrahedron 43 (1987) 2261-2272.
19 G. Briskot, K.Taraz, and H. Budzikiewicz, Siderophore vom Pyoverdin-Typ aus *Pseudomonas aeruginosa*, Z. Naturforsch. 41c (1985) 497-506.
20 Unpublished results from this laboratory.
21 M. Teintze, M.B. Hossain, C.L. Barnes, J. Leong, and D. van der Helm, Structure of Ferric Pseudobactin, a Siderophore from a Plant Growth Promoting *Pseudomonas*, Biochem. 20 (1981) 6446-6457.
22 Ch.-Ch. Yang and J. Leong, Structure of Pseudobactin 7SR1, a Siderophore from a Plant-Deleterious *Pseudomonas*, Biochem. 23 (1984) 3534-3540.
23 S. Wendenbaum, P. Demange, A. Dell, J.M. Meyer, and M.A. Abdallah, The Structure of Pyoverdine Pa, the Siderophore of *Pseudomonas aeruginosa*, Tetrahedron Lett. 24 (1983) 4877-4880.

24 P. Demange, S. Wendenbaum, A. Bateman, A. Dell, and M.A. Abdallah in: G. Winkelmann, D. van der Helm, and J.B. Neilands (Eds.), Ion Transport in Microbes, Plants and Animals, VCH, Weinheim, 1987, p. 167.
25 G. Pulverer, Erregerspektrum und Antibiotikaresistenz, Rhein. Ärztebl. (1972) 118-120.
26 M.N. Schroth and J.G.Hancock, Disease-Suppresive Soil and Root-Colonizing Bacteria, Science 216 (1982) 1376-1381.
27 J.W. Kloepper, J. Leong, M. Teintze, and M.N. Schroth, Enhanced Plant Growth by Siderophores Produced by Plant Growth-Promoting Rhizobacteria, Nature 286 (1980) 885-886.
28 P.E. Powell, G.R. Cline, C.P.P. Reid, and P.J. Szaniszlav, Occurance of Hydroxamate Siderophore Ion Chelators in Soils, Nature 287 (1980) 833-834.
29 E. Cowling, B. Krahl-Urban, and Ch. Schimansky, Ursachen von Waldschäden - Wissenschaftliche Hypothesen zur Erklärung der Ursachen, Österr. Chemie-Zeitschr. 88 (1987) 228-233.
30 Y.S. Cody and D.C. Gross, Characterization of Pyoverdin PSS, the Fluorescent Siderophore Produced by *Pseudomonas syringae pv. syringae*, Appl. Env. Microbiol., (1987) 928-934.
31 L. Torres, J.E. Pérez-Ortin, V. Tordera, and J.P. Beltrán, Isolation and Characterization of an Fe(II)-Chelating Compound Produced by *Pseudomonas syringae*, Appl. Microbiol. (1986) 157-160.
32 St.B. Philson and M. Llinás, Siderochromes from *Pseudomonas fluorescens*, J. Biol. Chem. 257 (1982) 8081-8085.
33 J.D. Newkirk and F.H. Hulcher, Isolation and Properties of a Fluorescent Pigment from *Pseudomonas mildenbergii*, Arch. Biochem. Biophys. 134 (1969) 395-400.
34 J.S. Buyer, J.M. Wright, and J. Leong, Structure of Pseudobactin A 214, a Siderophore from a Bean-Deleterious *Pseudomonas*, Biochem. 25 (1986) 5492-5499.

Affilliation:

Prof. Dr. Herbert Budzikiewicz
Institut für Organische Chemie der Universität zu Köln,
Greinstr. 4, D-5000 Köln 41, West Germany

Structure of Amavadin, the Vanadium Compound of *Amanita Muscaria* and Selective Binding of Vanadium

E. BAYER (Affiliation see page 344)

Vanadium has always been considered as an essential trace element of living organism. However the bilogical funtions of vanadium remained obscure, and even in 1980 Macara (ref. 1) regarded vanadium as "an element in search of a role". Only recently the isolation of a vanadium dependent peroxidase from algae (ref. 2), and a vanadium nitrogenase from Acotobacter (ref. 3) has changed this view.

Other examples of vanadium occurrence in nature are the relatively high concentrations of vanadium in the blood cells of tunicates, and in the fly agaric (*Amanita muscaria*). In both cases the vanadium enrichment is especially interesting, and propably the most spectacular example of metal enrichment in nature.

Despite many attempts the structure of the vanadium compounds in tunicates is still unknown, and nothing was known about the structure of the vanadium compound of *Amanita muscaria*.

The pale blue vanadium compound from *Amanita muscaria*, which we named Amavadin can be separated from aqueous mushroom extracts by HPLC on reversed phase columns as is displayed in Fig. 1.

Amavadin shows the characteristic electron spin resonance spectrum of tetravalent vanadium which can be used to follow isolation. The ESR spectrum is the same before and after isolation (ref.4). Therefore it is clear that no binding to a protein is occurring. The molecular weight determination reveals that amavadin is a low molecular weight compound of approximately 420 ± 20 dalton.

The structure elucidation was difficult because standard methods could not be utilized. The paramagnetism of amavadin caused extreme peak broadening of nuclear magnetic resonance spectra. Due to the low volatility of the vanadium complex no useful mass spectra in EI, CI or FAB mode could be obtained.

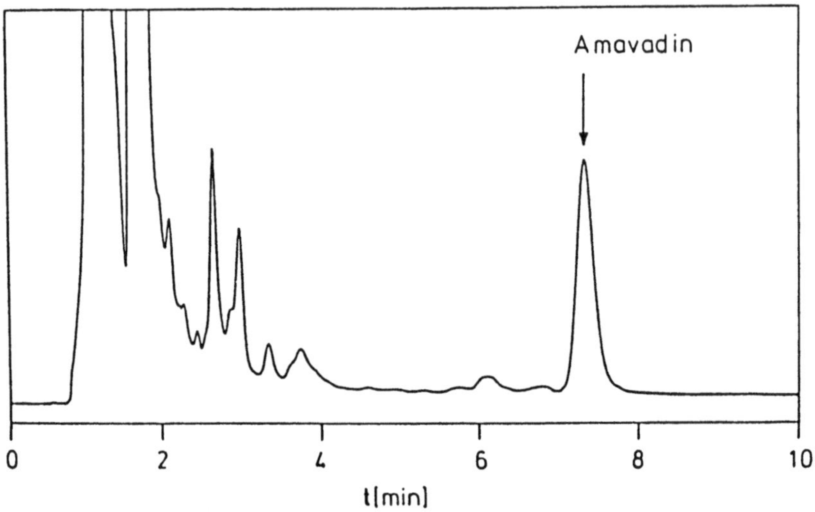

Fig.1 Separation of Amavadin by HPLC of mushroom extracts. Column: Latek ODS-1,5 um; 150x4 mm. Solvent 80% acetonitrile, 20% tetrabutylammoniumhydrogen sulfate. Flow 1 ml/min; pressure 1700 psi; temperature 23°C; detection: UV 235 nm.

All attempts to crystallize amavadin for x-ray structure determination have failed so far. It is also not possible to remove the vanadium without degradation or change of the ligand due to the extreme stability of the vanadium complex.

Tetravalent vanadium salts and coordination compounds normally occur in the vanadyl form VO^{2+} and a IR-band at 985 cm^{-1} in amavadin suggested that it is also a VO^{2} species.

The structure of the amavadin ligand had to be determined by chemical degradation (Fig. 2) and revealed a peptidic nature (ref. 5, 6, 7).

Strongly acidic and alcaline hydrolysis yielded L-alanine, pyruvic acid and acetaldehyde as only products, and indicated a deeper degradation.

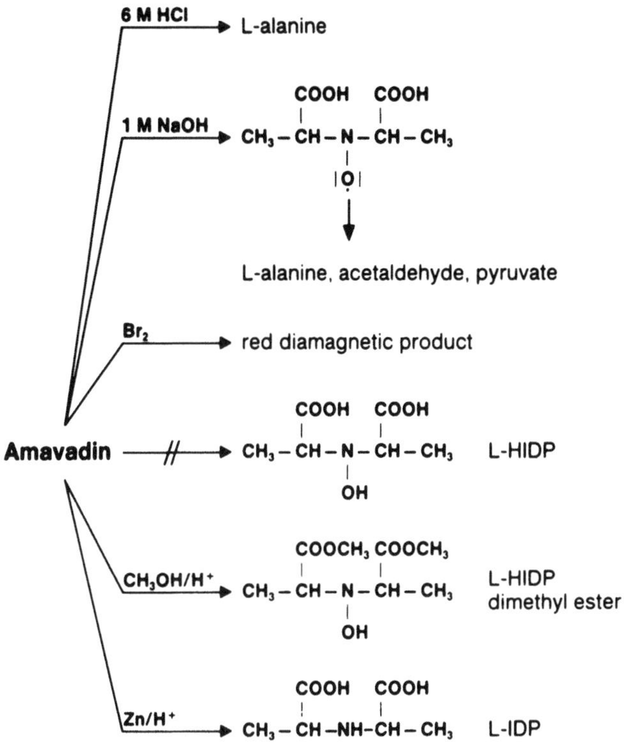

Fig. 2 Degradation reactions of amavadin

Under reducing conditions with zinc/HCl - -iminodipropionic acid (IDP) was isolated. IDP was already described in the literature by Karrer and Appenzeller (ref. 8), and occurs as RR and SS enantiomer and meso form. All forms were synthesized, and compared with the IDP obtained by degradation of amavadin. Since the amount of material was restricted a micromethod had to be devised for this purpose. Therefore capillary gas chromatography on Chirasil-Val (ref. 9, 10), a chiral polysiloxane peptide stationary phase was chosen. Very well resolved are the chiral forms of IDP as derivatization products with isopropyl isocyanate on Chirasil-L-Val, as is shown in Fig. 3. The IDP degradation product of amavadin is identical with SS-IDP.

IDP, which is closely related to the classical sequestering agent ethylenediamine tetraacetic acid (EDTA), belongs to a group of natural products ("opines") which can be found in crown gall tumors (ref. 11). IDP itself was found in mussels and squids where it occurred in the optically inactive meso form (ref. 12). However L-IDP is not the organic ligand of amavadin, because it does not produce amavadin when reacted with vanadylsulfate.

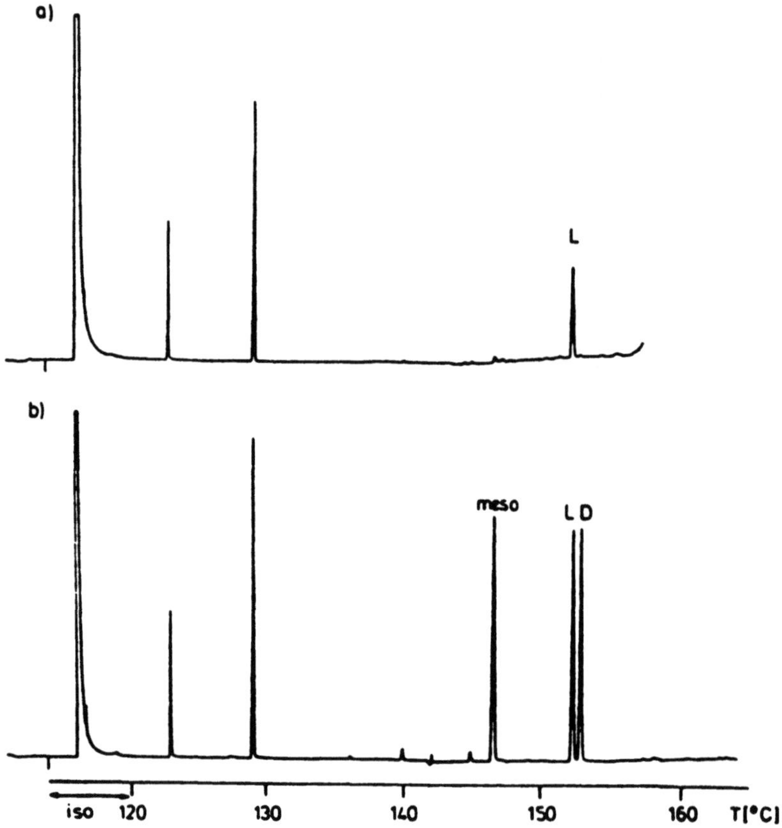

Fig. 3 Gas chromatogram of IDP from amavadin (a) and of meso-, L- and D-IDP after derivatization with isopropyl isocyanate

Upon treatment with diluted sodium hydroxide in air a change of the characteristic ESR spectrum of amavadin can be observed. First the signals disappear and subsequently a new spectrum occurs which can be assigned to a relatively stable nitroxyl radical, suggesting that the ligand of amavadin contains a NOH-group.

Finally by reacting amavadin with conc. sulfuric acid in methanol at elevated temperature a compound was isolated, which containd one oxygen more as IDP. The nmr and mass spectra suggested that the structure may be N-(LL-carboxyethyl)-N-hydroxy-L-alanine (L-HIDP). Since this compound was not yet described in the literature it had to be synthesized, including all stereoisomers (LL; DD; meso) in order to assign the structure.

Attempts to synthesize HIDP by direct oxidation of IDP failed. Preliminary experimentes indicated that an enzymatic oxidation by oxidase is possible, closing a possible biosynthetic link between the opines and amavadin.

For larger scale synthesis of this new and because of its vanadium selectivity extremely interesting sequestering agent the reaction of 2-bromopropionic acid has been chosen. L-HIDP can be obtained stereoselectively by condensation of hydroxylamine with D-bromopropionic acid (Fig. 4). The enantiomeric purity and comparison of educts and products again can be checked by chromatography on Chirasil-Val. L-HIDP from amavadin is identical in all properties with the synthesized L-HIDP.

Reaction of two molecules of HIDP with one equivalent of vanadylsulfate completes the total synthesis of amavadin (ref. 13), identical in properties with natural amavadin. Assuming the vanadyl group es central unit the structure in Fig. 5 was then assigned to amavadin.

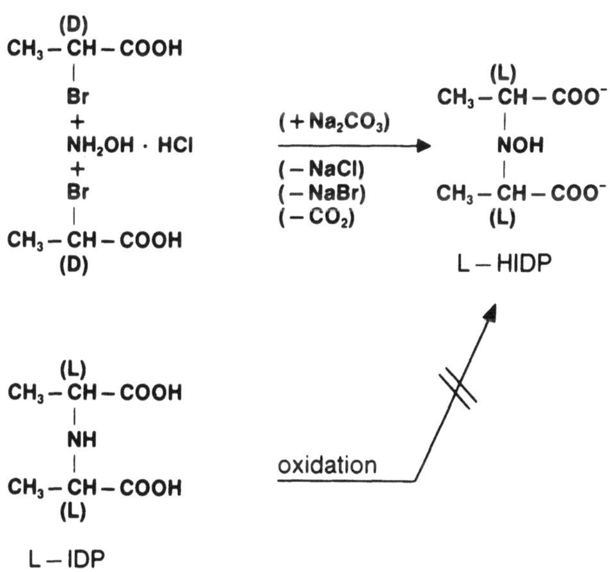

Fig. 4 Synthetic routes to HIDP

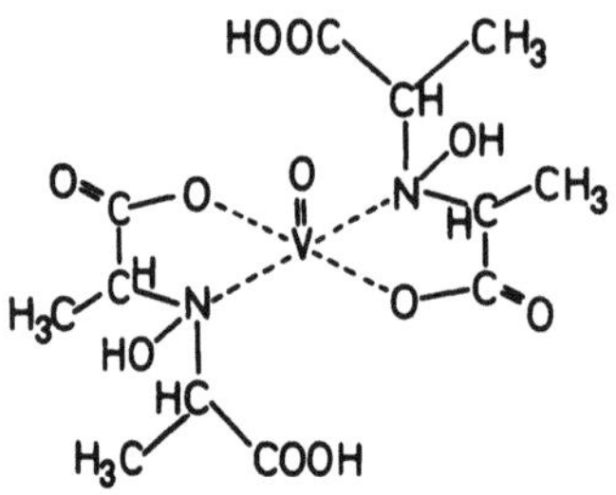

Fig. 5 First structural proposal for amavadin

The availability of larger quantities of the ligand of amavadin by synthesis prompted us to investigate in general the chelating properties of HIDP, and to compare it with the very well known sequestering capability of IDP.

Table 1 shows the log K values of IDP and HIDP for different metal ions, which have been investigated together with Anderegg (ref. 14). The stability of the 2:1-complex of HIDP is remarkable, and so far amavadin reveals itself as the most stable V(IV)-complex ever reported in literature.

TABLE 1

Logarithms of the stability constants $K_n=[ML_n]/[ML_{n-1}][L]$ for HIDP and IDP with different metal ions. Temperature: 298 K, ionic strength: 0.1 (from ref. 14).

Metal ion		Ligand	
		HIDP	IDP
Ca^{2+}	lg K_1	2.7	2.0
Ni^{2+}	lg K_1	5.7	7.6
	lg K_2	5.0	5.9
Cu^{2+}	lg K_1	9.2	10.6
	lg K_2	3.45	4.4
Zn^{2+}	lg K_1	5.05	6.6
	lg K_2	4.05	4.5
VO^{2+}	lg K_1	-.-	9.54
	lg K_1K_2	23.0	

It is more then a billion times more stable as the copper complex. Remarkable is also the favored formation of the 2:1 complex. Independent from the stoichiometric ratio of HIDP: vanadyl salt we never could observe a 1:1 complex, contrary to the observation of a step-wise complex formation in the case of other metals. This is in sharp contrast to the complex formation of IDP with vanadylions.

IDP forms only a 1:1 complex. The stability of this complex is much smaller as of amavadin. Additionally the vanadium selectivity disappears in the case of IDP. Copper forms a more stable chelate. So nature has created a unique vanadium selectivity only by inserting an oxygen into a NH-bond.

However if one looks at table 1 in more detail, it is difficult to understand why the vanadium complex of the N-hydroxy compound should be more stable than the vanadium compound of IDP on the basis of the structure proposed with a bond to nitrogen. Normaly one would assume that the nitrogen in a hydroxylamine derivative is less basic as in an amine, and therefore also the complex stability decreases. Table 1 teaches this for calcium, nickel, copper and zinc complexes of both ligands. With the exception of vanadium all other metal complexes of IDP are approximately by one order of magnitude more stable than those one of HIDP.

We therefore determined the complex stability constants of a larger variety of ligands related to HIDP, which are shown for the V(IV)- and Cu(II)-chelates in table 2. Only N-hydroxy-iminodiacetic acid forms a vanadium compound with similar stability as amavadin. All other ligands like the isomeric compound N-hydroxy-ß,ß'-iminodipropionic acid or the bidentate ligand N-hydroxy- -alanine form much weaker complexes.

This means that all ligands, especially also the NO-group as such, and not only the nitrogen as well as both carboxylic groups of HIDP are involved in the complex formation, and that the inner sphere of the coordination depicted in fig. 5 does not agree with the real structure.

Since no x-ray investigations could be obtained large-angle-X-ray scattering (LAXS) experiments were performed. They clearly indicated that no metal-ligand distance corresponding to a V=O distance (0.157-0.165 nm) was found, the shortest metal-ligand distance being larger than 0.19 nm. Addidionally the IR signal

TABLE 2

Complex formation constants ($K_n = ([ML_n]/[L][ML_{n-1}])$ for some amino and N-hydroxyamino acid metal complexes (T = 25°C, in 0.1 M KNO_3) (ref. 15).

Ligand	VO^{2+} lgK_1	VO^{2+} lgK_2	VO^{2+} lgK_3	Cu^{2+} lgK_1	Cu^{2+} lgK_2
HIDP	-	23.0*)		9.2	3.45
$HOOC-CH_2-N(OH)-CH_2-COOH$	-	21.9*)		8.45	4.10
$H_2N-O-CH_2-COOH$	$\tilde{n}_{max}<0.15$			5.02	
$H_2N-CH_2-CH_2-COOH$	<7.5			7.10	5.40
$HO-NH-CH_2-COOH$	6.4	6.2	5.1	Cu^{2+} reduction	
H_2N-CH_2-COOH	<6.6			8.46	6.83
$HO-N(CH_3)-CH_2-COOH$	6	5	4	Cu^{2+} reduction	
$H_3C-NH-CH_2-COOH$	<7.5			7.94	6.65
$HOOC-CH_2-CH(NHOH)-COOH$	7.24	5.4		6.54	4.3
$HOOC-CH_2-CH(NH_2)-COOH$	9.2			8.57	6.78
$HON(CH_2-CH_2-COOH)_2$	5.8	4.2		7.3	

*) lgK_1K_2

at 985 cm^{-1} does not appear in the analogous stable vanadium complex of N-hydroxyimino-diacetic acid. This excludes the presence of a vanadyl group and indicated a V(IV) cation without oxygen. The high acidity of amavadine (p_k = o) excludes also the free carboxylic groups. On the basis of LAXS and EXFAS investigations we have to assume that in amavadin the donating groups of the ligands are arranged like in fig. 6 depicted (ref. 15).

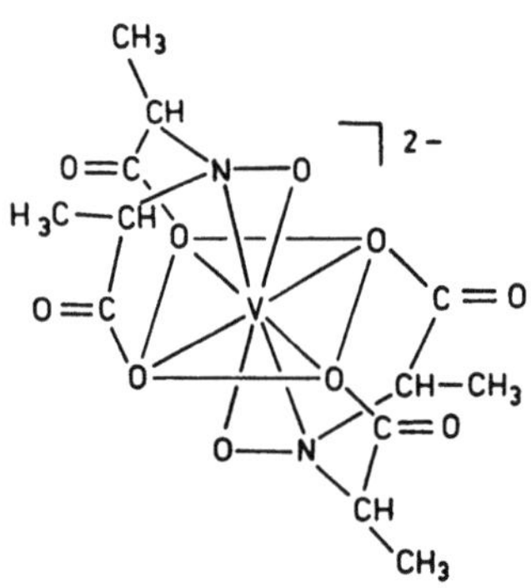

Fig. 6 Revised structural proposal for amavadin

The occurrence of amavadine seems to be restricted to a few Amanita species, as is shown in table 3 (ref. 16). However admirable is the architecture of the ligand to an unexpected and so far not observed vanadium selectivity. This example of tailor made selectivity of a chelating agent in nature would have been difficult to predict because it exceeds the imagination of a chemist.

TABLE 3

Amavadin and vanadium concentrations (mg/kg dry wt.) in a few air-dried mushrooms of the genus *Amanita* (C: cap, S: stalk, B: bulb) (from ref.16).

Species (origin)	part	amavadin	vanadium	V(% of total) in amavadin
A. muscaria (Finland)	C	127	85	19
	S,B	163	107	19
A. muscaria (Schönbuch)	C	119	77	19
	S	143	85	20
	B	369	213	21
A. muscaria, yellow variant (U.S.A.)	C	135	120	13
	S,B	278	218	16
A. regalis (Finland)	C	147	98	18
	S,B	202	122	12
A. velatipes (Michigan, U.S.A.)	C	195	257	9
	S,C	457	377	15

REFERENCES

1 I.G. Macara, Vanadium - an element in Search of a role, Trends biochem. Sci. 5 (1980) 92-94.

2 H. Vilter, Peroxidases from Phaeophyceae: a vanadium(V)-dependent peroxidase from *Ascophyllum nodosum*, Phytochemistry 23 (1984) 1387-1390.

3 R.L. Robson, R.R. Rady, T.H. Richardson, R.W. Miller, M. Hawkins and J.R. Postgate, The alternative nitrogenase of *Acotobacter chroococcum* is a vanadium enzyme, Nature 322 (1986) 388-390.

4 E. Bayer and H. Kneifel, Isolation of amavadin, a vanadium compound occuring in *Amanita muscaria*, Z. Naturforsch. 27b (1972) 207.

5 H. Kneifel and E. Bayer, Determination of the strucutre of the vanadium compound, amavadin, from fly agaric, Angew. Chem., Int.Ed.Engl. 12 (1973) 508.

6. P. Krauß, E. Bayer and H. Kneifel, Electron spin resonance investigations of amavadin, a natural product containing vanadium, Z. Naturforsch. 39b (1984) 829-832.
7. E. Koch, H. Kneifel and E. Bayer, Synthesis of new complexes: N-hydroxy- , '-iminodipropionic- and N-hydroxyiminodiacetic acid, Z. Naturforsch. 41b (19866) 359-362.
8. P. Karrer and R. Appenzeller, The stereoisomeric , '-iminodipropionic acids, Helv. Chim. Acta 25 (1942) 595-599.
9. H. Frank. G. Nicholson and E. Bayer, Chiral polysiloxanes for resolution of optical antipodes, Angew.Chem.Int. Ed.Engl. 17 (1978) 363-365.
10. E. Bayer, Chiral Recognition of Natural Products on optically active polysiloxanes, Z. Naturforsch. 38b (1983) 1281-1291.
11. J. Tempe, A. Petit, M. Holsters, M. von Montaga and J. Schell, The prosensitive step associated with transfer of the Ti plasmid during conjugation: possible relation to transformation in crown gall, Proc.Natl.Acad.Sci. U.S.A. 74 (1977) 2848-2849.
12. M. Sato, Y. Sato and Y. Tsuchiya, Studies on the extractives of molluscs - II. -iminodipropionic acid isolated from squid muscle extracts, Bull.Jpn.Soc.Sci.Fish. 43 (1977) 1441-1443.
13. H. Kneifel and E. Bayer, Stereochemistry and total synthesis of amavadin, the naturally occuring vanadium compound of *Amanita muscaria*, J.Am.Chem.Soc. 108 (1986) 3075-3077.
14. G. Anderegg, E. Koch and E. Bayer, N-Hydroxy- , '-iminodipropionic and N-hydroxy iminodiacetic acid as complexing agents and an example for selective coordination of the vanadyl ion in amavadin, Inorg.Chim.Acta 127 (1987) 183-188.
15. E. Bayer, E. Koch and G. Anderegg, Amavadin, an example for selective binding of vanadium in nature: studies of its complexation chemistry and a new structural proposal, Angew.Chem.Int.Ed.Engl. 26 (1987) 545-546.
16. E. Koch, H. Kneifel and E. Bayer, Occurrence of Amavadin in Mushrooms of the Genus *Amanita* , Z.Naturforsch. 42c (1987) 873-878.

Prof. Dr. Ernst Bayer
Institut für Organische chemie
Universität Tübingen
Auf der Morgenstelle 18
D 7400 Tübingen, FRG.

Micro-chemical Research Pattern and the Prospects of Its Industrial Development in the Developing Countries

SALIMUZZAMAN SIDDIQUI
H.E.J. Research Institute of Chemistry,
University of Karachi, Karachi-32, Pakistan.

In spite of the enormous work carried out in the chemistry of natural products, with particular reference to studies in the correlation of chemical structure and physiological activity, it still continues as an almost virigin sector of research. In this context, I would briefly deal with the mechanics of research and development with particular reference to the situation in the developing countries. First of all we have to take note of the fact that isolation of the chemical constituents of medicinal plants in micro quantities and elucidation of their structure with spectral evidence rarely lead to their industrial development.

I would at this point refer to the importance of chromatography and other new techniques which are being increasingly employed in the isolation of chemical constituents and reaction management. Without in any way wanting to minimise the importance of these techniques, I find that they are not always judiciously used and are mostly emloyed at the expense of orthodox procedures which continue to play an important role in the isolation of plant constituents. By way of illustration, I would here take the opportunity of citing a few examples from our work in the field.

In the traditional dietetics of the subcontinent highly nutritive and body building properties are attributed to Cicer arientum Linn, commonly known as Bengal gram or chana which contains arsenic and copper in micro quantities. In particular, chana soaked overnight in water, and sprouted chana, are considered as extremely nourshing, and are a regular item of diet for athletes and professional wrestlers. In view of these facts, a study of the chemical constituents of fresh, undried chana sprouts was undertaken and three crystalline products isolated from them, two of which proved to be isoflavones and another could be identified as asparagin [1]. The isolation of these products was greatly facilitated by excluding enzymic action and admixture with excessive quantities of non-crystallisable products through the device of extracting the uncrushed sprouts with organic solvents, using their outer membranes as a dialysing medium. On the other hand, when the sprouts were dried in shade, powdered and extracted with the same solvent, no trace of the crystalline constituents could be obtained from them.

On the alkaloidal side, I would quote the example of the Rauwolfia bases [2]. After a great deal of negative work extending over about two years, it was only on working up freshly collected and shade dried root material that nine new alkaloids were isolated, some of which have gained world wide importance in the treatment of cardiovascular diseases and mental ailments. Particular reference may be made on this occasion to the alkaloid ajmaline, which is being employed as the drug of choice in the treatment of cardiac arrhythmia and is being annually manufactured in tons by Kali Chemie under the commercial name Gilurhythmal.

During the same period (1930s) a white crystalline alkaloid, m.p. 362-63°C was isolated from the neutral fraction of the alcoholic extract in a yield of 0.1% [3]. Later, in 1952 Schlitller et al. reported the isolation of the same base under the name reserpine, which acquired global reputation as a hypotensive and sedative agent. As a result of long range clinical studies, carried out over a period of 10 years, Moyer et al., noted that its extended use results in heavy depressions bordering on suicidal tendencies and schizophrenia in about 50% of patients. On account of this complication, a large number of derivatives were prepared by different groups of workers with the object of eliminating its central sedative action. In this context, nitration studies carried out by us resulted in three isomers of nitro-reserpine, namely 1-, 12- and 9-nitroreserpine in 50, 12 and 3% yield, respectively [4]. Pharmacological studies on 1- and 12-nitroreserpine carried out by a leading German pharmaceutical firm, showed that their hypotensive activity is of the same order as that of reserpine without its toxic side effects like tremor ptosis and diarrhoea. 12-Nitroreserpine has much longer lasting hypotensive action than reserpine and showed better absorption through oral administration as compared with 1-nitroreserpine.

In the course of these studies it was noted that the experimental conditions in respect of the variables of concentration of reactants, temperature, duration, and the speed with which the reactants are brought together are highly crtitical. If these conditions are not meticulously observed, there is either failure of the reaction or formation of tarry material. For instance, if the optimum proportion of nitric acid (eight moles) in glacial acetic acid (Merck) is added slowly drop by drop with mechanical stirring at 20°C to a solution of reserpine in glacial acetic acid, reserpine nitrate crystallises out and there is no formation of a nitro derivative of the base. On the other hand, when under the same conditions the reaction mixture is quickly run through a burette without shaking, and allowed to stand for 4-5 min, the initial yellowish colour of the reaction mixture changes over to red and then deep red. On quenching the reaction at this stage by pouring into crushed ice, the three isomeric nitro derivatives are obtained in the yields mentioned above. It may be observed in this context that in the course of studies in its pharmacological studies over 30% of the substance had to be supplied.

It may be stated here that it took nearly 20 years before reserpine got exploited in the pharmaceutical industry in the Western countries and, ajmaline about 10 years longer to get established as an anti-arrhythmic drug, and its industrial exploitation is also restricted to the Western countries. This long time lag between research and developed relating to new therapeutic agents has been mainly due to lack of adequate facilities with us for interdisciplinary researches in the field. In the case of nitro-reserpines and nitro-ajmaline also we have got to note that their utilization would depend on the commercial considerations of the pharmaceutical industry of the developed countries.

The hectic pace at which modern science is advancing is broadly reflected in the fact that its literature references have a doubling period of about a decade, and the pace of this progress is being further accelerated with time. The phenomenal march of science thus holds out a challenge to our own and other developing countries which are late starters in the field, because if they want to catch up with scientific developments on the international level, their progress wil have to be at a vastly faster pace than in the advanced countries. The examples of Japan and Soviet Russia seem to indicate, that this need not be such a lost hope, and may be capable of achievement with wise and forceful planning on the part of the States and dedicated effort of their scientific communities.

It would be of interest to refer here to our work on Semecarpus anacardium Linn (vern. bhilawan). Owing to the medicinal importance attributed to it in traditional medicines, systematic investigation of the resinol occuring in the pericarp of its nuts, was undertaken in 1931, which resulted in the isolation of a phenolic substance (bhilawanol) through fractional distillation under reduced pressure. Studies in its structure revealed it as a catechol derivative with $C_{15}H_{27}$ unsaturated straight side chain, carrying an average of two double bonds. It was later established that it is closely related to uroshiol, obtained by Majima from the Japanese lacquer varnish, ki-urushi, (Rhus vernicifera) and differs from it only in the number and position of the double bonds in the side chain. These findings sbsequently led to the industrial development of the resinol for the production of stoving enamels, varnishes, and moulding compositions during the World War II. In a recent re-investigation of the petroleum ether insoluble, alcohol soluble component of bhilawan nuts, two dimeric flavonoids have also been isolated, the exact structure of which is under investigation.

The utter neglect of this potent drug in modern medical practice has been mainly due to its toxicity and vesicent action on external application or internal administration. Studies in its chemical constitution have to some extent made it possible to control its toxicity. Thus it was found that the free hydroxyl groups and the two double bonds in the side chain of bhilawanol are together responsible for its vesicent action. When both these active centres are eliminated as for instance in diacetyl or dimethyl-tetrahydrobhilawanol, a completely non-vesicent product results. Unfortunately, studies in the correlation of structure and activity in respect of its antirheumatic and a wide range of other therapeutic properties claimed for it have not so far been undertaken.

In view of the outstanding importance attributed to Melia azadirachta (neem) in folklore and traditional systems of medicine for the treatment of a wide range of human ailment, studies in its constituents were undertaken since the beginning of current century. However, the first two crystalline terpenoids nimbin and nimbinin were isolated in 1940 from the seed oil through solvent separation without involving prior hydrolysis [6]. More recent studies on the fresh, undried and uncrushed leaves, fruit 'twigs' stem bark and seed oil have resulted in the isolation and structure elucidation of around 50 new terpenoids [7,8], pharmacological studies on which are underway.

Further work on the various parts of the neem tree is aimed at the pilot plant establishment of new therapeutic agents as well as the constituents which are effective in controlling the target pests without undesirable environmental side effects.

Peganum harmala

In the course of basic studies in the germination metabolites of Peganum harmala seeds, it was discovered that the alkaloidal constituents are located in the husk and the kernel yields 20% of oil comparable in its constants and constituents to those of cotton seed oil [9]. However, the economic viability of harmal seeds as a new source of edible oil would largely depend on the utilization of alkaloids from the husk which constitutes 50% of the whole seed and contain ca. 7% of the alkaloidal bases. In view of this fact, intensive studies have been undertaken on the chemistry of the two main alkaloids of harmala, namely harmine, harmidine, and their derivatives, and a whole series of new potentially pharmacophoric derivatives of these bases have been prepared [10].

REFERENCES

1. S. Siddiqui, J. Sci. Ind. Res. (India), **4**, 68 (1945).

2. S. Siddiqui and R. H. Siddiqui, J. Indian Chem. Soc. **12**, 37 (1939).

3. S. Siddiqui, J. Indian Chem. Soc. **16**, 421 (1939).

4. S. Siddiqui and M. A. Khan, Pak. J. Sci. Ind. Res. **18**, 199 (1975).

5. S. Siddiqui, D. P. Bhal, R. P. Rastogi and V. N. Sharma, J. Sci. Ind. Res. (Inda) **8B**, 222 (1949).

6. S. Siddiqui, Current Science, **11**, 278 (1942).

7. S. Siddiqui, B. S. Siddiqui, S. Faizi and T. Mahmood, J. Nat. Prodcts **51**, 30 (1988).

8. I. Ara, B. S. Siddiqui, S. Faizi and S. Siddiqui, Phytochemistry (in press).

9. S. Siddiqui and N. Afza, Pak. J. Sci. Ind. Res. **21**, 46 (1978).

10. S. Siddiqui, B. S. Siddiqui and S. Begum, Z. Naturforsch., **40b**, 1747 (1985).

Isolation and Structural Studies on New Natural Products of Potential Biological Importance

Atta-ur-Rahman
H.E.J. Research Institute of Chemistry,
University of Karachi, Karachi-32, Pakistan.

INTRODUCTION

Large sections of populations living in rural areas of the Afro-Asian countries rely on the indigenous medicinal systems for the treatment of diseases. These are largely based on the use of roots, leaves, fruits and flowers of different plant materials. Isolation, structural and synthetic studies have accordingly been directed to the discovery of new natural products in many laboratories, including ours. Our interest is however not limited to traditional medicinal plants but we also pay attention to any new leads encountered in other natural sources. For instance marine plants and animals are an invaluable source of novel secondary metabolites many which possess a wide diversity of structures and pharmacological properties which match those obtained from terrestrial organisms. Some of the recent studies carried out by us are briefly presented here.

CHEMICAL CONSTITUENTS OF BUXUS PAPILOSA

Buxus papilosa (Buxaceae) is a shrub, widely distributed in the northern regions of Pakistan. This plant is a potential source of steroidal alkaloids. We have previously isolated over 40 new steroidal alkaloids from its leaves. The new alkaloids recently isolated are cyclobuxoviramine (1), buxafuranamide (2) (a novel alkaloid having an ether linkage between rings A and B of the steriodal base) and (+)-N-acetyl-N-demethylcyclomicrobuxeine (3). Temperature dependent ^{1}H-NMR studies on (3) indicate that it exists as a mixture of two conformers which equilibrate at higher temperatures. A new triterpenoid, buxatenone, has also been isolated from the roots of this plant. Buxatenone is structurally related to cyclobuxoviracine, an alkaloid reported by us previously (ref 1).

NEW BISINDOLE ALKALOIDS FROM PETCHIA CEYLANICA

Petchia ceylanica (Apocynaceae) is a plant found in the lowlands of Sri Lanka. A number of bisindole alkaloids having vincorine-type moieties linked together have been previously reported by us from this plant. As a result of further isolation and structural studies, two new bisindole alkaloids, ceylanine (5) and ceylanicine (6), have now been isolated. Ceylanine (5) represents the first bisindole alkaloid have a vincorine-type unit linked to a pleiocarpamine skeleton, through a bond between the benzene rings of the two moieties.

NEW ALKALOIDS FROM ALSTONIA MACROPHYLLA

Alstonia macrophylla Wall is a common plant in Sri Lanka. It is a rich source of monomeric and dimeric indole alkaloids. We have isolated three new alkaloids from this plant: 11-methoxygeissoschizine (7), 10,11-dimethoxygeissoschizine (8) and 16-hydroxy-N_b-demethylalstophylline (9). Geissoschizine is a key intermediate in the biosynthesis of indole alkaloids.

A NEW DIETERPENE FROM TINOSPORA MALABARICA

Tinospora malabarica (Miers) is cultivated throughout Pakistan. A novel furanoid bisnorditerpene, malabarolide (10), has been isolated from the fresh stems of this plant and its structure confirmed using X-ray crystallography and spectroscopic techniques. A number of 19-nor-clerodanes (ref. 2) have been reported previously, but malabarolide represents the first example of a 18,19-bisnorclerodane. A plausible biogenesis of malabarolide would involve oxidative removal of the 18-methyl group from a 19-norclerodane, or decarboxylation of a tinophyllol-type compound.

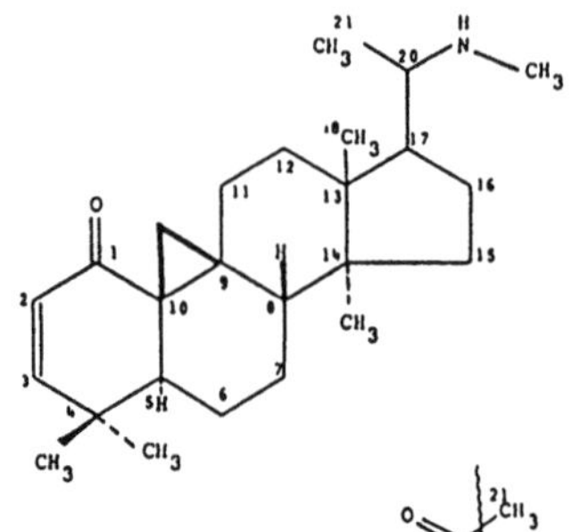

1

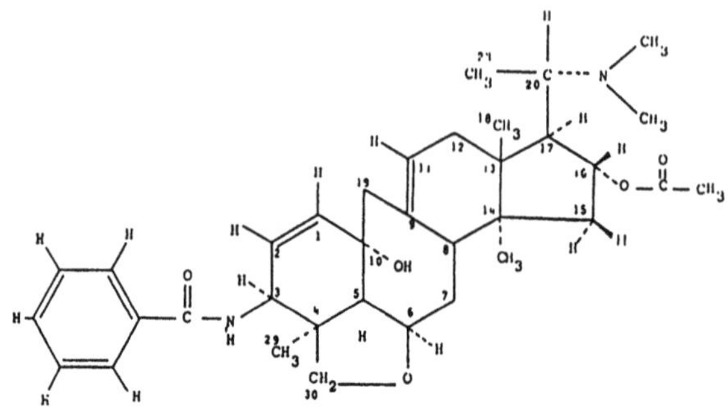

2

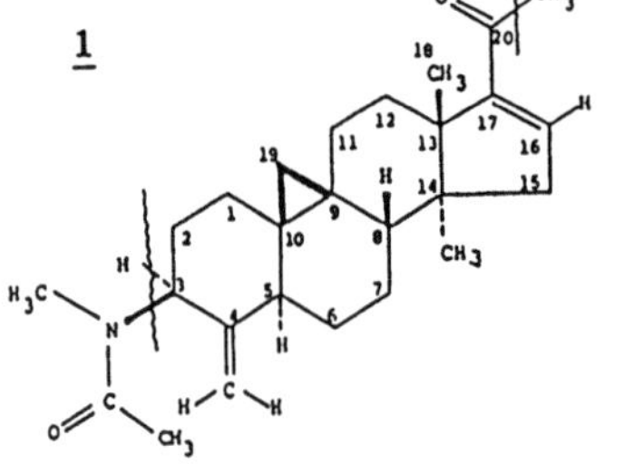

3

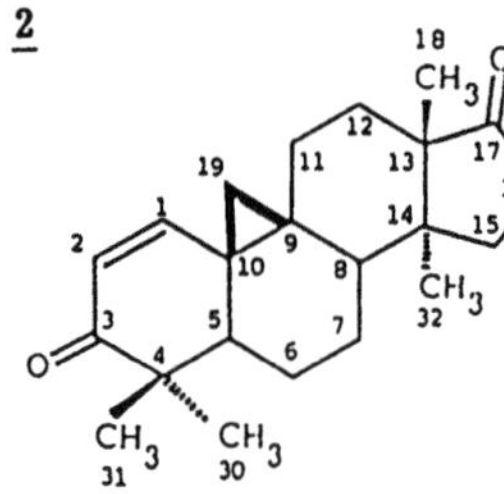

4

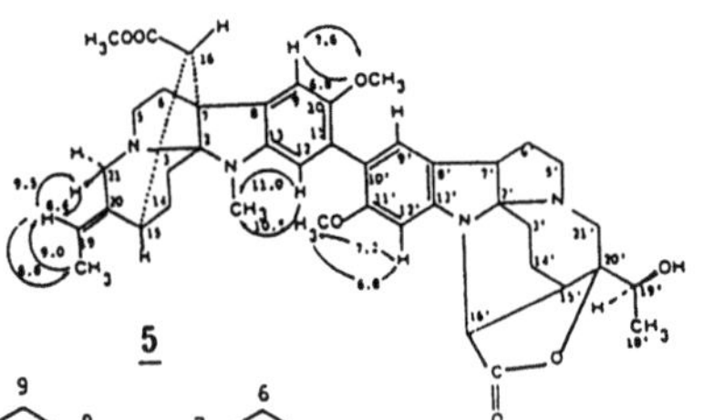

5

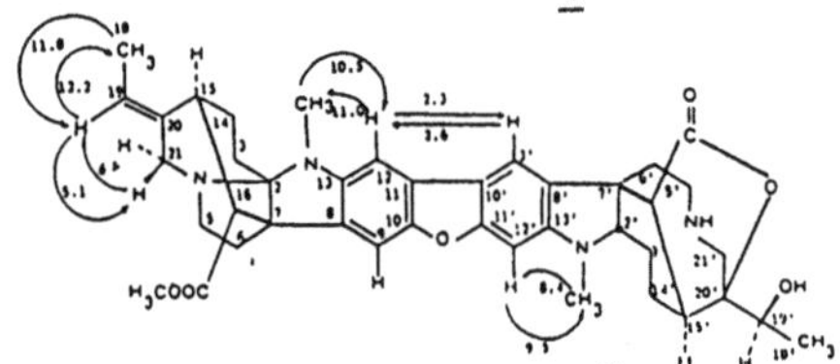

6

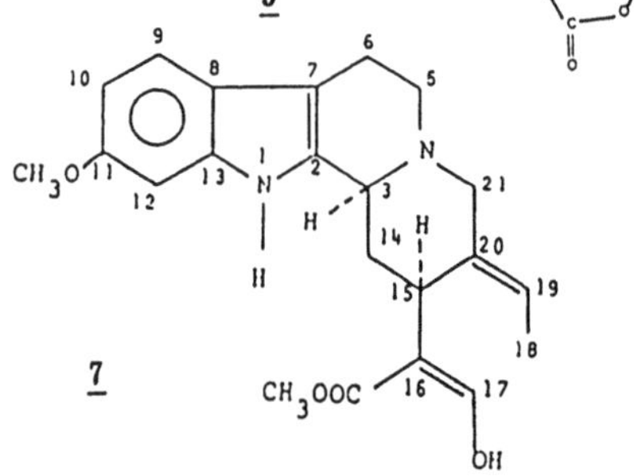

7

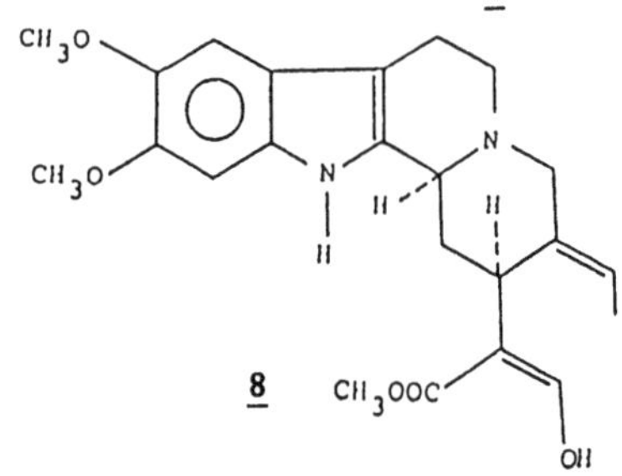

8

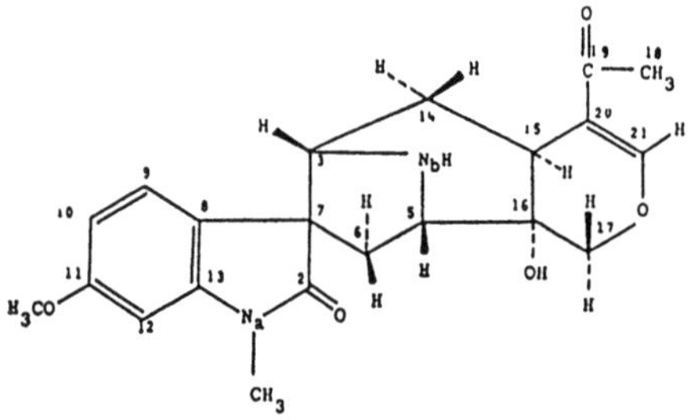

9

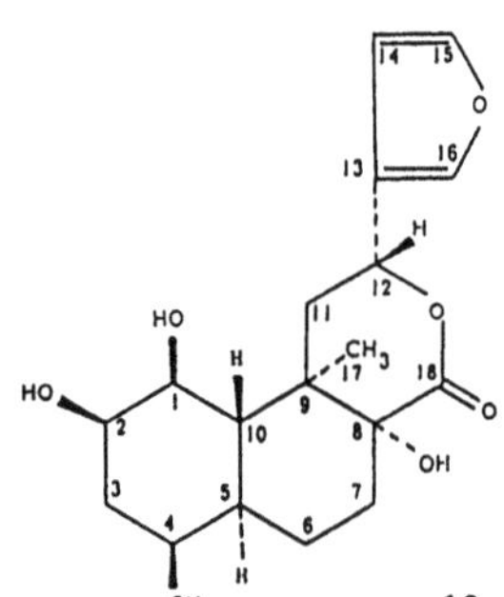

10

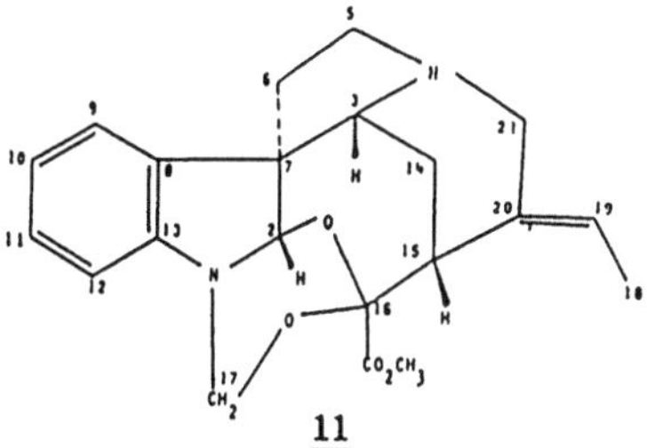

11

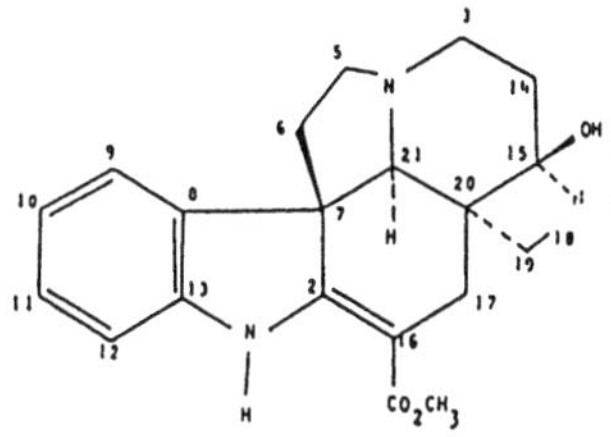

12

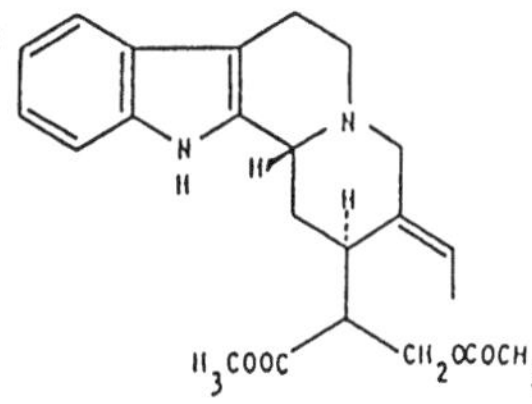

13

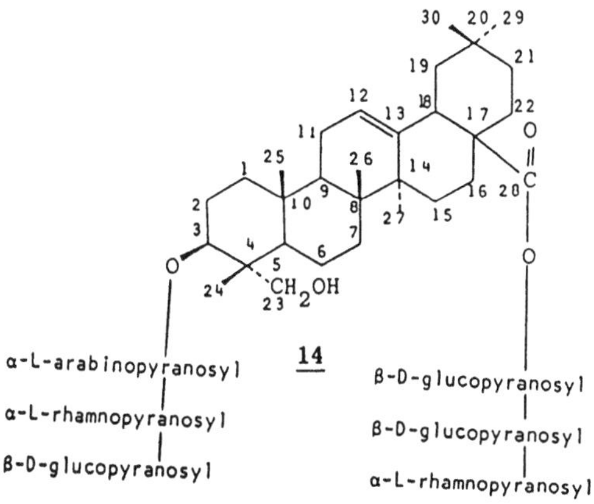

14

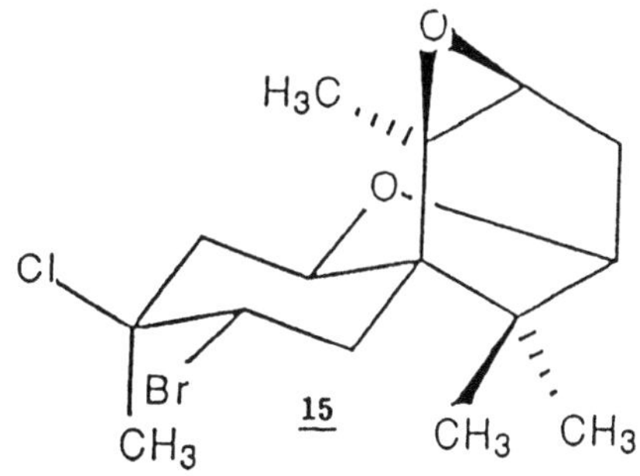

15

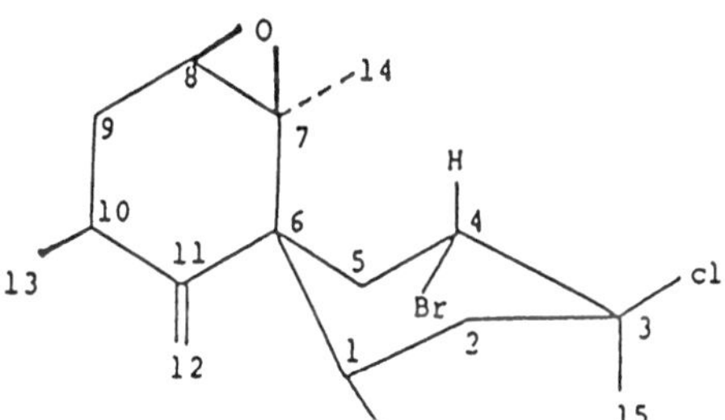

16

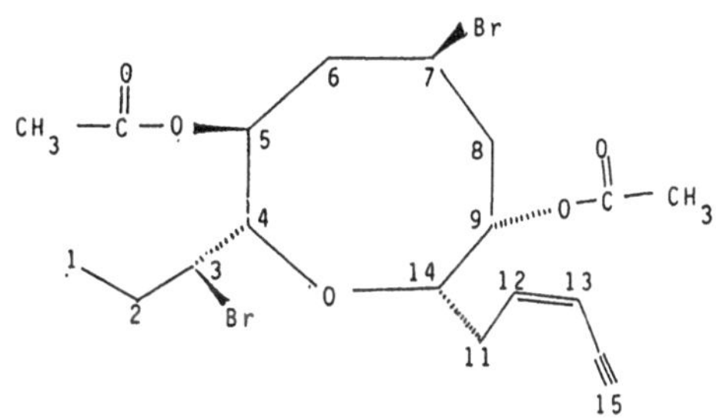

17

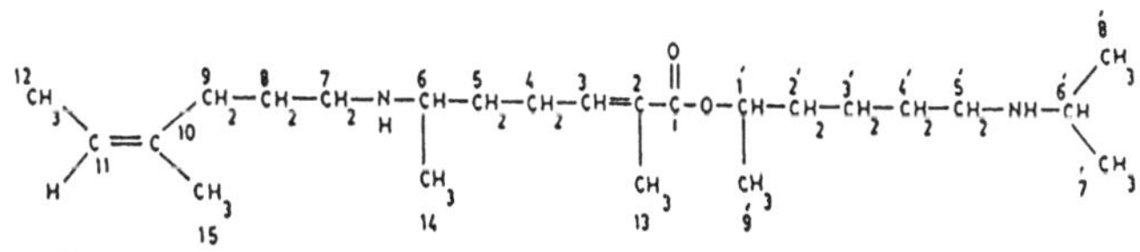

18

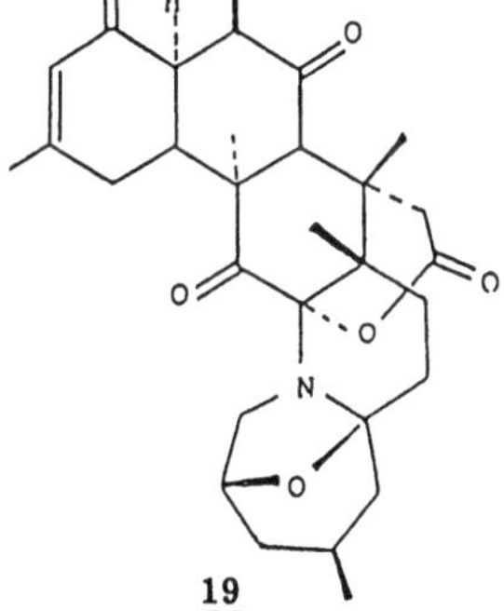

19

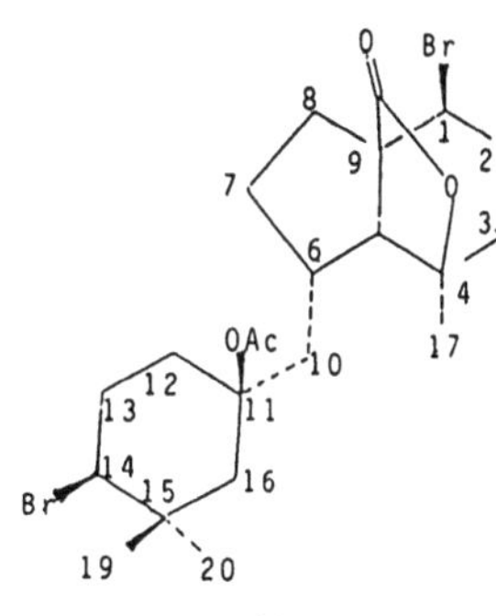

20

NEW ALKALOIDS FROM RHAZYA STRICTA

Rhazya stricta Decaisne (Apocynaceae) is a small shrub, growing profusely near Karacl It is used in the indigenous system of medicine as a bitter tonic, for soar throat and fever. We have previously reported twenty new alkaloids from it (ref.3). Recently we ha isolated three new indole alkaloids rhazizine (11), 15-β hydroxyvincadifformine (12) a bhimberine acetate (13).

A NEW SAPONIN FROM NIGELLA SATIVA

Nigella sativa Linn (Ranunculaceae) is an indigenous herbaceous plant. We have previous reported a number of alkaloids from it (ref.4). Recent studies on the chemical constituer of the seeds of this plant have led to the isolation of a new triterpenoidal saponin nigellosid (14). The structure of this saponin has been elucidated through spectroscopic studies a by chemical degradation.

CHEMICAL CONSTITUENTS OF LAURENCIA PINNATIFIDA

As a result of our studies on chemical constituents of marine plants and animals in t intertidal zone near Karachi, we have investigated red and brown algae and various mari animals. Laurencia pinnatifida belongs to the family Rhodomalaceae. Laurencia species a well known for the halogenated compounds of varied chemical structures and diverse biologi properties such as antimicrobial activity, cytotoxicity etc. We have isolated pinnataza (15), a new bridged sesquiterpene from Laurencia pinnatifida and its structure has be established through spectroscopic and X-ray crystallographic techniques. Another sesquite pene, pinnatifate (16) has also been isolated from this alga which is an acetylated prod of a previously known chamigrine alcohol (ref.5) Along with these sesquiterpeniods a n non-isopreniod acetylenic cyclic ether, pinnatifidine (17) has been isolated which represe the first diacetylated dibromo representative of this class.

A NEW LINEAR METABOLITE FROM STOCKEYIA INDICA

Stockeyia indica is a brown marine benthic alga, occurring in mid to lower littoral pools. novel linear metabolite (18) has been isolated and its structure has been established w the help of modern spectroscopic techniques.

NEW ALKALOID FROM ZOANTHID SPECIES

Zoanthids are small marine animals found growing as dense mats on intertidal rocks. Fr an unidentified species of zoanthid we have isolated a new alkaloid zoanthaminone (which belongs to a new class of alkaloids.

A NEW DITERPENOIDAL LACTONE FROM APLYSIA JULIANA

Aplysia juliana (sea hare) is a marine animal, belonging to the class of opisthobr molluscs. During the course of our studies on the chemical constituents of Aplysia juli we have isolated a new diterpenoidal lactone, agnasyl acetate (20).

Many of the compounds reported above have shown interesting anti-bacterial, anti-v and anti-cancer activities.

REFERENCES

1. Atta-ur-Rahman, M. I. Choudhary and (in part) M. Nisa, Phytochemistry. 24, 3082-3 (1985).
2. H. M. G. Al-Hamzi and G. A. Miana, J. Chem. Soc. Pakistan. 19, 125-142 (1987).
3. Atta-ur-Rahman and S. Malik, Phytochemistry. 26, 589-591 (1987) and referen therein.
4. Atta-ur-Rahman, S. Malik, He Cun-Heng and J. Clardy, Tetrahedron Letters. 2759-2762 (1985) and references therein.
5. M. L. Bittner, M. Silva, V. J. Paul and W. Fenical, Phytochemistry. 25, 987- (1985).

Isolation and Structural Studies on Chemical Constituents from Some Plants and Marine Organisms of Pakistan

Viqar Uddin Ahmad*, Mohammad Shaiq Ali, Nasreen Bano, Shaheen Bano, Ayjaz Fatima, Iqbal Fatima, Kaniz Fizza, Tanveer Ahmad Farooqui, M.Aslam Khan, Shaista Perveen, Sabiha Qazi, Tahir Rasheed, Ghahzala Rizwani, Viqar Sultana, K.Usmanghani and Shafiuddin.
H.E.J. Research Institute of Chemistry, University of Karachi, Karachi-32, Pakistan.

In the meeting held in January 1987 I presented before you [1] the work done in our group on the isolation and structure elucidation of new natural products. Today I would like to discuss the structure of new natural products which have been isolated by us in the last few years which were not included in my last lecture.

Saponins from Guaiacum officinale

In continuation of our work on new saponins from Guaiacum officinale, we have isolated eight new saponins named as guaianin A (**1**), guaianin A2 (**2**), guaianin B (**3**), guaianin C (**4**), guaianin D (**5**), guaianin E (**6**), guaiacin A (**7**), and guaiacin B (**8**). The structures of these saponins were determined through spectroscopic studies permethylation followed by hydrolysis as well as isolation of the genuine sapogenin through Smith degradation.

Pluchea arguta Boiss

Chemical investigation of Pluchea arguta led to the isolation of ten new sesquiterpenes of eudesmane type. These are arguticin (**9**), argutinin (**10**), pluchecin (**11**), odontin (**12**), odonticin (**13**), pluchecinin (**14**), 3'-chloro-2'-hydroxy-arguticinin (**15**), deacetoxy-3-chloro-2-hydroxy-arguticinin (**16**), dideacetoxy argutin (**17**), odontinin (**18**), and odonticinin (**19**). The structures of these compounds were determined with the help of spectroscopic studies particularly two dimensional nmr spectroscopy.

Crataeva nurvala Buch Ham

From this plant the new alkaloid cadabicine methyl ether (**20**) was isolated and correlated with the cadabicine.

Prosopis juliflora

This plant contains alkaloids which show strong antibacterial and antifungal activity. A new minor alkaloid isolated by us is juliflorinine(**21**), an isomers of the already known alkaloid juliflorine (juliprosopine).

Cocculus hirsutus

We have isolated a new isoquinoline alkaloid cohirsine (**22**) from this plant, the sturcture of which was determined through spectroscopic studies.

Primula denticulata
The plants belonging to Primulaceae are known to contain saponins which are responsible for their medicinal properties From P. denticulata we isolated a new saponin denticulatin (**23**). Its structure was determined through acidic hydrolysis before as well as after permethylation and also through spectroscopic studies.

Caralluma tuberculata
From this edible vegetable we were able to isolate two new saponins, named as caratuberside A(**24**) and caratuberside B (**25**). Their aglycones belong to pregnane series of steroids.

Marine Natural Products
During our work on active metabolites from starfish we have isolated from Astropecten indicus three novel steroidal glycosides named as indicoside A (**26**), B (**27**) and C (**28**). The red alga Laurencia pinnatifida yielded three new halogenated sesquiterpenoids pinnatificine (**29**), pinnatifidone (**30**), and 1ß,4ß,10ß-tribromo-3-chloro-7(14)-ene-α-chamigrene (**31**). Compounds **29** and **30** are new additions in the list of compounds having ether linkage between C-1 and C-10.

References.
1. Viqar Uddin Ahmad, S. Bano, N. Bano, A. Fatima, I. Fatima, M. Shafi, S. Perveen, S. Qazi, K. Fizza, T. Rasheed, F. V. Mohammad, A. Nasir, V. Sultana, A. Amber, S. Arif, "Isolation and Structure Elucidation of Natural Products, Atta-ur-Rahman (Editor), "Recent Advances in Natural Products Chemistry, Karachi, 1987

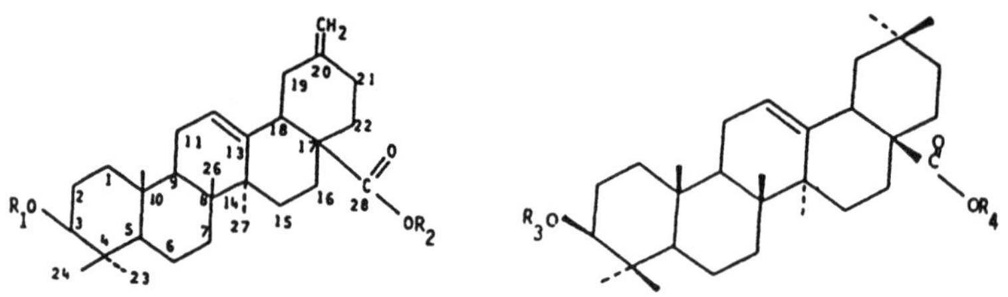

1.R_1=α-L-rhamnopyranosyl(1→2)-α-L-arabinopyranosyl
R_2=H
2.R_1=α-L-rhamnopyranosyl(1→2)-α-L-arabinopyranosyl
R_2=ß-D-glucopyranosyl
3.R_1=α-L-rhamnopyranosyl(1→3)-ß- D-glucopyranosyl(1→3)-α-L-arabinopyranosyl
R_2=ß-D-glucopyranosyl
4.R_3=α-L-rhamnopyranosyl(1→ 3)-ß-D-glucopyranosyl (1→3)-α-L-arabinopyranosyl
R_4=ß-D-glucopyranosyl
5.R_1=α-L-rhamnopyranosyl (1→3)-α-L-rhamnopyranosyl (1→2)-α-L-arabinopyranosyl (3 1)-ß-D-glucopyransoyl
R_2=H
6.R_1=α-L-rhamnopyranosyl (1→3)-α-L-rhamnopyranosyl (1→2)-α-L-arabinopyranosyl (3 1)-ß-D-glucopyranosyl
R_2=ß-D-glucopyranosyl
7.R_1=ß-D-glucopyranosyl (1→ 3)-ß-L-arabinopyranosyl
R_2=ß-D-glucopyranosyl
8.R_3 = ß-D-glucopyranosyl (1→ 3)-α-L-arabinopyranosyl
R_4=ß-D-glucopyranosyl

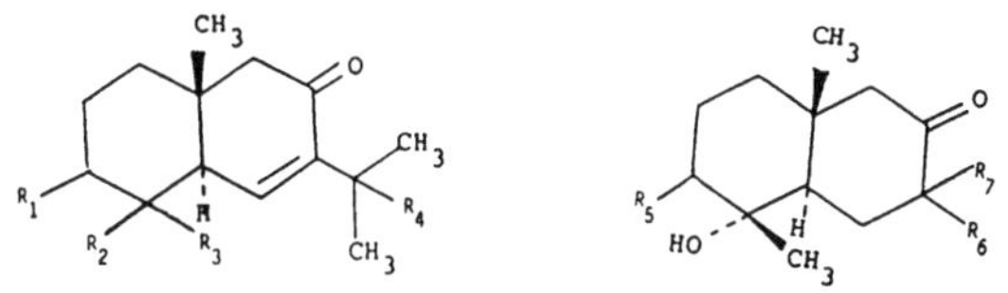

9. R_1=α-O-C(=O)-$\underbrace{\text{C}(\text{CH}_3)\text{-C}(\text{H})}_{\text{O}}$-$CH_3$, R_2=α-CH_3, R_3=β-OAc, R_4=β-OH

10. R_1=α-O-C(=O)-$\underbrace{\text{C}(\text{CH}_3)\text{-C}(\text{H})}_{\text{O}}$-$CH_3$, R_2=α-CH_3, R_3=β-OAc, R_4=β-OOH

11. R_1=β-O-C(=O)-C(CH_3)=C(H)—CH_3, R_2=α-CH_3, R_3=β-OH, R_4=β-OOH.

12. R_5=β-OH, R_6= —C(=CH_2)——CH_3, R_7=α-OOH

13. R_1=β-O-C(=O)-C(CH_3)(OH)-C(H)(OH)-CH_3. R_2=α-OH R_3=β-OH, R_4=β-OOH

14. R_5=β-O-C(=O)-$\underbrace{\text{C}(\text{CH}_3)\text{-C}(\text{H})}_{\text{O}}$-$CH_3$, R_6=β-C(=CH_2)— $COOCH_3$. R_7=α-H

15. R_1=α-O-C(=O)-C(CH_3)(OH)-C(H)(Cl)-CH_3. R_2=α-CH_3. R_3=β-OAc, R_4=β-OH

16. R_1=α-O-C(=O)-C(CH_3)(OH)-C(H)(Cl)-CH_3, R_2=α-CH_3, R_3=β-OH, R_4=β-OH

17. R_5=α-O-C(=O)-C(CH_3)(OH)-C(H)(OH)-CH_3, R_6, R_7= =C(CH_3)(CH_3)

18. R_1=α-O-C(=O)-C(CH_3)(OAc)-C(H)(OH)-CH_3. R_2=β-CH_3, R_3=α-OAc, R_4=β-OH

19. R_1=α-O-C(=O)-C(CH_3)(OH)-C(H)(OH)-CH_3, R_2=α-CH_3, R_3=β-OH, R_4=β-OH

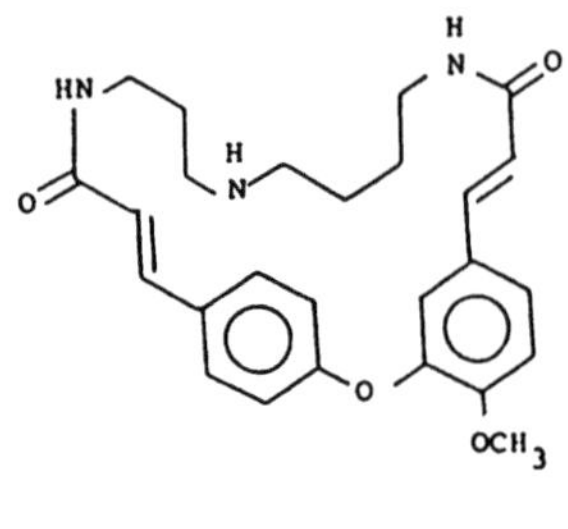

20

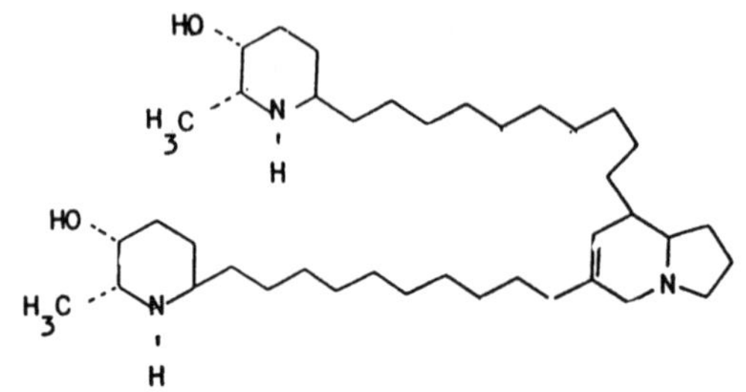

21

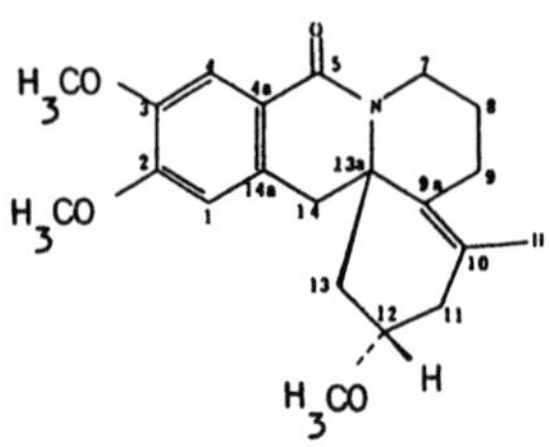

22

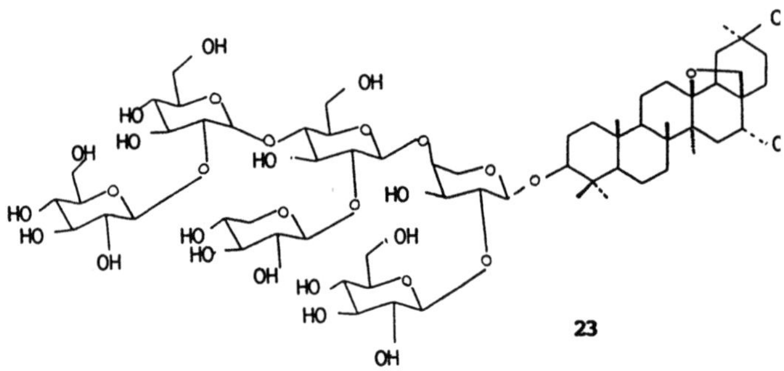

23

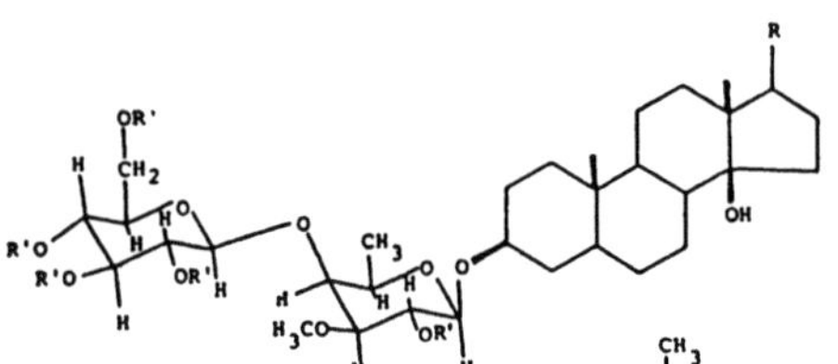

24 = R= $-\overset{CH_3}{\overset{|}{C}}=O$, R'= H

25 = R= $-\overset{CH_3}{\overset{|}{C}H}-OH$, R'= H

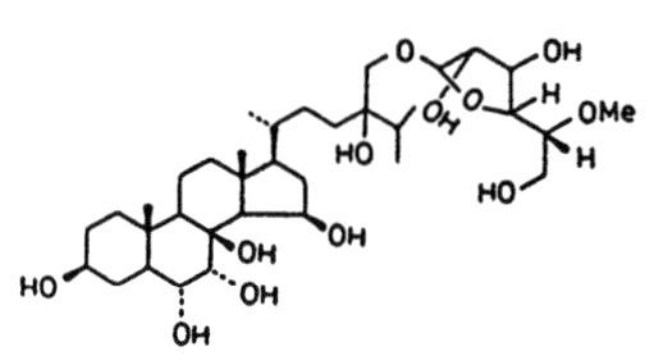

26

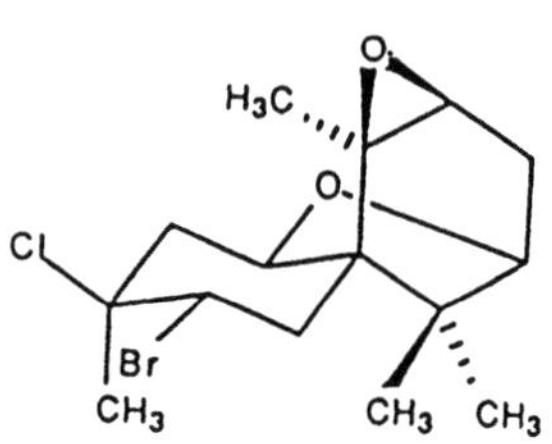

29

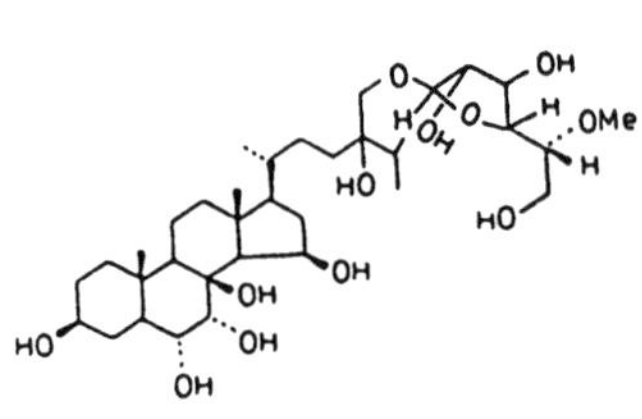

27

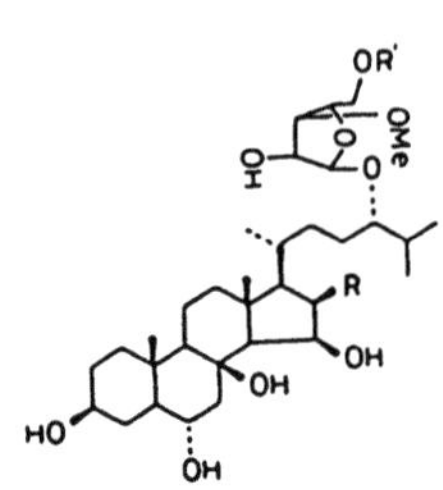

28

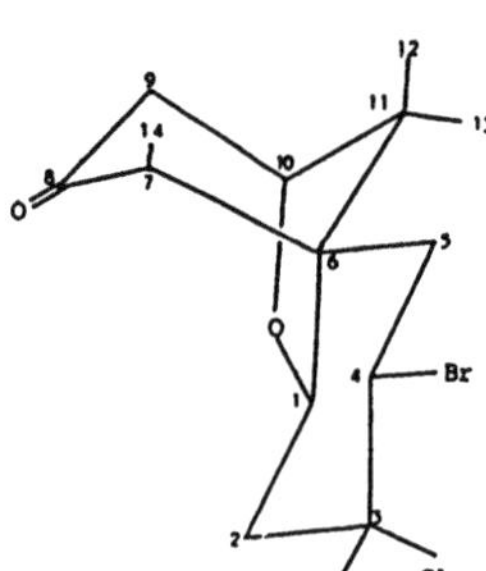

30

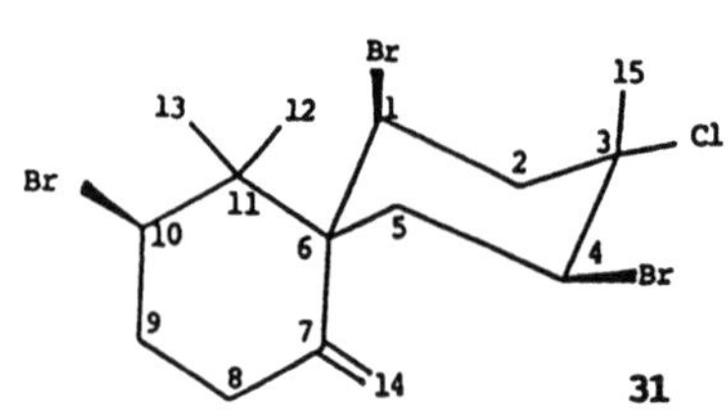

31

Subject Index

MIX
Papier aus verantwortungsvollen Quellen
Paper from responsible sources
FSC® C105338

If you have any concerns about our products,
you can contact us on
ProductSafety@springernature.com

In case Publisher is established outside the EU,
the EU authorized representative is:
Springer Nature Customer Service Center GmbH
Europaplatz 3, 69115 Heidelberg, Germany

Printed by Libri Plureos GmbH
in Hamburg, Germany